Lecture Notes in Physics

363

V. Ruždjak E. Tandberg-Hanssen (Eds.)

Dynamics of Quiescent Prominences

Proceedings of the No. 117 Colloquium
of the International Astronomical Union
Hvar, SR Croatia, Yugoslavia 1989

Springer-Verlag
Berlin Heidelberg GmbH

Editors
Vladimir Ruždjak
Hvar Observatory
58 450 Hvar, Yugoslavia

Einar Tandberg-Hanssen
Space Science Laboratory, NASA, MSFC
Huntsville, AL 35812, USA

ISBN 978-3-662-14499-2 ISBN 978-3-540-46293-4 (eBook)
DOI 10.1007/978-3-540-46293-4

Originally published by Springer-Verlag Berlin Heidelberg New York in 1990
Softcover reprint of the hardcover 1st edition 1990

Bookbindung: J. Schäffer GmbH & Co. KG., Grünstadt

PREFACE

Informal discussions starting in 1986 among a number of scientists working in the field of prominence physics revealed a need for a broad discussion on the understanding of the dynamic nature of quiescent prominences. A successful, albeit more limited, workshop on the dynamics and structure of solar prominences was held in Palma Mallorca in 1987, and it is more than ten years since the last IAU Colloquium on prominences was held in Oslo.

Considerable progress in our understanding of the nature of quiescent prominences has been made since the Oslo meeting, and it was felt the time was ripe to gather observational scientists, data analysts and modelers, and theoreticians and conduct an in-depth discussion. The idea for an IAU Colloquium proposed by Croatian solar astronomers, following the initial discussions, was supported by Commisions 10 and 12 and approved by IAU. The meeting was held in the pleasant medieval town of Hvar on the island of the same name, Croatia, Yugoslavia, and the Local Organizing Committee consisted of V.Ruždjak (Chairman), K.Brajša, R.Brajša, M.Malarić, D.Plačko-Vršnak, and B.Vršnak.

The Scientific Organizing Committee included E.Tandberg-Hanssen (Chairman), O.Engvold, J.Kleczek, J.L.Leroy, M.Machado, H.Morozhenko, E.Priest, and V.Ruždjak.

89 participants from 23 countries contributed to a very successful meeting by delivering 11 invited talks, 49 contributed papers, and 16 posters and, not least, by participating in the numerous valuable discussions that followed the various contributions. We thank them all for their participation.

We want to thank IAU for financial support and we are indebted to Hvar Observatory, Faculty of Geodesy, University of Zagreb for both financial and logistic support.

The majority of the contributed and poster papers are published in full in a special issue of the Hvar Observatory Bulletin, 1989, Vol. 13 No 1.

ZAGREB AND HUNTSVILLE
JANUARY 1990

V. RUŽDJAK

E. TANDBERG-HANSSEN

TABLE OF CONTENTS

1 G.A. Porfirjeva
2 V. Ruždjak
3 M.G. Rovira
4 B. Schmieder
5 E. Tandberg-Hanssen
6 F. Chiuderi-Drago
7 D. Plačko-Vršnak
8 E.A. Makarova
9 L. Geondjan
10 V. Bommier
11 I.B. Voloshina
12 L.N. Kurochka
13 B. Vršnak
14 M. Malarić
15 A.I. Kiryukhina
16 V. Vujnović
17 K.F. Novikova
18 A.F.A.A. Galal
19 M.I. Gigolashvili
20 S.F. Martin
21 V.V. Zharkova
22 G.M. Simnett
23 H. Balthasar
24 H.-J. Schober
25 A. Kovacs
26 J. Klacka
27 K. Brajša
28 R. Brajša
29 V. Gaizauskas
30 J.L. Ballester
31 C. Ridgway
32 U. Anzer
33 T. Hirayama
34 P. Ambrož
35 L. Dezso
36 S. Pohjolainen
37 S. Urpo
38 T. Nikiforova
39 A.N.Sedjikina
40 E.G. Rudnikova
41 M. Gnevyshev
42 I.S. Kim
43 Y.N. Redcoborody
44 Gu Xiao-Ma
45 Z.B. Korobova
46 E.R. Priest
47 B.S. Nagabhushana
48 T.A. Darvann
49 P.J.J. De Bruyne
50 E. Jensen
51 C. Chiuderi
52 Fang Cheng
53 B. Rompolt
54 J.-L. Leroy
55 S. Koutchmy
56 D. Novocky
57 V.I. Kulidzanishvili
58 P. Heinzel
59 E. Landi Degl´Innocenti
60 M. Molodensky
61 P. Mein
62 P. Kotrč
63 J. Deliyannis
64 C.D.C. Steele
65 G.T. Buyukliev
66 Z. Mouradian
67 P. Demoulin
68 O. Engvold
69 J.-C. Vial
70 U. Degenhardt
71 E. Wiehr
72 V.Yu. Klepikov
73 R. van der Linden
74 V. Rušin
75 J. Kubota
76 P. Gouttebroze
77 L. Hejna
78 F. Farnik
79 A.I. Poland

Not appearing on the group photo

T. Fiedler
T.G. Forbes
G.V. Jakunina
S. Jurač
K.R. Lang
J.C. Noens
J.P. Rozelot
N.D. Tromans
G. van Hoven
J.B. Zirker

CONDITIONS FOR THE FORMATION OF PROMINENCES AS INFERRED FROM OPTICAL OBSERVATIONS

Sara F. Martin
Big Bear Solar Observatory
Solar Astronomy 264-33
California Institute of Technology
Pasadena, CA, USA

Abstract

In the optical region of the electromagnetic spectrum, the conditions most frequently associated with the formation of prominences are: (1) the existence of opposite polarity photospheric magnetic fields on opposing sides of a prominence, (2) a coronal arcade that connects the magnetic fields on opposing sides of a prominence, (3) a transverse magnetic field configuration in the chromospheric and photospheric polarity inversion zones that is approximately perpendicular to the direction of maximum magnetic field gradient between adjacent patches of opposite polarity line-of-sight magnetic flux, (4) in active regions or decaying active regions, the alignment of chromospheric fibrils in a polarity inversion zone approximately parallel to the transverse magnetic field component and parallel to the long axis of the future prominence, (5) the long-term (hours to days) converging flow of small patches of opposite polarity magnetic flux towards a common polarity inversion zone, and (6) the cancellation of encountering patches of magnetic flux of opposite polarity at a photospheric polarity inversion boundary (interpreted as the transport of magnetic flux upwards or downwards through the photosphere). Because these are observed conditions found from magnetograms and filtergrams at various wavelengths, they do not necessarily represent independent physical conditions. Although none of these conditions have proven to be individually sufficient for prominence formation, a combination of 3 of these conditions might prove to be both necessary and sufficient. The following hypothesis is offered for study and evaluation: condition (2) and the combination of conditions (5) and (6), if dynamically maintained for a sufficient length of time, will invariably result in the formation of a prominence.

1.0 INTRODUCTION

This review discusses conditions observable in the optical region of the spectrum that the author considers to be essential clues to the formation of the magnetic field structure of prominences. Conceived ways of filling the magnetic field structure with mass and other physical properties of prominences are not covered in detail; other reviews thoroughly discuss most possibilities (Tandberg-Hanssen, 1974; Hirayama, 1985; Malherbe 1987, 1988; Zirker 1989; Priest, 1989).

The prominences discussed herein include only those that would correspond to 'filaments' in monochromatic observations of the solar disk. This means that the words filament and prominence can be used interchangeably when considering the physics of these phenomena. When describing observations, the term 'prominences' is used according to the historical definition to refer to phenomena that extend above at the limb and 'filaments' is used to describe the same phenomena when observed against the disk. Due to the prevalence of observations of the disk, the term 'filament' is used more often than 'prominence'. However, in general discussion, when referring to both limb and disk observations, the term 'prominences' is used, to refer to both filaments and prominences. 'Prominences' in this review includes all categories: quiescent prominences, both large-scale and small-scale, and prominences of all scales in active regions and between active regions. This review covers the growth or extension of existing prominences and the reformation of prominences after eruption but excludes any general discussion of prominence eruptions which are the topic of another review in this volume. Specifically excluded in this review are transient phenomena such as surges, loops and arches which accompany flares and are sometimes included in the general category of active prominences when observed at the limb.

2.0 CONDITIONS ASSOCIATED WITH THE FORMATION OF FILAMENTS AND PROMINENCES

2.1 Sites of Formation with Respect to Line-of-sight Magnetic Fields

Upon completion of the first magnetograph made for systematically producing spatial images of solar magnetic fields, Babcock and Babcock (1955) soon found that large quiescent prominences existed at the boundaries between large areas of photospheric magnetic fields of opposite polarity. After a number of other authors (Howard, 1959; Avignon et al., 1964, Howard and Harvey, 1964; Martres et al., 1966) found that prominences in active regions, outside of active regions and in between active regions also occurred between opposite polarity magnetic fields, Smith (1968) concluded that the condition of opposite polarity fields on opposing sides of a prominence was an invariable condition for the existence of prominences. Because the magnetic fields are always present before the prominences are seen, it is assured that this is a necessary condition for their formation (Martin 1973). Hereafter this first condition will be referred to as the '+/-' condition.

Throughout this paper the area between large-scale fields of opposite polarity will be often called the 'polarity inversion zone'. Under specific conditions to be described, the polarity inversion zone will be designated as the 'filament channel'.

The site of a filament within a large-scale polarity inversion zone is shown in Fig. 1 in an H-alpha filtergram and drawn on a corresponding magnetogram from the Big Bear Solar Observatory. The filament is in two segments and both parts lie in between the large-scale fields of opposite polarity. In the magnetogram, areas of negative and positive polarity are respectively denoted by black and white immediately outside of the contours. The contours are introduced to increase the visual dynamic range by programming the display to reverse color each time the saturation level is reached during the integration of successive video frames containing the magnetic signal.

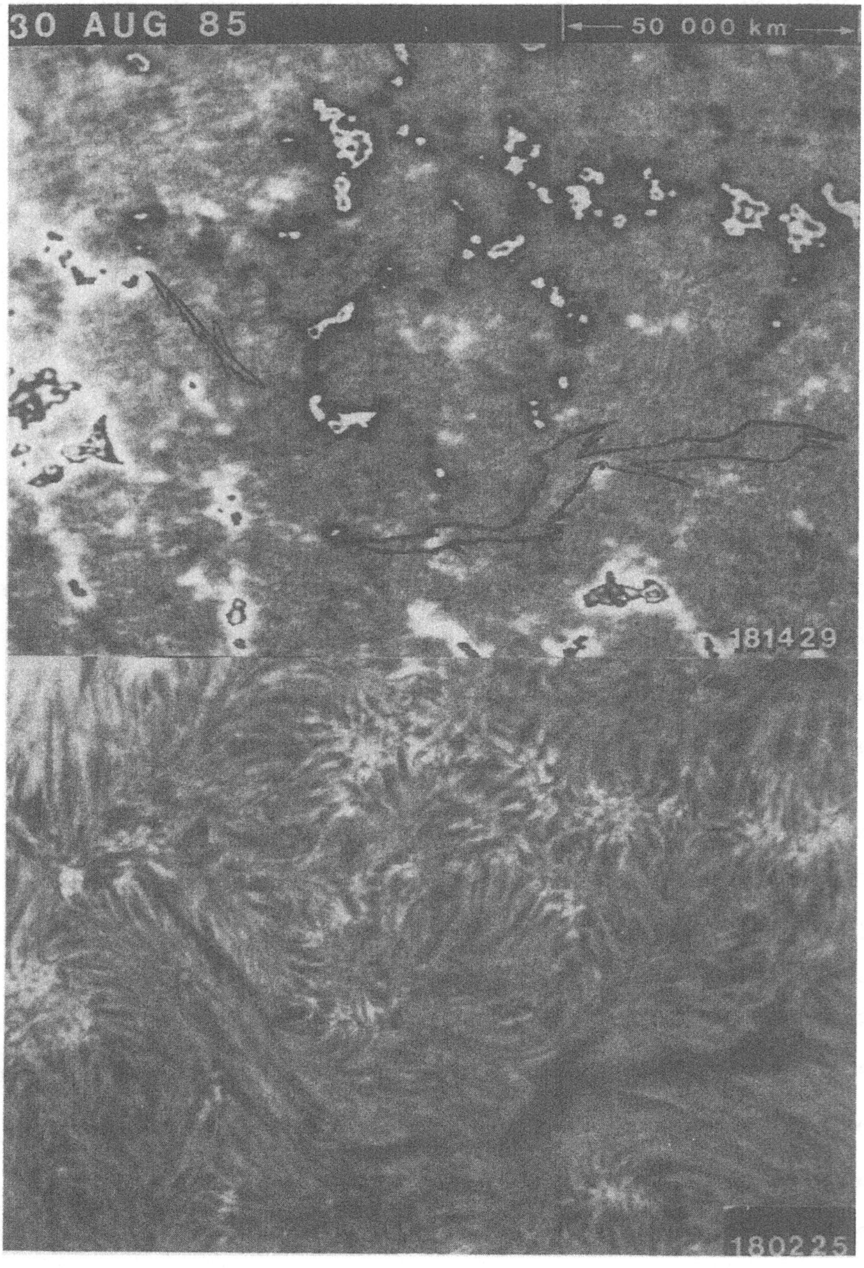

Figure 1. An outline of the filament shown in the lower half is superposed on a videomagnetogram of the line-of-sight component in the upper half. The filament occurs between large-scale magnetic fields of opposite polarity but some small-scale magnetic fields lie under or very close to the base of the filament. The polarity of the magnetic field is determined by the color immediately surrounding the contours. Negative polarity is black and positive polarity is white. The contours are introduced to extend the visually dynamic range in the magnetograms. Each additional contour within a feature signifies an increase in magnetic strength by a factor of 2 over the nearest outer contour. These images were recorded at the Big Bear Solar Observatory.

The +/- condition only applies to the large-scale fields. It is seen in Fig. 1 that amidst these large-scale fields, there are small-scale magnetic fields of both polarities. These small-scale fields do not obviously affect the filament except when they are in the polarity inversion zone below the filament. The importance of the polarity inversion zone can be seen in the corresponding dynamics of filaments and the magnetic fields below or close to filaments. Filaments slowly adjust their shape as the polarity inversion zone changes, becoming narrow where the polarity inversion zone is narrow and wider where the polarity inversion zone is wide (Bruzek and Durrant, 1977).

The converse of this first necessary condition for the existence of prominences is not true. It is recognized that not all boundaries between opposite polarity photospheric magnetic fields are sites of promiences. For example, note the gap between the two segments of the filament in Fig. 1. Hence, the condition of +/- fields is not a sufficient condition for prominence formation. This is also illustrated in Fig. 2 which shows two boundaries between opposite-polarity, line-of-sight magnetic fields and corresponding H-alpha images below. In one case a large filament is present; in the other no major filament has formed. The differences in these two circumstances are discussed further in Section 2.4.

2.2 Magnetic Field Arcades Overlying Prominences

A second condition associated with quiescent prominences, well-known from eclipse observations, is the existence of a closed arcade of magnetic field lines which overlies a prominence and connects the opposite polarity magnetic fields on opposing sides of the

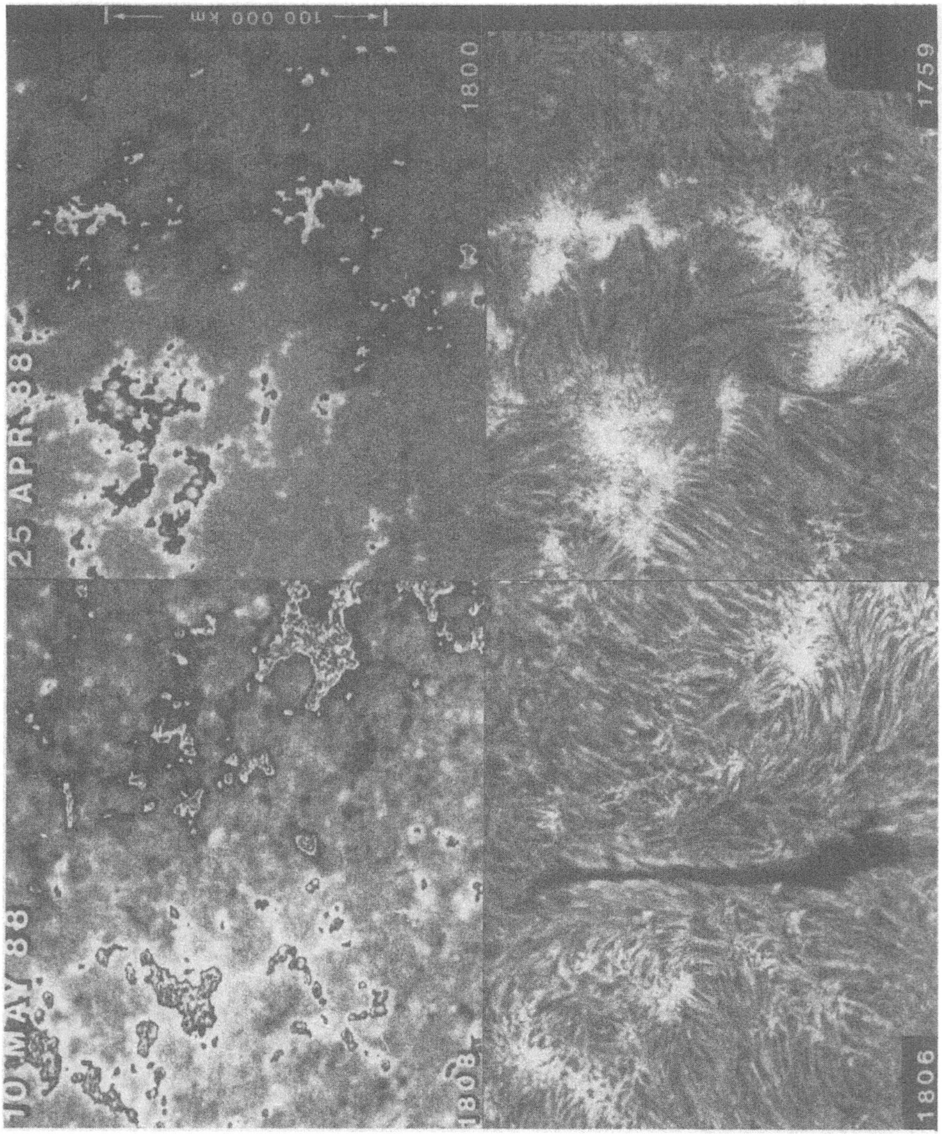

Figure 2. Videomagnetograms and H-alpha images from Big Bear Solar Observatory showing, on the left, a polarity inversion zone which is associated with a filament and, on the right, a major polarity inversion zone without a major filament. Filaments do not form in polarity inversion zones where the fibrils appear to connect opposite polarity magnetic fields. Filaments are observed to form where fibril-like structures divide rather than join closely-spaced magnetic fields of opposite polarity.

prominence. The arcade is visible in white-light (von Klüber, 1961; LeRoy and Servajean, 1966; Kawaguchi, 1967; Newkirk, 1971), in X-rays (Vaiana et al., 1973; McIntosh et al., 1976; Serio et al. 1978), and in EUV images (Schmahl et al., 1982). Many eclipse observations in white light and monochromatic light reveal a dark cavity, signifying reduced electron density, between the prominence and the overlying arcade (Waldmeier, 1941, 1970 and references therein). The arcade is often faint and hence, the cavity was discovered before the arcade. Observations of the corona aboard Skylab (McQueen, Sime and Picat, 1983) also have revealed the presence of coronal 'voids' above the cavity and the arcade. When a prominence is viewed from one of its ends at the limb, the arcade has been called a 'helmet' and there often is often one or more narrow, coronal streamers extending nearly radially from the top of the 'helmet'. The coronal voids are long, narrow zones, that lie high above the helmet, adjacent and parallel to the coronal streamers.

Tandberg-Hanssen (1979) shows an example in which the helmet and streamer enclose two adjacent arcades, each overlying a prominence. Ionized particles are trapped in these magnetic loops of the arcade and the density in these structures is of the order of 10^{-9} cm^{-3} (Davis and Krieger, 1982). The X-ray arcades over quiescent prominences are always weak and diffuse (McIntosh et al., 1976). Over active region filaments, the X-ray arcades are much brighter.

The angle between the long axis of a prominence and the overlying arcade can vary over a wide range. If this angle is small, the arcade as a whole is sometimes described as being 'highly sheared' with respect to the long axis of a prominence (Webb and Zirin, 1981).

From full-disk X-ray images of the sun, it is now recognized that closed arcades of field lines do exist between most, but not all, adjacent magnetic fields of opposite polarity (Serio et al., 1978; Davis and Krieger, 1982). The most noteworthy exceptions are the magnetic fields within and next to coronal holes (Webb et al., 1978). In this context it is important to note that the base of coronal holes correspond to areas at the photosphere where unipolar network magnetic fields exist (Vaiana et al., 1973) rather than to mixed-polarity network fields as defined by Giovanelli (1982). The unipolar fields within coronal holes are open to the interplanetary space and do not connect to adjacent fields of opposite polarity (Bohlin, 1977; Levine, 1977). If the second condition of a closed arcade overlying adjacent +/- magnetic fields, is truly a necessary condition for the formation and existence of prominences, we would expect to find few or no prominences at polarity inversion boundaries which are close to the boundaries of coronal holes. In addition, we should expect that the growth or expansion of a coronal hole towards a polarity inversion boundary containing a prominence, would eventually result in the eruption or dissipation of the prominence. This expectation has been verified by Webb et al., (1978). They found that coronal 'transients', most of which are associated with erupting prominences, have a significantly higher rate of occurrence in the vicinity of coronal holes than at other locations on the sun. Such findings strongly indicate that an overlying arcade of closed field lines probably is a necessary condition for the formation and sustenance of a prominence. The importance of the arcade to prominence sustenance is further substantiated by the association of their eruption with coronal mass ejections. When the corona is distended or opened above a prominence during the early stage of a coronal mass ejection, the prominence is invariably observed to ascend and follow the expanding corona outward into the interplanetary medium. From this association it is deduced that the coronal arcade is a magnetic enclosure that prevents the eruption of an underlying prominence and thereby contributes to its sustenance.

From X-ray observations, it is also now known that magnetic arcades can exist without an underlying prominence (McIntosh et al., 1976). Thus, the second necessary condition for prominence formation, is not a sufficient condition.

2.3 Conditions from Transverse Magnetic Field Measurements

Due to current observational limitations, the measurement of the transverse component of solar magnetic fields has only been done successfully in active regions and there is a scarcity of published examples showing the direction of the photospheric field relative to the filaments above (Ding et al., 1987 (Fig. 5 and 7). Although the spatial resolution is not high, the transverse fields appear to be parallel to the long axis of filaments.

From the association of the sites of filaments and some flares as well as the associations of erupting filaments with flares, it is known that the magnetic configurations which apply to flares, also apply to filaments. Moore, Hagyard, and Davis (1987), Hagyard, Venkatakrishnan, and Smith (1989) and Venkatakrishnan, Hagyard, and Hathaway (1989) have shown that flares are associated with transverse magnetic field configurations that have the maximum degree of 'magnetic shear'. Use of the word 'shear' in this case refers to the magnetic field configuration rather than to motion. Magnetic shear means that the direction of the transverse photospheric field along the polarity inversion zone where flares occur (and also filaments) is at a large angle with respect to the direction that a potential magnetic field would have at that location as viewed against the disk.

Fig. 3 is a magnetogram from the Marshall Space Flight Center Observatory used here to illustrate the transverse configurations that are and are not associated with the sites of filaments. This magnetogram is also illustrated in Venkatakrishnan, Hagyard and Hathaway (1989) which discusses flare positions with respect to magnetic field configurations. The arrows show the direction and magnitude of the transverse component of the magnetic field while contours show the line-of-sight component. The outermost contour of the negative polarity fields are shown by a broader line than the positive polarity fields. The polarity inversion zone close to the center of the illustration, and labelled A, is an example of a

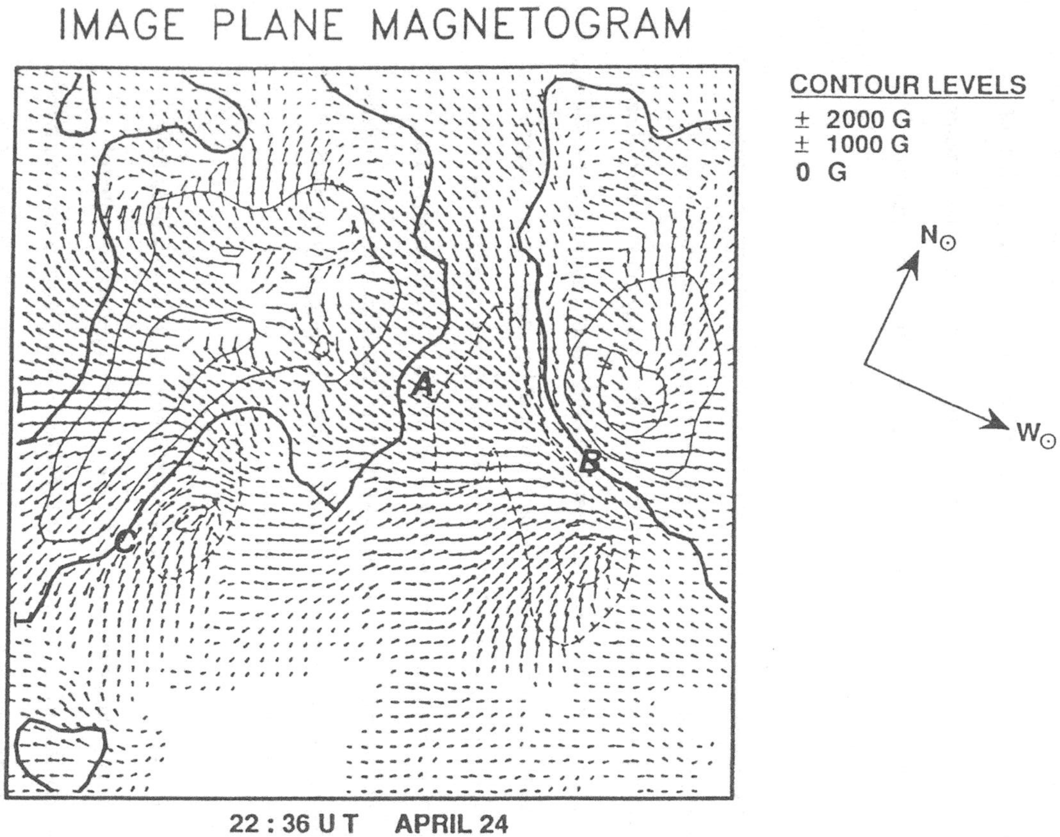

Figure 3. This magnetogram from Marshall Space Flight Center depicts the line-of-sight magnetic field component as contours and the transverse component by small-scale arrows. Areas of negative polarity are enclosed by broad, continuous lines and areas of positive polarity by thinner, dashed lines. Location A, between the broad line to the left and the thinner, dashed line to the right, corresponds to a line-of-sight polarity inversion zone where a filament cannot form. B and C are sites where filaments can form. The filament that formed at site B is shown in Kurokawa et al. (1984).

location where no filament can form; the arrows directly cut across the contours of the line-of-sight component inferring that the transverse component directly connects the opposite polarity line-of-sight fields. In contrast, the polarity inversion zone to the right of center, labelled B, is an example of the configuration where a filament can exist. The arrows run parallel to the contours of the line-of-sight component rather than crossing the contours. A filament formed in the lower part of this polarity inversion zone on the previous day and already erupted with a flare prior to the time of this magnetogram (Kurokawa et al., 1987)

These sites where filaments can and cannot form should be compared to the sites with and without filaments in H-alpha filtergrams as shown in Fig. 2. Where the fibrils, like the arrows of the transverse component, cross the polarity inversion zone directly from one polarity to the opposite, there is no filament (right side of Fig. 2). Where there is a filament (left side of Fig. 2), the fibrils are directed along the long axis of the polarity inversion zone and parallel to the filament.

Fig. 3 also has a polarity inversion zone to the lower left of center, labelled C, in which the fibrils only at the left end of the polarity inversion zone are aligned appropriately for a filament. The other end of the polarity inversion zone shows the configuration, like area C, where a filament cannot form.

Due to the paucity of high quality transverse magnetograms at the present time, it is still important to use the more abundant H-alpha images as a proxies for chromospheric magnetograms (Foukal, 1971; Smith 1971; Zirin, 1972). By this means, much has been learned about the local conditions of the magnetic field prior to, during, and after the formation of filaments. The next section is devoted to the discussion of conditions relating to filament formation, growth, and reformation as seen in H-alpha filtergrams, and the interpretation of this information in terms of the magnetic field direction in and near filaments.

2.4 Conditions of Prominence Formation from Ha Filtergrams

2.4.1 Circumstances of Aligned Fibrils

In a study of the chromospheric conditions associated with the formation of filaments, Smith (1968) observed that before the filaments formed in or near active regions, the chromospheric fibrils, associated with the opposite polarity magnetic fields adjacent to the polarity inversion zone, would invariably avoid becoming connected to the fibrils of opposite polarity. Instead they would sharply curve into the polarity inversion zone. In the middle of the polarity inversion zone, the fibrils would thus be aligned with the direction that was to become of the long axis of the filament. Adjacent fibrils along a polarity inversion boundary would in this way create a long path in which fibrils would be aligned approximately parallel to each other; in a long polarity inversion zone, many fibrils would be aligned approximately end to end. (Further discussion and examples can be found in Prata (1971), Foukal (1971), Martin (1973), Rompolt and Bogdan (1986).)

The direction of the aligned fibrils provides an easy means of distinguishing between photospheric polarity inversion zones which are and are not associated with overlying filaments as illustrated in the right side of Figure 2. Filaments cannot form where the fibrils appear to directly link opposite polarities. This circumstance is seen over most of the polarity inversion zone in the right side of Figure 2. Only in the lower part of the polarity inversion zone is there a location where a small filament has formed. It is seen at the site where positive and negative magnetic fields are close together. Note that the fibrils immediately adjacent to the small filament are approximately parallel with the long axis of the filament; neither the fibrils nor the filament reveal any structural connection to the opposite polarity magnetic fields that are close together on both sides of the filament. This is a general property of fibrils and filaments which can be seen in most of the remaining illustrations in this paper. This is important information for modelling filaments because it suggests that the magnetic fields from one side a filament do not take a short or obvious route through the filament to the nearest opposite polarity magnetic fields on the other side of the filament.

Even under conditions of excellent image quality, as seen in Figure 4, the fibrils do not obviously connect areas of opposite polarity across or through the filament. Additionally, the mass motions seen in time-lapse films, even under the best conditions also do not show any obvious connection between the opposite polarity fields on both sides of the filament. With rare exceptions, usually such connecting structures are only seen when flare loops suddenly develop often accompanying the eruption of a filament. The relevant aspects of the association of prominences and flares is discussed further in Section 2.4.3 on prominence reformation.

The same long filament in the left in Fig. 2 is shown 50 minutes earlier in Fig. 4. Although the general shape of the filament is the same in the two images, almost all of the fine structure of the filament has completely changed. In the time-lapse films, one sees that the fine structure of the filament is constantly changing. Small changes are seen in most of the fine structure within a few minutes. In 20 to 30 minutes, the fine structure of the entire filament has changed. There is continuous mass flow along the long axis of the filament coincident with the appearance and disappearance of the fine structure. This can be interpreted either as a recycling of the filament mass or as a process of continuous formation of new strands of the filament as previous ones disappear.

Rompolt and Bogdan (1986) suggest that the alignment of fibrils, along rather than across the polarity inversion zone, can be a consequence of the relative motions of the footpoints of the opposite polarity magnetic fields on opposing sides of a polarity inversion zone. This is an

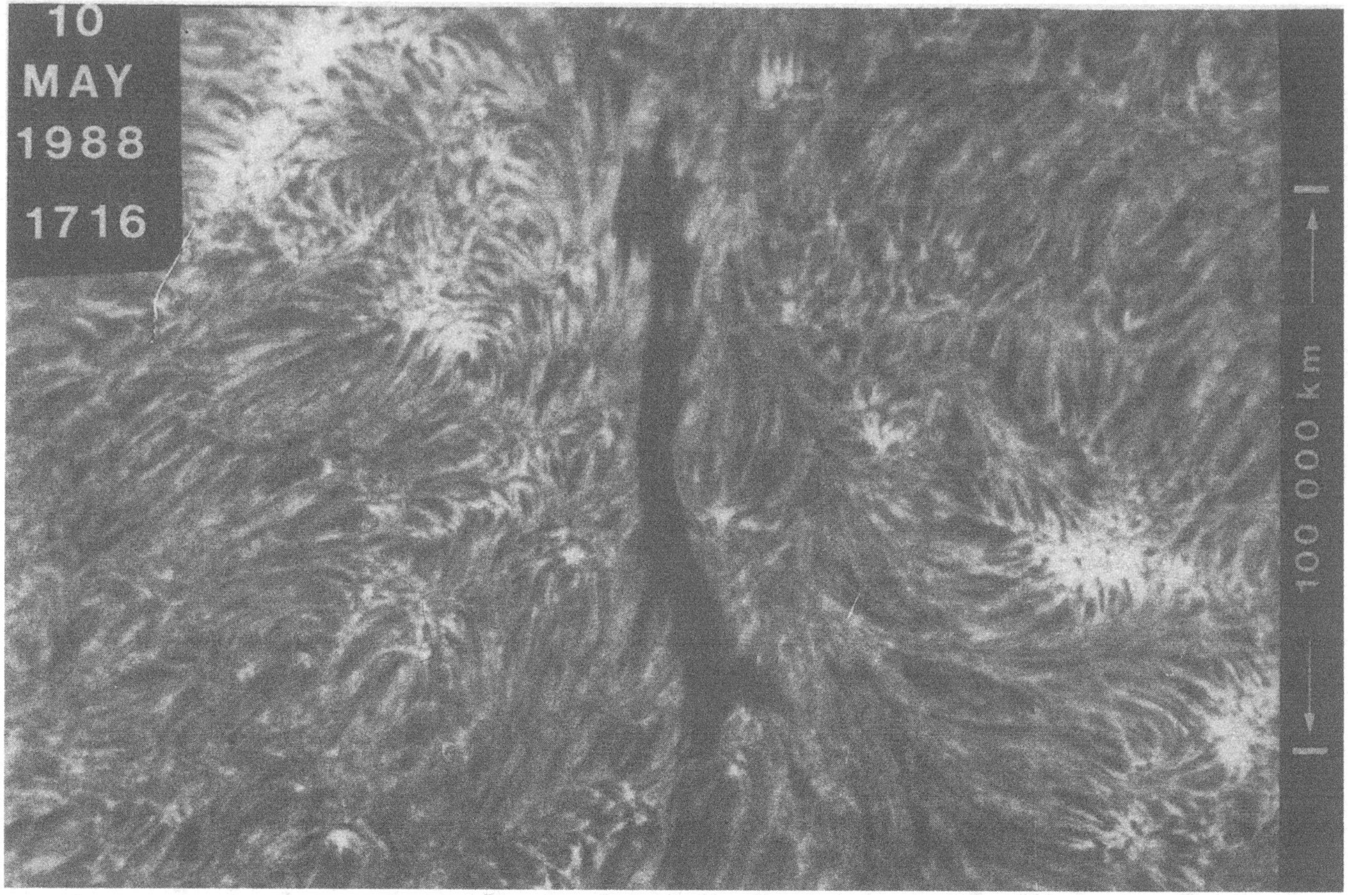

Figure 4. The direction of the fine structure within this filament can be compared with the direction of the fine structure of the adjacent chromosphere. Chromospheric fibrils below and adjacent to the filament lie nearly parallel to the fine structure within the filament and are approximately parallel to the long axis of the filament.

important point because authors sometimes discuss the circumstance of relative anti-parallel motions (also often called shearing motions) and ignore the circumstance of parallel motion which does not necessarily imply shearing motion. Different circumstances might result in different fibril patterns as clearly illustrated by Rompolt and Bogdan (1986). Fig. 5 is a composite of their illustrations depicting the two extreme cases of anti-parallel motion and parallel motion. In the anti-parallel case, the fibrils are curved in opposite directions on opposing sides of the polarity inversion zone. In the parallel case, the fibrils are curved in the same direction. (A case of parallel motion, resulting in the lengthening of the polarity inversion zone, is published in Martin, Livi and Wang, 1985). In both situations, the fibrils are evidence of a path of mostly horizontal magnetic field where filaments can form. The anti-parallel case implies that the magnetic field along the filament is unidirectional but in the parallel case, it seems that the magnetic field on opposing sides of a filament could be in opposite directions, if only temporarily. The lower left end of polarity inversion zone C in Figure 3 shows an interesting small zone of apparent anti-parallel, transverse fields. Gary and Hagyard (1990) discuss methods for resolving the 180 degree amiguity in this and other cases.

There are specific circumstances in which filaments or parts of filaments show no association with underlying fibrils. In general, extremely high filaments, surrounded by photospheric magnetic fields of low flux density, show little or no association with the underlying fibrils. An example is in Figure 6. Even for low-lying filaments, the condition of aligned fibrils must be considered a dominant rather than a necessary condition since exceptions are occasionally found where a filament is seen in projection against a plage where no fibrils are seen (Martin 1973). Also the condition of aligned fibrils is not obvious for small-scale filaments on the quiet sun. Either the associated magnetic fields are too weak or the polarity inversion boundary is too short for this condition aligned fibrils to become apparent. An arrow points to an example of such a small-scale filament in Fig. 6.

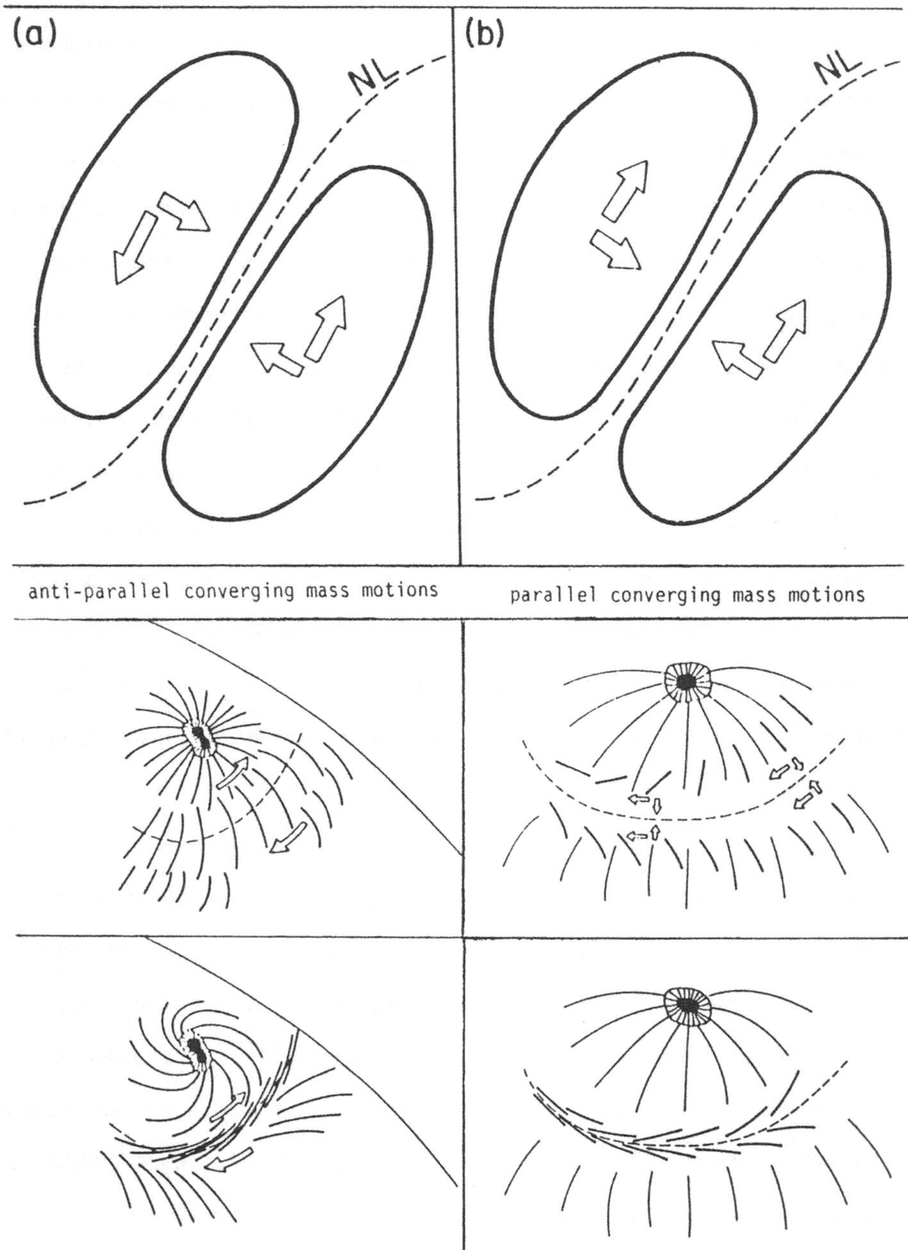

Figure 5. This diagram is a composite of diagrams in Rompolt and Bogdan (1986). It illustrates how the chromospheric fibrils can change direction as a consequence of motions of the photospheric magnetic fields related to the fibrils. Components of motions parallel and anti-parallel to the polarity inversion zone (NL), along with converging motion, can both result in the alignment of fibrils along the polarity inversion zone.

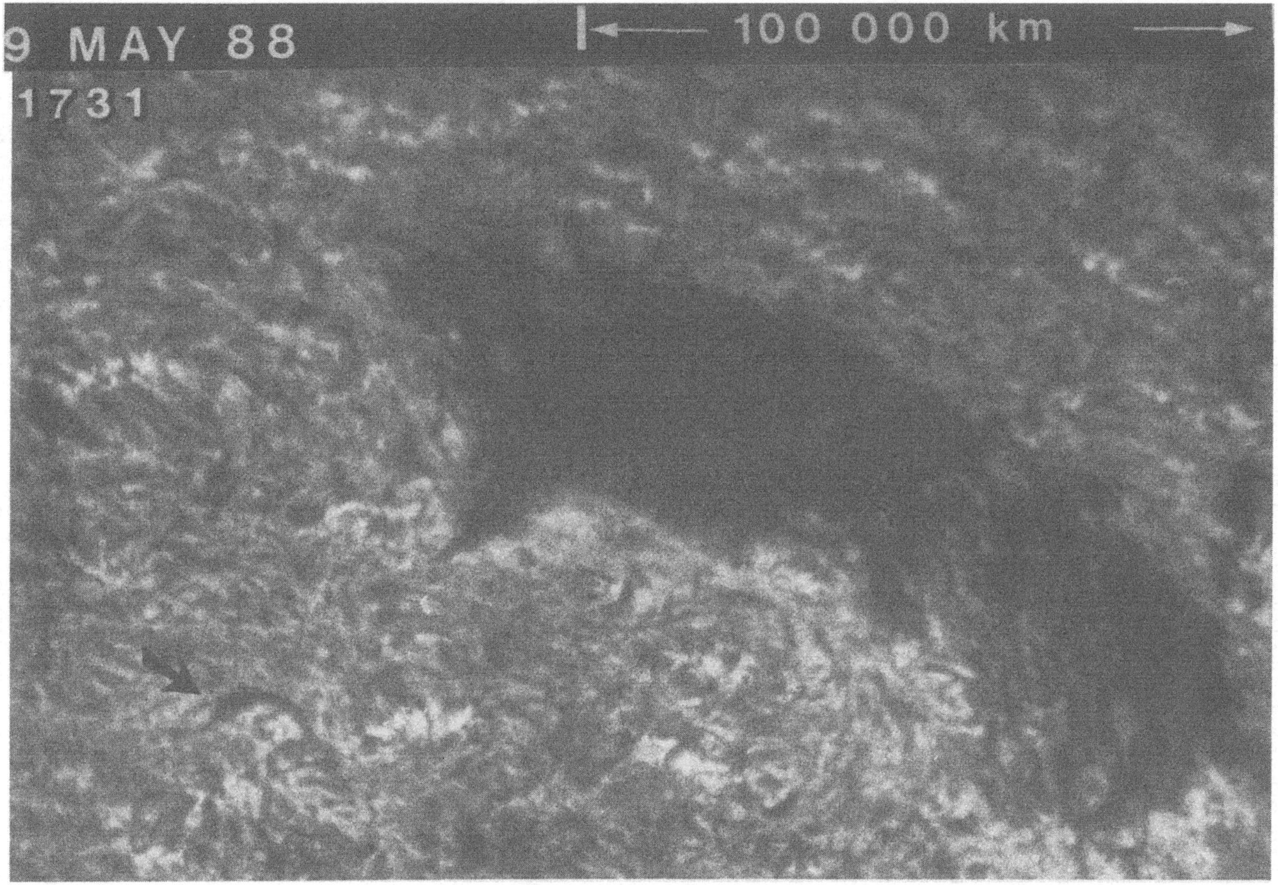

Figure 6. A large-scale filament and a small-scale filament (arrow points to it) related to polarity inversion zones between very weak magnetic fields. In such very weak fields, where there is also no obvious plage, the fibrils do not show the organized patterns that can be seen in Figures 1, 2, and 4.

Another circumstance where fibril alignment is often not seen, is at the polarity inversion boundaries that develop when a new active region (bipolar magnetic field) develops in a pre-existing active region or complex of active regions of high flux density. In these cases, sometimes only a single fibril separates (does not join) regions of opposite polarity. In one case, Gaizauskas (1990) observed the formation of a filament as a thickening and darkening of the previous fibril structure. Gaizauskas described the filament as an extra accumulation of mass at the site of a fibril.

In circumstances where it commonly occurs, the reorientation of fibrils prior to filament formation provides an important clue about filament formation: a special non-potential magnetic field configuration at the polarity inversion zone in the chromosphere and photosphere must develop.

2.4.2 The Filament Channel

The path in the chromosphere below a filament is called the 'filament channel'. A filament channel corresponds to a path of fibrils aligned along rather than across a polarity inversion zone. A filament channel does not necessarily have a filament above it at all times. For example, in Fig. 1 there is a gap in the filament; this is a temporary gap; on the previous day, the filament also occupied this part of the filament channel (Fig. 3, Martin 1986).

Filament channels correspond to polarity inversion zones but not all polarity inversion zones correspond to filament channels. For example in the right side of Fig. 2, most of the polarity inversion zone, excepting the lower part, has neither a filament channel nor a filament. A long filament would not be expected to form at this location unless the fibrils extending between the opposite polarity fields would first change direction by about 90 degrees such that they no longer suggest a direct connection between the opposite polarity fields.

Filament channels are often recognizable in H-alpha filtergrams taken in the wings of the Ha line out to approximately .5A from the center of the line. In active regions observed in the wing of the line, the filament channel is usually darker than the average adjacent background where there is plage. This is probably due to the presence of mostly horizontal fields and the general absence of line-of-sight magnetic fields in the filament channel. Because there is also little or no line-of-sight fields in the channels of quiescent filaments, there is a marked reduction of spicules in the filament channel compared to the surrounding area.

The width of the filament channel is variable in space and time. The width at any given location apparently depends on the magnetic flux density adjacent to the channel. The widest channels are associated with sites of low magnetic field gradient; narrow channels correspond to sites of high magnetic field gradient. Where the magnetic field gradients are extremely high in active regions, the filament channel is only a line, sometimes referred to as the 'neutral line' or '0 line' when referring to magnetograms of only the line-of-sight component. At the boundaries of active regions and in between adjacent active regions, the development of the filament channel precedes appearance of filament mass in H-alpha.

2.4.3 Conditions of Prominence Reformation

It is well known that solar flares occur in association with the eruption of filaments. It is important to recognize that flare loops and filaments cannot co-exist in precisely the same space and that each has distinctly different spatial associations to the overall magnetic field geometry in the vicinity of a polarity inversion zone. Occasionally flare loops develop above filaments. More often, the prominence erupts and then flare loops temporarily exist in the space previously occupied by the filament (Martin, 1979). As a flare decays, the loops form at successively higher altitudes; a filament can then develop under the flare loops in the same location as the previous filament.

Tang (1986) has further shown that many filaments, at the time of eruption, have divided into two sections or layers. In these cases, only the upper section erupts; the lower layer remains unchanged in direction. Such cases show that the basic magnetic field geometry necessary for the reformation of a filament still exists after an eruption. This is consistent with many observations showing that, in active regions, the reformation of a new filament is often seen to begin within less than an hour after an eruption and during the decay of solar flares. (See recent examples in Martin, 1989; Gaizauskas, 1990.) On the quiet sun, the reformation of quiescent prominences is typically slower; involving many hours or several days (d'Azambuja and d'Azambuja, 1948). The average times between quiescent eruptions were found to be 5-8 days (Serio et al., 1978). Such observations indicate three important pieces of information about prominence formation: (1) the conditions for prominence formation are not destroyed by eruption, (2) the formation of prominences is probably a continuous process, and (3) the process of formation is more rapid at polarity inversion zones bounded by areas of high flux density than by low flux density.

2.5 The Condition of Converging Magnetic Fields

Recent observations of magnetic fields under and adjacent to filaments (Martin, Livi and Wang, 1985; Martin, 1986; Hermans and Martin, 1986; examples in the paper), reveal a common denominator among filament sites; they occur where small, apparent fragments of magnetic field of opposite polarity slowly flow and converge toward a common boundary. Because large-scale solar magnetic fields tend to expand with time (Leighton, 1964; Stenflo, 1972; Mosher, 1977; Howard and LaBonte, 1981), converging flows of opposite polarity fields are expected under many circumstances. Converging patches of magnetic fields of opposite polarity are commonly found between active regions, at the borders of growing active regions, along the polarity inversion zones within active regions, between the decaying active regions and between their opposite polarity remnants near the solar poles. Prominences often form at all of these various boundaries where small, opposite polarity patches of magnetic flux encounter one another. Hence converging flows might be a necessary condition for prominence formation.

Diverging flows of opposite polarity patches are probably rare. The possibility of diverging magnetic fields are seldom mentioned in the literature except in terms of global solar patterns of magnetic flux. There is an absence of information on whether prominences are ever related to diverging magnetic fields.

2.6 The Condition of Cancelling Magnetic Fields

Shelke and Pande (1983) and Maksimov and Ermakova (1986) observed that quiescent prominences tend to form where the magnetic field gradients across polarity inversion zones diminish from day to day as seen in full disk magnetograms or in synoptic maps made from full-disk magnetograms. Magnetic field gradients across a polarity inversion zone can decrease either because of diverging flows or because the magnetic flux disappears in the polarity inversion zone. Because divergence flows are rare, one should suspect that the observed lowering of the average magnetic field gradient along large-scale polarity inversion zones, might be due to the disappearance of magnetic flux. Mention of the expected, long term (day-to-day or month-to-month) reduction in magnetic flux around prominences has been made by Mosher (1977) and Zwaan (1978) who anticipated that line-of-sight magnetic fields must somehow disappear under prominences. Zwaan further speculated that magnetic reconnection near prominences could be a prerequisite for magnetic flux to be removed via the downward transport of field lines through the photosphere.

Direct, confirming observations of the cancellation of opposite polarity magnetic fields under a filament and prior to filament formation were first made by Martin, Livi and Wang (1985). They defined the term 'cancellation' to be 'the mutual loss of magnetic flux of opposite polarity at a common boundary as seen in magnetograms of the line-of-sight component'. Cancellation was observed at the polarity inversion zone for 2 days prior to the formation of the filament mentioned in Martin, Livi and Wang (1985) and described more completely by Martin (1986). Converging flows and cancellation of opposite polarity magnetic flux were the only dynamic magnetic field changes observed under the filament and during its formation.

An example of the formation of a small quiescent prominence was also illustrated by Martin (1986). Again the formation was accompanied by convergence and cancellation of magnetic fields immediately below the prominence.

Small-scale cancelling magnetic fields on the quiet sun are often present due to the formation and decay of numerous small active regions and ephemeral active regions. Hermans and Martin (1986) showed that the small-scale filaments form and erupt on the quiet sun at rates of hundreds per day. They found the formation of these filaments to be related to the occurrence of cancelling magnetic features and to be analogous to large quiescent filaments.

The cancellation of magnetic fields under filaments in decaying active regions has now been observed many times in videomagnetograms taken at the Big Bear Solar Observatory . High rates of cancellation, in areas of moderate to low flux density, tend to be accompanied relatively soon by the eruption of a filament and an accompanying flare. If cancellation continues, the immediate reformation of the filament is expected and subsequent eruptions and reformations of the filament along the same polarity inversion boundary are both possible and likely. Conversely, low rates of cancellation are associated with long-lived filaments.

3.0 AN EXAMPLE OF THE FORMATION, DEVELOPMENT, ERUPTION AND REFORMATION OF A FILAMENT

Figures 7 through 13 illustrate a 7-day sequence of H-alpha filtergrams and line-of-sight magnetograms. In this series, the H-alpha filtergrams depict various aspects of the formation of a filament: its growth in length, its eruption with a flare, its reformation, and its further growth in length. The magnetograms show the convergence and cancellation of magnetic flux throughout the entire sequence.

The long-term formation and day-to-day evolution of this filament is shown in Fig. 7. Consecutive images are displayed in columns from left to right. In the first image on 27 April 1988, there is no filament. The area of bright plage in the middle of the image is divided into two major sections by a series of fibrils called 'field transition arches' because they appear to join opposite polarities when compared with magnetograms showing the line-of-sight component. The filament begins to form on the next day, 28 April, at the lower end of the bright plage. It is a conspicuous filament by the following day. Until 2 May the filament continues to grow in length at both ends. On 2 and 3 May, a short adjoining filament develops in the plage at the upper end of the large filament.

Fig. 8 shows the formation of the filament on 28 April 1988. In the first 2 frames on the left, the newly developing filament appears like a single fibril. However, by 2112, the small filament is darker and longer than the fibril structures in the adjacent plage.

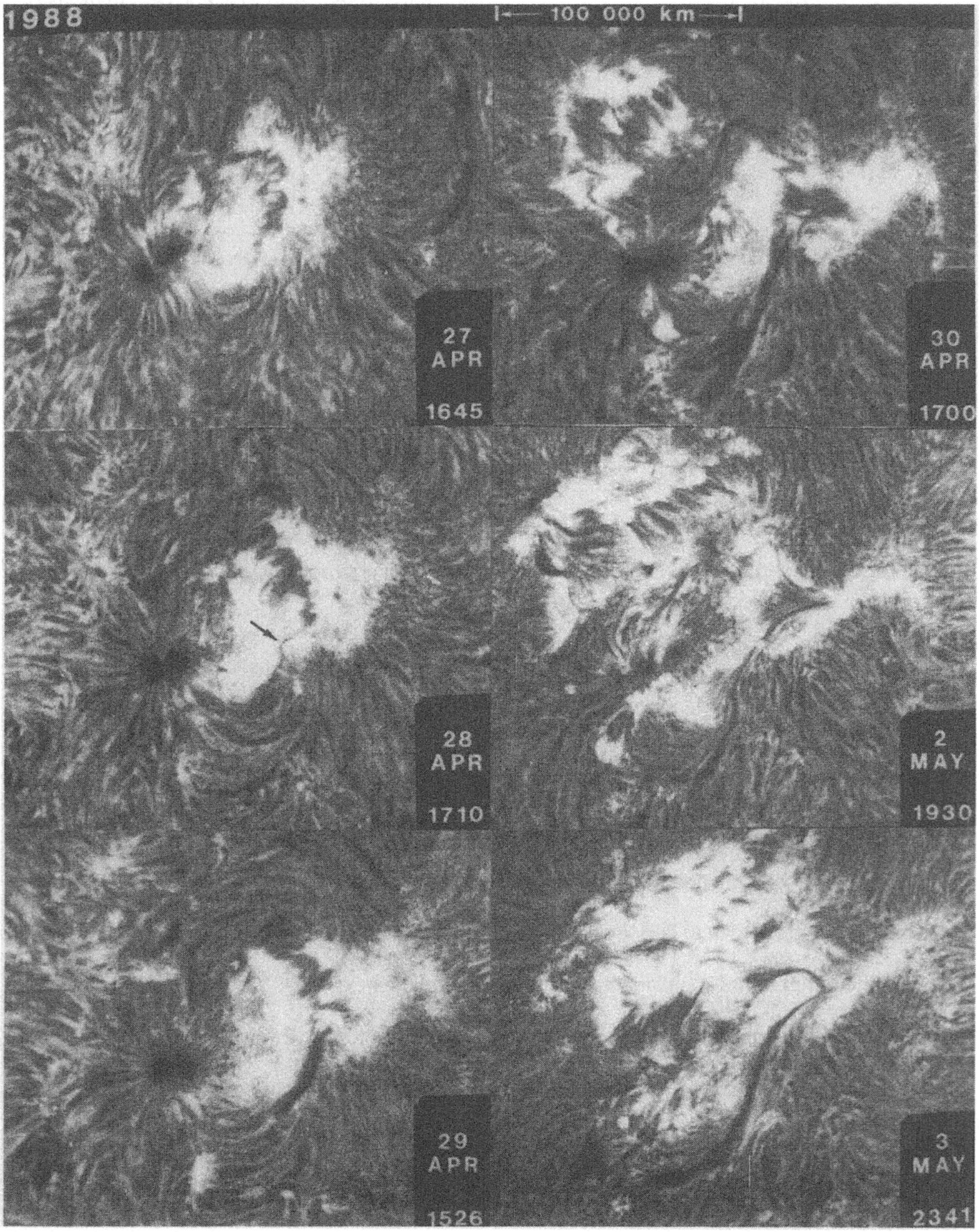

Figure 7. Day-to-day sequence showing the formation and evolution of a filament. The filament begins to form on 28 April and extends initially along the border of the active region following the direction of the chromospheric fibrils. As the field transition arches disappear in the middle of the bright plage, the filament slowly extends into the active region and by 3 May, joins another newly-formed small filament.

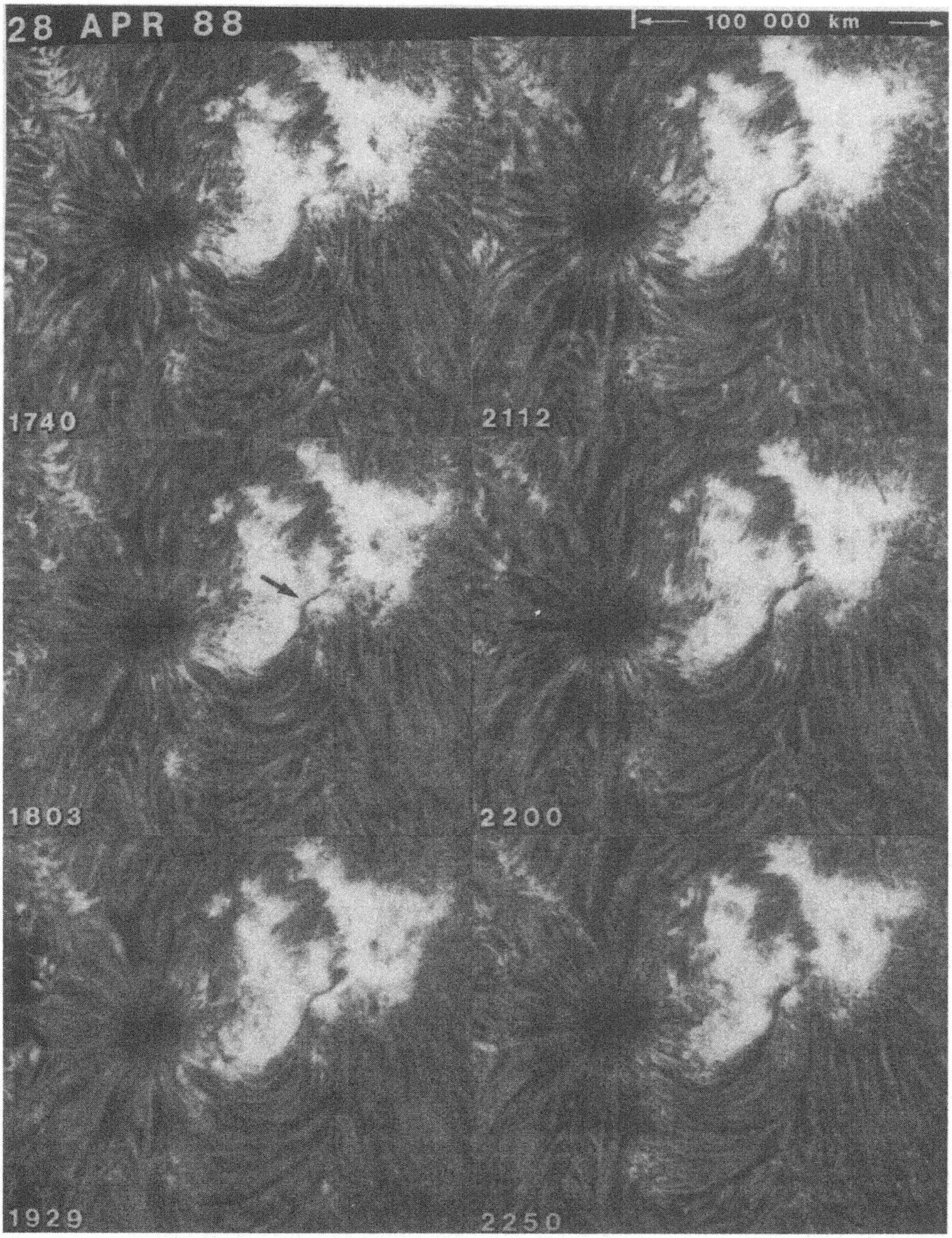

Figure 8. The slow formation of the filament begins by 1803. It extends upward a small distance into the plage by 1929 but thereafter it grows in the opposite direction along the border of the bright plage where its direction is already defined by the pre-filament fibrils.

Fig. 9 shows some of the same development in the filtergrams as in Fig. 8 in comparison with magnetograms. The magnitude of the magnetic flux is represented in two forms: gray-scale and contours. Shades from medium gray to white are positive polarity and shades from medium gray to black represent negative polarity. When the digitizer reaches saturation, the color is programmed to reverse; white becomes black and black becomes white. This change introduces the apparent contours in the images. Increasing numbers of contours represent increasing magnetic flux. The polarity is still represented by the color, black or white, outside of the lowest contour.

The arrows on the magnetograms in Figure 9 point to the small negative polarity patches of magnetic flux that are seen to cancel with adjacent positive polarity magnetic flux. At 1628, the first small filament structure already is aligned along the polarity inversion zone and has one of its endpoints between the patches marked N_0 and the positive polarity field to the left; the other apparent endpoint is between N_1 and the positive polarity field to the left. By 2029, N_0 has split into several smaller patches which have spread parallel to the polarity inversion zone. Thus, the polarity inversion zone has elongated toward the lower part of the frame. Patch N_1 steadily becomes smaller throughout the day because it is cancelling with the neighboring positive flux which initially shows a local maximum with a second level contour. As the two patches cancel, the reduction in the positive flux is distinguished by the slow disappearance of the positive-polarity, second-level contour. The cancellation of the negative polarity is seen from the concurrent disappearance of the second level and then the first level contours of N_1. This example of cancellation is typical. When the resolution of magnetograms is good enough to reveal fine structure, the cancellation is then identifiable only in small patches of magnetic flux which are close together. It is seen that the fragments of N_0 also decrease in area (flux) throughout the day; N_0 loses its only first level contour. The decrease in the positive flux is not conspicuous next to N_0 in this short

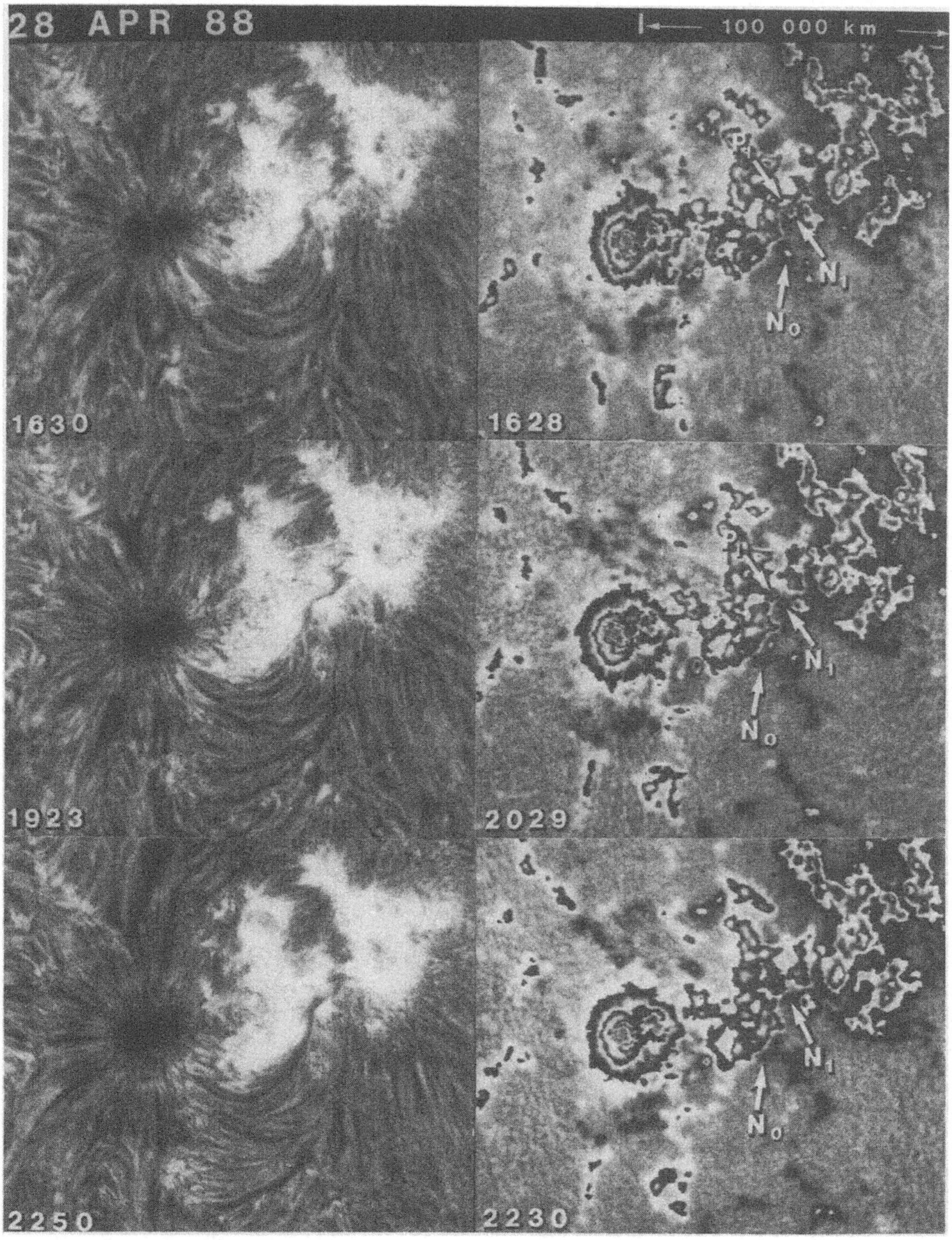

Figure 9. The magnetograms on the right show where the filament devlops with respect to the line-of-sight magnetic fields. The filament grows both northward and southward along the polarity inversion zone starting between the magnetic field patches labelled P_1 and N_1 which cancel throughout the day. Another cancellation site exists between N_0 and the adjacent positive polarity field. The filament grows most in the direction of this cancelling site.

interval because the percent of change is too small to be obvious. From studying many such examples of cancellation over longer time intervals, we know that both polarities always decrease simultaneously although the decrease in flux is sometimes only conspicuous in the smaller patches. To summarize, Fig. 9 illustrates the filament becoming longer and darker as the underlying patches of magnetic flux converge and cancel. The several interpretations of cancellation are discussed in Zwaan (1985, 1987).

Figure 10, left side, shows the continued growth of the filament during the interval from 28-30 April. It is seen that the filament extends towards the lower parts of the frames where the polarity inversion zone is broad and the fibrils are already aligned parallel to the long axis of the filament. In the magnetograms on the right, the magnetic flux is seen to change along the polarity inversion zone. Between the first two magnetograms on 28 and 29 April, the negative polarity patches that were adjacent to the positive polarity flux on 28 April, completely cancel and are replaced by other patches of flux that have migrated towards the polarity inversion zone. These patches also cancel; the result is an average localized broadening of that part of the polarity inversion zone. The lower end of the filament grows much more than the upper end and it is seen that the filament tends to be narrow where the magnetic field gradient between opposite polarities is high and broad where the magnetic field gradient between opposite polarities is low. However, the magnetic field gradient is not static; it is constantly changing everywhere along the polarity inversion zone due to both the convergence and cancellation of magnetic flux.

In the first frame in Fig. 11, it is seen that the fully developed filament still has approximately the same configuration that it had on 30 April as seen in Fig. 7 and Fig. 10; however, it is darker, and on average, broader along its whole length; these are signs of impending instability. Indeed by 1555, the eruption of this filament is underway. Although a small component of motion is seen to the lower right during the eruption, most of the mass of

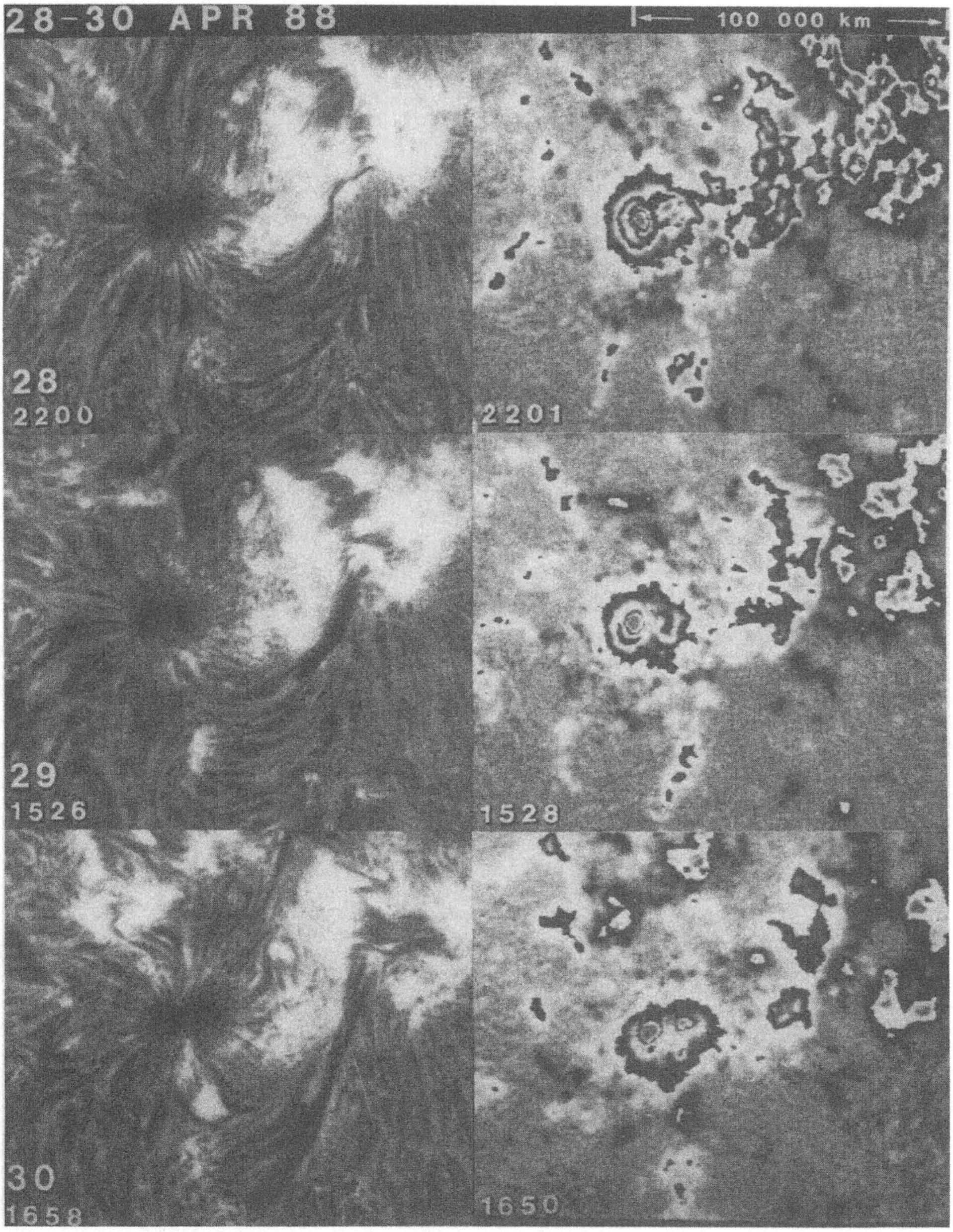

Figure 10. The filament grows wide where it is bounded by the strong fields related to plage on the left and weak background fields on the right. On these successive days, the new cancellation sites develop beneath the growing filament because patches of magnetic flux migrate toward the polarity inversion zone as the older cancellation sites disappear.

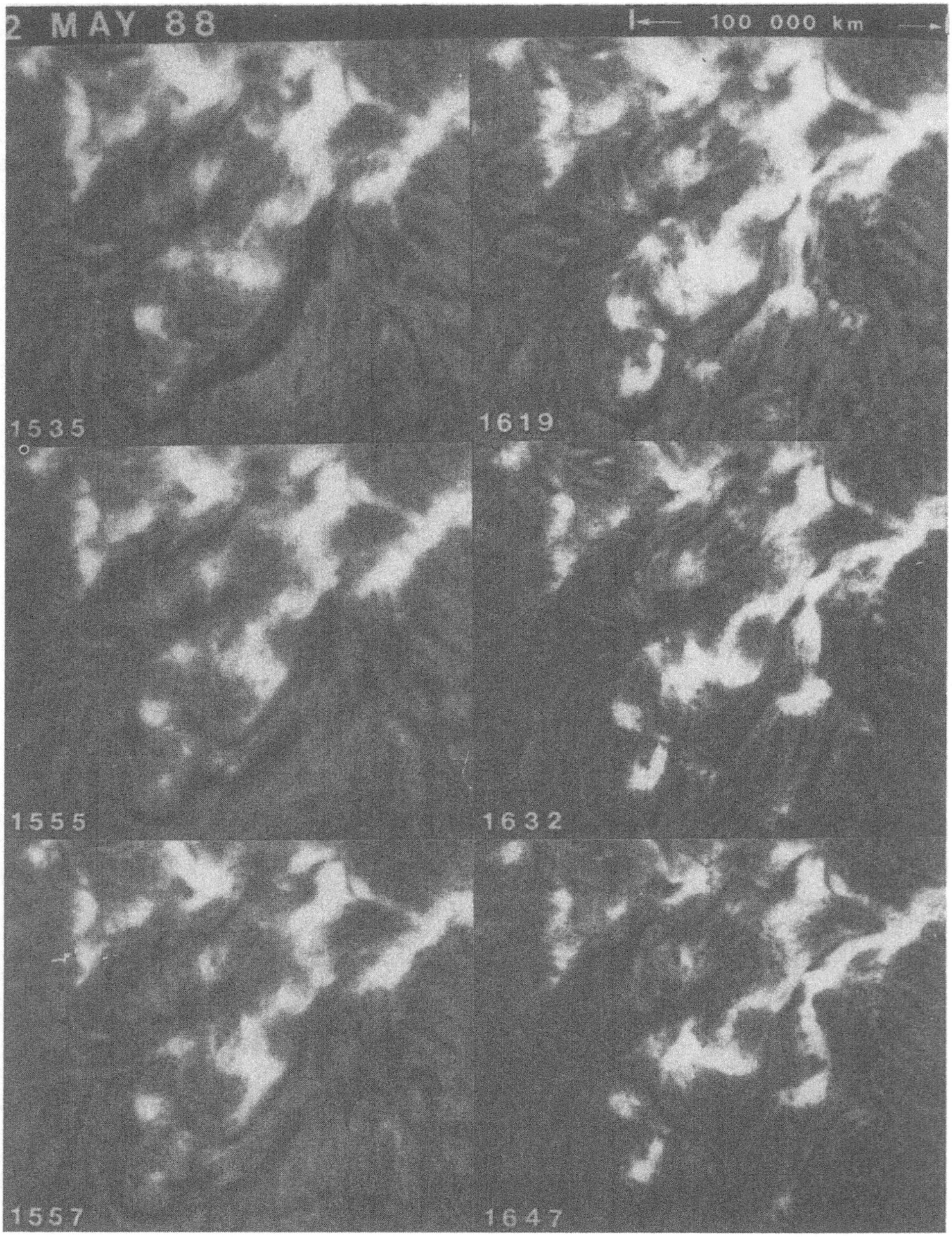

Figure 11. The fully developed filament becomes unstable and erupts outward. Due to its high line-of-sight velocity component, the filament is Doppler-shifted out of the 1/4 Angstrom passband of the H-alpha filter. A small component of motion in the plane of the sky can be seen towards the bottom of the frames until 1557. A flare is develops on the chromosphere on both sides of the site of the erupted filament. The reformation of the filament already begins during the flare.

the filament is moving outward in the line of sight as confirmed by observations in the blue wing of the line (not shown here). The chromospheric part of the associated flare is seen to develop rapidly between 1557 and 1619. It is also seen that a low, narrow part of the filament does not erupt. Many similar examples, in which only the upper part of a filament erupts, have been discussed and illustrated by Tang (1986). The lower part of the filament immediately begins to rebuild even before the flare ends.

During flares, there are no rapid changes in the line-of-sight component of the observed photospheric magnetic flux (Livi et al. 1989). Also, there are no obvious short-term changes in the rate of convergence and cancellation before, during, or after flares. H-alpha observations of the eruption of filaments, the formation of flare loops, and other associated events indicate that dramatic changes in the magnetic fields do take place during solar flares but it is now apparent that these changes take place in the corona where routine magnetic field measurements cannot yet be made.

The reformation of the filament is illustrated in Fig. 12 throughout a two-day interval. By 3 May, the filament is fully redeveloped. It is a matter of definition and interpretation whether one chooses to call this process, the reformation of a filament or the building of a new filament at the location of a previous filament. I shall call it 'reformation'; the data indicate that the process of formation simply continues unabated throughout the eruption and flare. The reformation is seen as the spontaneous appearance of long narrow threads which appear to flow as they become visible. The flowing motion is in the direction of the long axis of the filament and the adjacent fibrils. The appearance of new filament threads continues until the filament is entirely rebuilt.

Fig 13 illustrates the changes that take place in the line-of-sight magnetograms during the reformation of the filament. On 2 May, there is only one conspicuous cancellation site; in

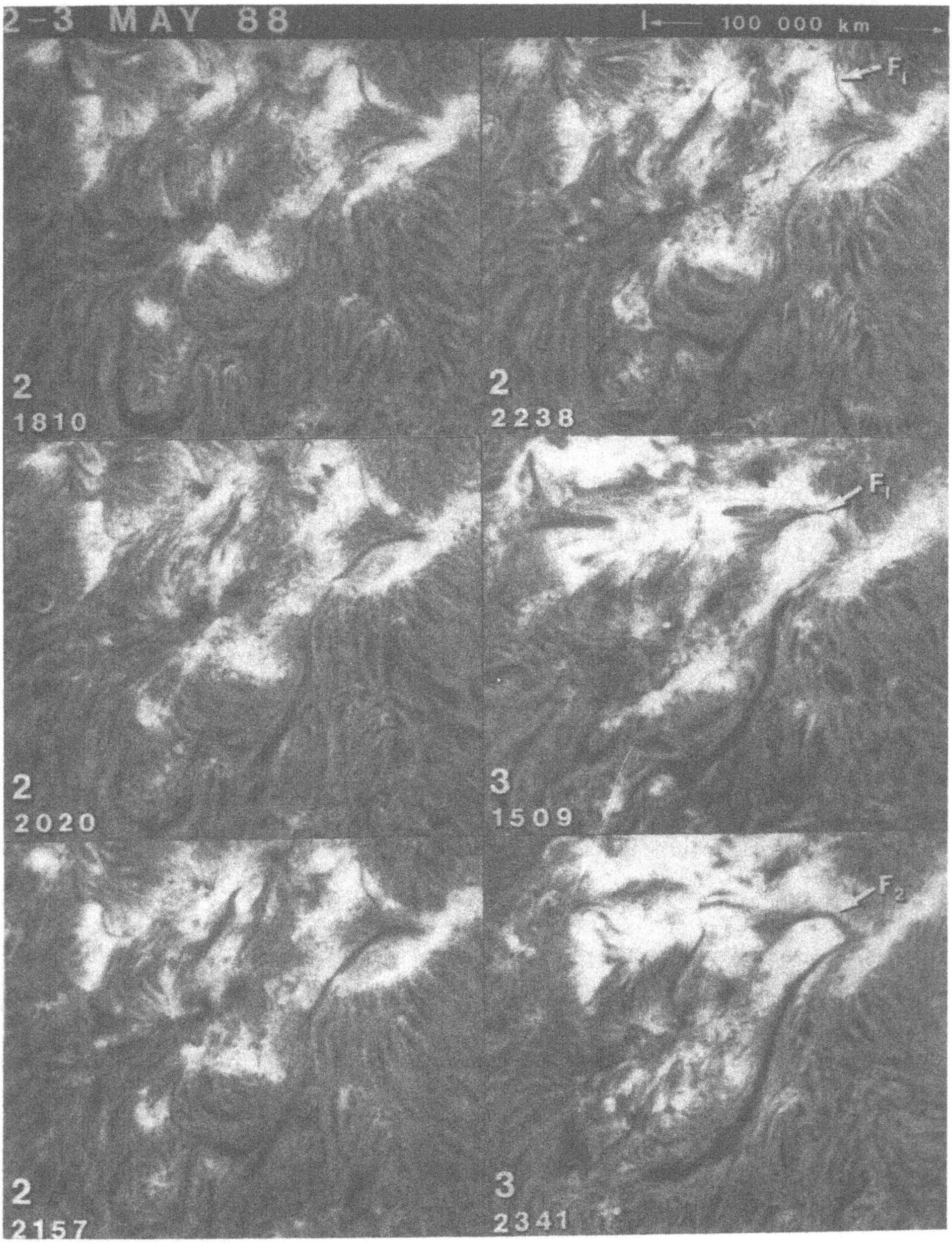

Figure 12. Following the flare in Figure 11, the filament reformation is more rapid than its initial formation. Within one day, by 3 May, 1509 (on the right), the filament has completely reformed. It joins a newly-formed filament, F_2, in the plage. Small filaments F_1 and F_2 form where magnetic flux from a new, adjacent active region is converging upon the plage of the pre-existing active region.

the H-alpha filtergrams, it corresponds to the plage at both sides of the narrow, upper end of the filament. Cancellation is seen at this site from 2258 on 2 May until 2341 on 3 May. On 3 May, magnetic flux of opposite polarity migrates together to form a new cancellation site in the polarity inversion zone above the upper end of the filament. During this time, the filament grows a new segment which extends from the previous end of the filament to the new cancellation site.

In addition a new small, adjoining filament, F2, forms above the upper end of the long filament. The formation of F2 is accompanied by cancellation at the upper part of the frame and by the migration of positive polarity flux to the upper right toward the negative flux. This migration brings additional opposite-polarity flux into contact where it can cancel. The formation of the filament is thus accompanied by convergence, cancellation and a steepening of the magnetic field gradient at the polarity inversion zone where the filament forms.

In Figure 13, more images of the reformation are shown. The migration of flux seen in the magnetograms can also be seen in H-alpha plage. Referring back to Fig. 10, a new active region is seen to develop in the upper left part of the images. During 2 and 3 May (Fig. 12 and 13), the new active region is continuing to grow. The expansion of the magnetic field of the new active region toward the right, possibly accelerates the cancellation and convergence of the magnetic flux above and to the left of the small, new filament. It is probably not a fortuitous circumstance that filaments form rapidly in active regions, where convergence and cancellation rates are high, and form more slowly between decaying active regions where convergence and cancellation sites are fewer in number.

The cancellation of magnetic flux has now been observed so often before filaments form, during their lifetimes and before their reformation that I propose that cancellation is a necessary condition for prominence formation. However, various interpretations of cancellation need to

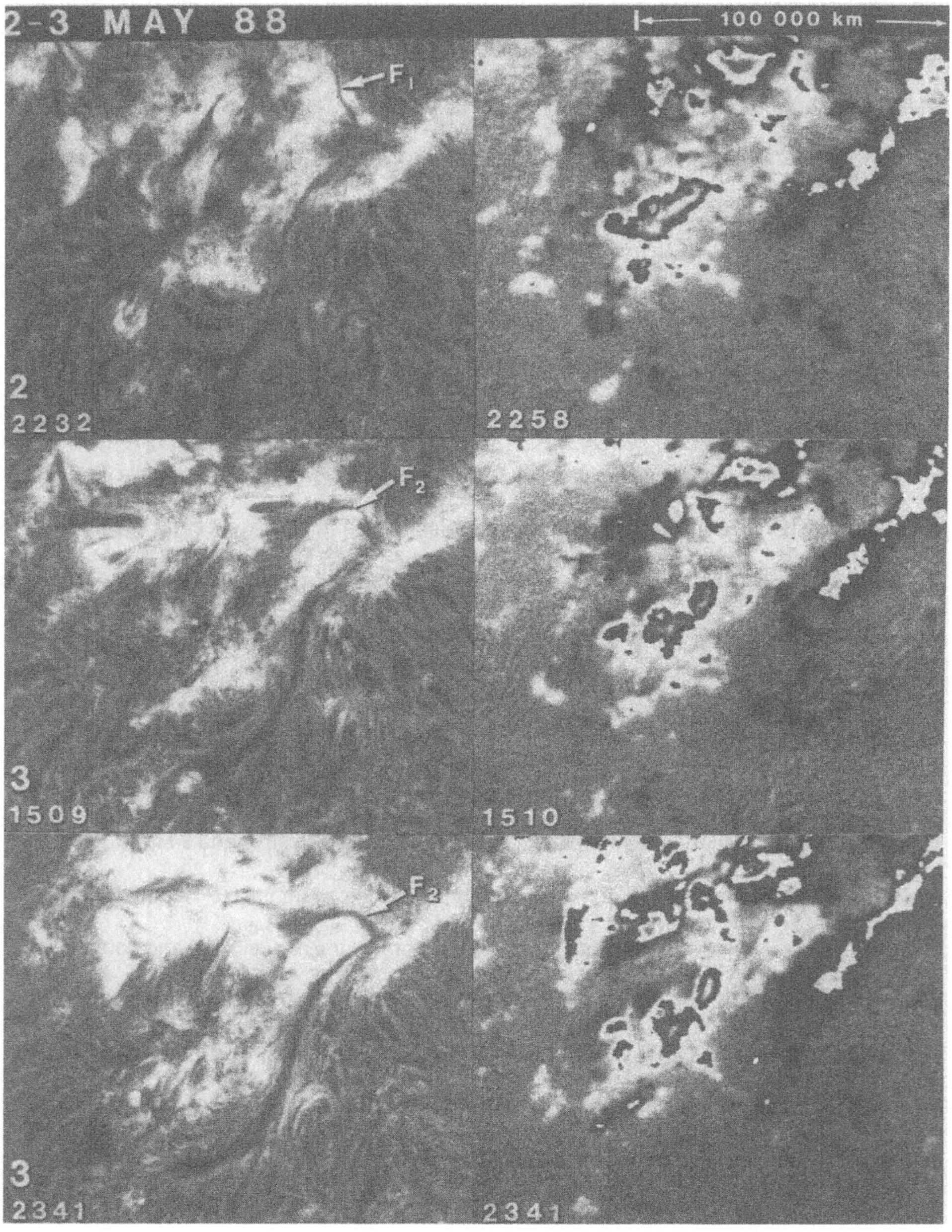

Figure 13. Small filament, F_1, disappears because its polarity inversion zone disappears due both to cancellation and migration of the magnetic flux on both sides of it. The new filament, F_2, develops along the nearby, changing polarity inversion zone. The arrow points to the primary cancellation site associated with the formation of F_2.

be considered in order to understand of the physical significance of this phenomenon to the formation of filaments. The several ways that magnetic flux can disappear via transport through the photosphere are discussed and illustrated in Zwaan (1985, 1987).

4.0 DISCUSSION

4.1 Necessary and Sufficient Conditions for Prominence Formation

To simplify the discussion, the six conditions discussed in the previous sections will be abbreviated to two words as follows, with the subsequent words in parentheses to be understood throughout this discussion:

(1) opposite polarities (on opposing sides of a prominence referring to the line-of-sight component)

(2) overlying arcade (in the corona rooted in areas of dominantly unipolar, unipolar magnetic flux of opposite polarity)

(3) transverse fields (in the photosphere below a prominence having an extreme non-potential configuration in the polarity inversion zone that is approximately perpendicular to the maximum magnetic field gradient between adjacent patches of opposite polarity, line-of-sight field)

(4) aligned fibrils (in the chromosphere, parallel to the transverse magnetic field component and parallel to the long axis of a prominence)

(5) converging fields (of opposite polarity toward one another, line-of-sight fields)

(6) cancelling fields (in the line-of-sight magnetic component only)

These six conditions are found from several types of observations and are not all independent conditions. The condition of opposite polarities (1) is implicit in the condition of the overlying arcade (2). The conditions of transverse fields (3) and aligned fibrils (4) are synonymous conditions seen respectively in H-alpha filtergrams and magnetograms. These two conditions, (3) and (4), are probably consequences of either or both of the dynamic conditions: convergence (5) and cancellation (6). It is clear from the observations that cancellation cannot occur unless convergence first takes place or unless new magnetic fields of opposite polarity grow in juxtaposition.

Considering all of the above interrelationships, none of these conditions by themselves are sufficient for prominence formation. The necessary conditions are might be the following combination:

(1) an overlying arcade

(2) convergence and

(3) cancellation

The existence of the overlying arcade can be viewed as a relatively static condition since such arcades can be very long-lived. However, convergence and cancellation are dynamic conditions; many patches of magnetic flux of opposite polarity can encounter one another and cancel completely both during the formation and continued life of a prominence. I hypothesize that, under the environment of an overlying arcade, the existence of a prominence depends on rates of convergence and cancellation as well the duration of the convergence and cancellation. The minimum rate of convergence and cancellation for sustaining a prominence might be very low because quiescent prominences, especially polar crown prominences, exist where the flux density is very low and the number of cancelling features are only a few per day. Quantitative measures of cancellation under various circumstances are needed to determine if there are such suggested thresholds for prominence formation and sustenance.

The magnetic field geometry of a prominence and its environment, as deduced from all of the observations cited in this paper, is depicted in the schematic drawing in Fig. 14. In this depiction, the prominence does not share magnetic field lines with the overlying arcade. Some of the magnetic field lines previously belonging to the inner part of the arcade have been reconfigured into the magnetic field lines within the prominence possibly by magnetic reconnection. It has not been observed how this happens; this is a deduction from the observations of small patches of converging and cancelling magnetic flux. Two recent models of prominences suggest different ways that this reconfiguring might take place (van Ballegooijen and Martens, 1989; Kuijpers, 1990). In both models, the assumed magnetic field reconnection is a continuous process that sporadically occurs along a polarity inversion zone. Under quiescent prominences, as depicted in Fig. 14, there are relatively few sites of cancellation at any given time. However, because of the cancellation, the flux density of the line-of-sight component is less under and adjacent to the prominence than at the footpoints of the majority of the arcade.

If the overlying arcade, convergence, and cancellation are all required for prominence survival, then it would follow that the removal or cessation of any of these conditions would result in the destruction of a prominence. The destruction of a prominence might occur by several possible means: (1) the draining of all of the mass out of a prominence if the convergence and cancellation rates are too low, (2) eruption, if the overlying field is removed or reconfigured due to reconnection with external magnetic fields on the sun or (3) the build-up of magnetic fields or electric currents in a prominence to the extent that the overlying arcade becomes unstable, expands outward and becomes a coronal mass ejection. The commonplace erupting prominence might be a consequence of either (2) or (3) above. These ideas on the formation, sustenance, and destruction are open to testing by current observational techniques with existing instrumentation.

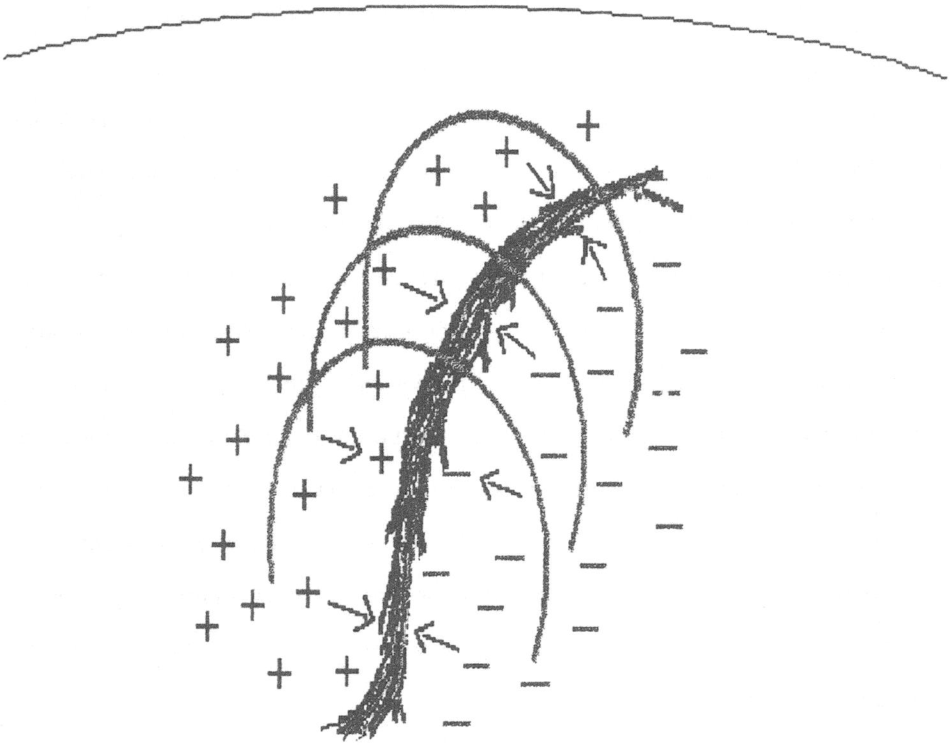

Figure 14. This schematic represents the geometry of a filament with respect to the arcade that connects the adjacent, unipolar magnetic fields of opposite polarity. The magnetic field in the filament is primarily transverse and is directed mostly along the filament at a large angle with respect to the overlying arcade. At present, there is no evidence that the magnetic field lines in a filament are directly connected to the overlying arcade. This paper presents evidence that the magnetic field in filaments is closely associated with the convergence and cancellation of small, discrete fragments of magnetic flux under or immediately adjacent to the filament. The convergence is depicted by the arrows. Cancellation is represented only by the reduced magnetic flux density in the polarity inversion zone.

4.2 Shearing Motion - A Clarification

A discussion of prominence formation in the context of present-day research would not be complete without some mention of a possible role or lack of role for shearing motion. Shear or shearing motion can be defined as motion of the patches of one polarity along a common boundary with opposite polarity fields. The shearing component of motion can be anti-parallel or parallel (Fig. 5) but implies that the motion of the footpoints of one polarity is either greater or lesser than the adjacent polarity or in opposite directions. Both large-scale and small-scale shearing motion have been invoked in some models of prominences. In Figs. 6-12, it is shown that small-scale shearing motions sometimes accompany the converging flow of opposite polarity fields toward a common polarity inversion boundary. However, in these observations, the shearing motion, in general, is secondary to the converging component.

Significant shearing motion of the photospheric magnetic field has been observed in some circumstances and not necessarily in conjunction with the formation of a filament (Kurokawa 1987; Zirin and Wang, 1989). Gaizauskas (1990), however, has shown an example of a filament formation in an active region in which shearing motions were present. Such motions might have important consequences, especially to flare occurrence or to the rate of formation of filaments. However, in some clear examples of filament formation (Martin, Livi and Wang 1985; Martin 1986 and examples in this paper) large shearing motions were not observed.

A distinction needs to be made between the usage of the term 'shear' to refer to a velocity pattern, as above, and another usage of the word 'shear' to refer to solar magnetic field configurations. All filaments, as demonstrated in this paper, (and flares - Moore, Hagyard, and Davis, 1987) are associated with a large-scale magnetic field configuration that is often described as a 'sheared' configuration. This usage of the term 'shear' is ambiguous for two reasons. First, it can imply either of two magnetic field geometries: (1) situations in which the field lines between opposite polarities develop an S-shaped kink or (2) situations in which a large angle exists between field lines in the vicinity of prominences and and those of the overlying arcade. Secondly, the term 'shear', when used in the context of magnetic field configurations, is confusing because the so-called 'sheared configuration' in most of the observed circumstances cited above, is not a consequence of large-scale shearing motion. Additionally, small-scale shearing motions alone can not result in the observed large-scale 'sheared configuration'. To account for the prominence configuration small-scale shearing motions have to be accompanied by a substantial converging component of motion and possibly other magnetic field dynamics such as magnetic reconnection as suggested in the theoretical models of Van Ballegooijen and Martens (1989) and Kuijpers (1990).

One suggested way to avoid this confusion of terms would be to use the term 'shear' to only refer to the specific type of velocity pattern that is defined in other disciplines such fluid dynamics or geophysics. Except when quoting earlier papers, other terminology can be used to describe magnetic field configurations.

5.0 CONCLUSIONS

Prominences form above polarity inversion zones (as seen in magnetograms of the line-of-sight component) only where all of the following observational conditions are met:

(1) an arcade of coronal magnetic fields joins the adjacent opposite polarity photospheric magnetic fields, and

(2) small patches of opposite polarity fields flow into juxtaposition at discrete locations along the polarity inversion zone, and

(3) the encountering, small patches of opposite polarity fields cancel (disappear concurrently at their mutual boundaries)

These conditions applyy to the entire spectrum of prominences: small-scale prominences which form and erupt within a few hours on the quiet sun, active region prominences, prominences between active regions, quiescent prominences, and polar crown prominences.

It is hypothesized that the above three conditions are necessary and sufficient for prominence formation. Conditions (2) and (3) somehow result in the formation of a transverse field that is approximately perpendicular to the direction of convergence. Does this happen via magnetic reconnection?

The direction of the transverse component of the magnetic field is parallel to the fibril structure of the chromosphere. The fibrils where prominences form always divide rather than join adjacent magnetic fields of opposite polarity. However, prominences do not form from the fibril structure; instead they are a dynamic composite of fine structures that form and decay as seen superposed against the background fibril structure.

New observational and theoretical works are needed to understand all of the physical processess involved in the formation and maintainance of and prominences.

Acknowledgements

The author gratefully acknowledges support from the Air Force Office of Scientific Research under Contract AFOSR-87-0023. The acquisition of H-alpha filtergrams and magnetograms at the Big Bear Solar Observatory is made possible by NASA grant NGL05002034 and NSF grant ATM-8513577.

REFERENCES

Avignon, Y., Martres, M.., Pick, M.: 1964, *Ann. Astrophys.*, **27**, 23.

Babcock, H.W. and Babcock, H.D.: 1955, *Astrophys. J.* **121**, 349.

Bohlin, J.D.: 1977, in Zirker (ed.), *Coronal Holes and High Speed Wind Streams*, Colorado Associated University Press, p. 59-60.

Bruzek, A. and Durant, C.: 1977, *Illustrated Glossary for Solar-Terrestrial Physics*, p. 63-65.

Davis, J.M. and Krieger, A.S.: 1982, *Solar Phys.* **80**, 295.

d'Azambuja, L. and d'Azambuja, M.: 1948, Annales de l'Observatoire de Paris, **Tome VI.**

Ding, Y. J., Hagyard, M.J., DeLoach, A.C., Hong, Q.F., and Liu, X.P.: 1987, *Solar Phys.* **109**, 307.

Foukal, P.: 1971a, *Solar Phys.* **19**, 59.

Foukal, P.: 1971b, *Solar Phys.* **20**, 298.

Gaizauskas, V. 1990, this volume.

Gary, G.A. and Hagyard, M.J.: 1990, *Solar Phys.* in press.

Giovanelli, R.G.: 1982, *Solar Phys.* **77**, 27.

Hagyard, M., Venkatakrishnan, P. and Smith, J.B. Jr.: 1990, *Astrophys. J. Sup.,* in press.

Hermans, L.M. and Martin, S.F.: 1986, in A. Poland (ed.), *Coronal and Prominence Plasmas*, NASA Conference Publication 2442, p.369.

Hirayama, T.: 1985, *Solar Phys.* **100**, 415.

Howard, R.F.: 1959, *Astrophys. J.*, **130, 193.**

Howard, R.F. and Harvey, J.W.: 1964, *Astrophys. J.*, **139**, 1328.

Howard, R. and LaBonte, B.J.: 1981, *Solar Phys.* **74**, 131.

Kawaguchi, I.: 1967, *Solar Phys.* **1**, 420.

Kopp, R.A. and Pneuman, G.W: 1976, *Solar Phys.* **50**, 85.

Kurokawa, H., Hanaoka, Y., Shibata, K. and Uchida, Y.: 1987, *Solar Phys.* **108**, 251.

Kuijpers, J.: 1990, to be published in *Plasma Phenomena in the Solar Atmosphere*, 1989 Cargese workshop, eds. M.A. Dubois and D. Gresillon.

Leighton, R.B.: 1964, *Astropys. J.* **140**, 1547.

Leroy, J.L. and Servajean, R.: 1966, *Ann. Astrophys.* **29**, *263.*

Levine, R.H.: 1977, in Zirker (ed.), *Coronal Holes and High Speed Wind Streams*, Colorado Associated University Press, p. 103-143.

Livi, S.H., Martin, S., Wang, H. and Ai, G.: *Solar Phys.* **121**, 197.

Malherbe, J.-M.: 1987, Doctoral Thesis, University of Paris VII.

Malherbe, J.-M.: 1989, E.R. Priest (ed.), *Dynamics and Structures of Quiescent Solar Prominences*, Kluwer Academic Publishers, p. 115.

McIntosh, P.S., Krieger, A.S., Nolte, J.T., and Vaiana, G.: 1976, *Solar Phys.* **49**, 57.

McQueen, R.M., Sime, D.G. and Picat, J.-P.: *Solar Phys.* **83**, 103.

Maksimov, V.P. and Ermakova, L.V.:1986, Contributions Skalnate Pleso, **15**, 65.

Martin, S.F.: 1973, *Solar Phys.* **31**, 3.

Martin, S.F.: 1979, *Solar Phys.* **64,** 165.

Martin, S.F.: 1988, *Solar Phys.* **117**, 243.

Martin, S.F.: 1990, in J.O. Stenflo (ed.) *Solar Photosphere: Structure, Convection and Magnetic Fields,* I.A.U., p. 129.

Martin, S.F.: 1986, in A. Poland (ed.), *Coronal and Prominence Plasmas*, NASA Conference Publication 2442, p. 73.

Martin, S.F., Livi, S.H.B., and Wang, J.:1985, *Australian J. Physics,* **38**, 929

Martres, M.J., Michard, R., Soru-Iscovici, I.: 1966, *Ann. Astrophys.* **29**, 245.

Mosher, J.M.: 1977, Ph. D. Thesis, California Institute of Technology.

Newkirk, G. Jr.: 1971, in Macris (ed.), *Physics of the Solar Corona,* p. 73.

Prata, S.W.: 1971, *Solar Phys.* **20**, 310.

Priest, E.R.: 1989, (ed.), *Dynamics and Structures of Quiescent Solar Prominences*, Kluwer Academic Publishers

Rompolt, B. and Bogdan, T.:1986, in A. Poland (ed.), *Coronal and Prominence Plasmas*, NASA Conference Publication 2442, p. 81.

Schmahl, E.J., Mouradian, Z. Soru-Escaut, I. and Martres, M.J. 1982, *Solar Phys.* **81**, 91.

Serio, S., Vaiana, G.S., Godoli, G., Motta, S., Pirronello, V. and Zappala, R.A.: 1978, *Solar Phys.* **59**, 65.

Shelke, R.N. and Pande, M.C.: 1983, *Bull. Astr. Soc. India*, **11**, 327.

Smith, S.F.: 1968, in Kiepenheuer (ed.) *Structure and Development of Solar Active Regions*, IAU Symp. **35**, p. 267.

Smith, S.F.: 1971, (ed.) Howard, *Solar Magnetic Fields,* IAU Symposium, p. 323.

Stenflo, J.O.: 1972, *Solar Phys.* **23**, 307.

Tandberg-Hanssen, E.: 1974, *Solar Prominences,* D. Reidel Pub. Co., p. 78-106.

Tandberg-Hanssen, E.: 1979, (ed.) E. Jensen, P. Maltby, and F.Q. Orrall, *Physics of Solar Prominences,* IAU Colloq. No. 44, p. 139.

Tang, F.: 1986, *Solar Phys.* **105**, 399.

Vaiana, G.S., Krieger, A.S. and Timothy, A.F.: 1973, *Solar Phys.* **32**, 81.

van Ballegooijen, A.A. and Martens, P.C.H.: 1989, *Astrophys. J.* **343**, 971.

Venkatakrishnan, P., Hagyard, M., and Hathaway, D.H.: 1989, *Solar Phys.* **122**, 215.

von Kluber, H.: 1932, *Z. Astrophys.* **4**, 1.

von Kluber, H.: 1961, *Mon Not. Roy. Astron. Soc.* **123**, 61.

Waldmeier, M.: 1941, *Ergebnisse und Probleme der Sonnenforschung,* Leipzig, p. 234.

Waldmeier, M.: 1970, *Solar Physics,* **15**, 167.

Webb, D.F., McIntosh, P.S., Nolte, J.T., and Solodyna, C.V.: 1978, *Solar Phys.* **58**, 389.

Webb, D.F. and Zirin, H.: 1981, *Solar Phys.* **69**, 99.

Zirin, H.: 1972, *Solar Phys.* **22**, 34.

Zirin, H. and Wang, H.: 1989, *Solar Phys.* **119**, 245.

Zirker, J.B.: 1989, *Solar Phys.* **119**, 341.

Zwaan, C.: 1978, *Solar Phys.* **60**, 213.

Zwaan, C.: 1985, in R. Muller (ed.), *Lecture Notes in Physics,* Springer Verlag, Berlin, Vol. **233**, p. 263.

Zwaan, C.: 1987, *Ann. Rev. Astron. Astrophys.* **25, 83.**

Discussion

Priest: What is the difference between a fibril and a small filament?

Martin: Filaments (by definition) are structures wether large or small that divide opposite polarity field in the same sense that a valley is a division between mountain peaks. Fibrils, according to our present usage of this term are the H_α structures that compose the horizontal fine structure of the chromosphere in general. The fibrils in general lie parallel to the direction of the magnetic fields in the chromosphere. Many fibrils extend horizontally from plage clusters. Small filaments are possibly special sets of fibrils in a specific magnetic field configuration.

Priest: I prefer your option (c), namely reconnection submergence as a model for cancellation of magnetic features for the following reasons: (i) when opposite polarity photospheric fragments approach it is natural theoretically (because of the coronal Alfven time) that their field lines should easily reconnect in the corona where the magnetic field dominates
(ii) before and during cancellation one would expect a small release of energy near the overlying location of reconnection, and this would naturally explain the observed occurrence of an X-ray brignt point or He 10830 dark point
(iii) It is very hard for coronal fields to lift up very dense photospheric material as in your option (b) and this would certainly be observed as an upflow of dense photospheric material to the overlying atmosphere.

Possible objections to (c) are: how to provide the filament mass? (easily by sucking or injecting up from the chromosphere or by condensation from the corona); how to explain the lack of a transverse structure if the field lines cross from one polarity to another? (This exists also in (B). By adding a strong longitudinal field component, and in any case not all field lines, especially the transitory ones here, would show up as fibrils).

Martin: Regarding item (i) in the above comment- although we have found in our study with D. Webb of X-ray bright points that the bright points are more closely related to cancelling features then to ephemeral active regions, there is not necessarily a direct relationship between the mechanism of cancellation and the mechanism that produces the X-ray emission. You might be correct saying that configuration (c) is represantative of a way of creating a bright point by reconnection- but it is also the same configuration of reconnection that is commonly associated with flares. Both X-ray bright points and flares have properties quite different from cancellation. If cancellation is a consequence of magnetic reconnection, it more likely takes place in a medium more dense than the corona because the process is very slow in contrast to flares and flaring bright points.

Forbes: What evidence do you have that flux cancellation is not simple submergence of a loop as you show in case (a) of your figure?

Martin: The evidence comes from observations not directly related to prominences. Ephemeral active regions are new small bipoles in which the opposite polarities are connected by obvious fibrils or arch filaments. Isolated ephemeral regions never decay by cancelling themselves. That is, the opposite polarities do not come back together unless one pole cancels with an external fragment of flux. Also the connecting fibrils between opposite polarities of an ephemeral region do not shorten and gradually disappear as one would expect if submergence were taking place. Ephemeral regions usually disappear by cancelling and merging with external magnetic fields. Secondly, cancellation occurs frequently between some of the elementary bipoles within growing active regions. It is unlikely that submergence and emergence are taking place simultaneously in a growing active region.

Van Hoven: an alternative interpretation of the function of the field reconnecton near the photosphere is that it merely increases the length of the field lines, and thus decreases the dominant parallel (electron) conductive heat input to the pre-prominence volume. Then the radiative instability will go to start where the density is highest near the bottom of these field dips, as they rise.

Martin: Thank you for your comment. Alternative interpretations should be considered.

Hirayama: Do truely quiescent prominences develop as active region filaments as you described?

Martin: Yes, an example is published in "Coronal Prominences and Plasmas". Additionally F. Tang has published a paper showing that more quiescent prominences form in the polarity inversion zones between active regions than in active regions. The same processes of convergence and cancellation take place in the boundaries between very weak remnants of multiple active regions. The only differences are that less flux is converging and cancels and the formations takes much longer.

Anzer: If prominences are formed by subphotospheric reconnection, then the dense prominence material would have to be pulled through the photosphere and the prominence would first appear below the chromospheric structures. When it then moves further up it would strongly perturb the existing fibril structure. Is there any observational evidence for this?

Martin: Yes, first the fine structure within prominences and the fibrils below prominences are continuosely changing. Secondly, some filaments are bright at their base. The cause of this brightening is not definitively known. Possibly it is excitation triggered by the reconnection process.

Zirker: Please comment on the magnetic structure near the footpoints of quiescent, growing prominences.

Martin: The footpoints most likely lie at vertices of supergranulation cells.
The vertices between supergranule cells are observed sites where network magnetic field and intranetwork magnetic fields converge. If opposite polarity magnetic fields encounter one another, "cancellation" occurs. More work needs to be done to see if the feet of prominences change position as new cells are born and old ones die.

PROMINENCE MAGNETIC FIELD OBSERVATIONS

Iraida S. Kim
Sternberg State Astronomical Institute, Moscow State University
Universitetskij prosp. 13, Moscow 119899, U.S.S.R.

1. Introduction

The tendency of the complex researches of the solar activity manifestations, such as spots, flocculai, filaments (prominences), flares, coronal structures, etc., is widely recognized nowadays. However, not a single solar structure displays such a close relation with the others and appears so evident indicator of the solar large-scale magnetic field as prominences do (Babcock and Babcock, 1955; Mc Intosh, 1972).

Extremely diverse in shape, life time, relations with the active regions, prominences remain enigmatic solar structures for us. In some cases using numerical modeling, we can describe some events, but we are still practically unable to foretell with confidence prominence formation, its duration in time, or destruction. Today nobody has doubts that the main parameter determining such a phenomenon as prominence is a magnetic field. That's why so many theoretical and experimental papers are devoted to prominence magnetic fields. I would specify prominences as indicators of a specific, still unknown to us configuration of the magnetic field in the solar atmosphere.

2. Instrumental achievements

The history of prominence magnetic research is based on the photoelectric method of determining the Sun's longitudinal magnetic field using the Zeeman diagnostics (Babcock, 1953). Further successes in the theory and technology made it possible to create the Stokes polarimeters. I won't dwell in detail on the methods of the solar magnetographic studies that can be easily found in reviews by Beckers (1968), Grigoriev (1977), O En Den (1978), Leroy (1979), Stenflo (1985), Kim (1985), Leroy (1988).

Now I'll deal with the problem of determining the magnetic field in the solar atmosphere and, in particular, in prominences. There is a trend nowadays to speak of indirect and direct methods of determining prominence magnetic fields. The indirect methods are based on getting the magnetic field data either from knots movements or Hα-fibrils directions using filtergrams. The direct methods are usually those determining the magnetic fields using polarization diagnostics.

2.1. *Indirect methods*

An indirect method requires knowledge of velocity and density, as well as the assumption on the dominance of the magnetic field energy over the kinetic one. This method makes it possible to determine the lower limit of the magnetic field strength. Velocity is determined accurately enough using prominence filtergrams. Density can be determined less accurately, but fortunately, the field strength is proportional to the density square root. According to Idlis et al. (1956) prominence fields do not exceed 10 G. Further successful developments of this method by Ballester and Kleczek (1983) and Ballester (1984) has provided data on the magnetic fields of a number of active prominences.

The obvious advantage of this method is that it does not require sophisticated equipment. Its disadvantage is the necessity to use additional assumptions on the prominence matter density. Evidently the method can be applied only to prominences exhibiting rapid movements (loops, surges, eruptions, sprays, active region filaments). Its successful implementation for quiescent prominences is doubtful.

We can expect much from solving the inverse problem of determining the magnetic field in the solar atmosphere using the known field in the chromosphere level. The field normal component distribution is known from the magnetograms. The transverse field is identified by the Hα-fibriles orientation (Kulikova et al.,1986; Molodensky and Filippov,1988).

2.2. *Direct methods*

Magnetographic studies require both special instruments developments and use of rather complicated methods of polarization measurement analysis. Contemporary direct methods of prominence magnetic field determination are based on analysis of circular (the Zeeman diagnostics) and linear (the Hanle diagnostics) polarization measurements.

The Zeeman diagnostics considered classic has been in use since the beginning of the '60. The pioneer measurements of prominence magnetic fields by Zirin and Severny (1961) and Ioshpa(1962) were made using Babcock's type magnetographs. Here I'd like to stress again the fact that prominence magnetic field measuring is highly specific. Firstly, the Zeeman splitting is extremely small compared with the half width (FWHM) of the emission line profile used (Hα, Hβ, D_3, etc.). Thus, for the D3 line (FWHM<0.4 A) the Zeeman splitting for the 10 G field will be 1.8×10^{-4} A. Secondly, prominence brightness averages 10^{-4} of the photosphere brightness. That's why it becomes exceedingly important to take into account the "parasitic" polarization.

Babcock's magnetograph two fixed slits did not allow to take into account the variations of brightness, half widths as well as doppler velocities existing even in quiescent prominences. As magnetograph feeding optics coelostat mirrors with considerable instrumental polarization (Milovanov, 1974; Bashkirtsev, 1976; Backman and Pflug, 1983) were often employed. So measurements were carried out only for more bright prominences. A number of these restriction have been overcome in the Climax magnetograph designed specifically to measure the solar atmosphere fields (Lee et al., 1965; Lee et al., 1969). The two main merits of this device are the employment of a coronagraph as a feeding optics and a special system providing for constant comparison of the Zeeman splitting with the predetermined shift of the line used. In the '60 American astronomers using this magnetograph made vast magnetographic studies of prominences that have not lost their importance even today.

The Zeeman diagnostics provides for rather simple interpretation of observations in most cases. Its merits are obvious while studying prominences with fields exceeding 15 G. It should be noted that the noise level of most longitudinal field magnetographs used to study prominences corresponds 1 G with the integration time of several minutes and the space resolution of 5″ (seconds of arc). When measuring the transverse component the accuracy of such instrument is 50 G. Then we should not forget about the non-local thermodynamic equilibrium (Lamb, 1970). Due to the partial excitation of the prominence emission lines by the anisotropic radiation field of the underlying photosphere, coherences between the Zeeman sublevels can be induced. The atomic polarization will be particularly important for the Hydrogen and neutral helium lines, whose fine structure is comparable to the Zeeman splitting by the weak prominence magnetic field (Landi Degl'Innocenti, 1982).

By the middle of the '70 a lot of observational data on the longitudinal magnetic field have been obtained by Rust (1966), Harvey(1969), Tandberg-Hanssen (1970), Smolkov and Bashkirtsev (1973). According to the data of American astronomers the field strength in quiescent prominences averaged 5-15 G and in active ones - 80 G. Measurements of the Soviet astronomers showed quiescent prominence fields up to 100 G, active ones - 1000 G and sometimes - 10 000 G in "dashes" (Shpitalnaja and Vjalshin, 1970).

These discrepancies made their French colleagues use the new method based on **the Hanle diagnostics.** In 1929 Ohman (1929) noted that the emission in prominence lines must be linear polarized due to the directivity of the radiation exciting field. For purely resonant scattering the polarization vector is tangential to the Sun's limb, and the polarization degree increases with the height. A non-vertical field results in decreasing the polarization degree and rotating the polarization direction, that is the Hanle effect. The emission line polarization observed by a number of authors. Hyder (1965) and Thiessen (1951) tried to evaluate the magnetic field strength using linear polarization measurements. However, the usage of polarization measurements for magnetographic purposes has became feasible thanks to the progress in the quantum theory of the polarized radiation (House, 1970; Stenflo, 1976; Bommier and Sahal-Breshot, 1978;). I will not dwell on this method in detail because I do not think I can do it better than the French and the Italian astronomers have done (Leroy, 1979; Sahal-Breshot, 1981; Bommier et al., 1985; Landi Degl'Innocenti, 1986). I would like just to express my admiration at skilful interpretation of the polarization measurements given by these astronomers.

Today we do not discuss the possibility of using the Hanle diagnostics for magnetographic studies as 15 years ago. Here I just note that the Hanle diagnostics of observations at the Pic du Midi (France) and later at the Sacramento Peak (the USA) observatories was based on looking for the best correlation between the estimated and the observed parameters. Today we consider this method as direct despite it requires a lot of assumptions and calculations.

The Hanle diagnostics provides for determining the field vector and is more sensitive compared with the Zeeman diagnostics when dealing with the fields of only several G. But sometimes it does not provide for an acceptable interpretation of the data obtained. The field vector may be determined in points with the height (h) more than 15″. The Hanle diagnostics may be applied to quiescents since the Hanle effect is characterized by the field critical strength (B_c) depending on the atomic parameters. Thus, B_c is 3 and 8 G for the main and the "red" components of the D_3 line respectively. One must keep in mind an existence of non-magnetic depolarization (collisions and accidental anisotropy of the radiation field), doppler velocities in prominences when using emission lines having the powerful Fraunhofer lines.

2.3. Historical remarks

The prominences longitudinal magnetic field measurements carried out in the '70 have constituted an important stage in prominence researches providing observational potential for the decade to come. However, at the same time these very studies have proved the necessity to get reliable statistical data corresponding to long duration observation periods carried out using the same type of instrument.

In the '70 several groups in the USA, France and the USSR have undertaken programs to create special polarimeters and obtain the magnetic characteristics of an "average" prominence. Reports on developing such polarimeters appeared in the '70 (Rust, 1972; Ratier, 1975; O En Den et al.,1976). Later we've come to know valuable results obtained by the French astronomers (Leroy, 1979 and 1988), in which they used the Hanle "wide band" diagnostics. With small strength fields the polarization parameters are approximately constant along the line profile making it possible to use radiation from the whole line. Circular polarization in this case is zero.

The full scale Hanle diagnostics implementation requires simultaneous observations in two lines at least. Such observations have became feasible after a Stokes polarimeter has been developed at the Boulder and the Sacramento Peak observatories (Baur et. al.,1981). This instrument registers Stokes profiles providing for the Hanle "narrow-band" diagnostics (House and Smartt,1982). Two components of the D_3 line (the central one consisting of five lines and the "red" triplet) are used. Both "wide band" and "narrow band" Hanle diagnostics are implemented for fields not exceeding 30 G.

In 1972 under Prof. G.M.Nikolsky (the Solar Activity Laboratory of IZMIRAN, the USSR) an elaboration of a magnetograph scanning along the line profile was started (O En Den et al., 1977). Later the program of magnetographic studies was successfully conducted together with the specialists of the Astrophysical Institute of Paris (Drs S.Koutchmy and G.Stellmacher). After Prof. G.M.Nikolsky premature death in 1982 (on September 28, 1989 he would have celebrated his 60th birthday), the magnetograph was added with the doppler velocity channel. Nowadays this Soviet-French instrument is known in our country as Nikolsky's magnetograph (Nikolsky et al., 1984a).

2.4. Nikolsky's magnetograph

To separate the "parasitic" polarization from the prominence signal and to exclude the effects of variations of brightness, line half widths and doppler velocities, the circular polarization has to be recorded along the whole line profile. Our magnetograph was designed for measuring of magnetic fields in the solar atmosphere. The 53 cm Nikolsky coronagraph (Gnevyshev et al., 1967) feeds the magnetograph attached to the coronagraph polar tube.

A piezo-electrically scanned Fabry-Perot interferometer (FPI) and a lithium niobate crystal are used as a dispersing and an electrooptical analyzer respectively. A special system is used both for the FPI mirrors justing and for the simultaneous record of prominence emission line and doppler velocities using reference spectral line (Kim et al.,1984). Two photomultipliers are used as detectors of prominence and reference lines signals. During the I and V Stokes profiles record a prominence picture with the magnetograph "pin-hole" location is photographed. The electronic circuits was described in detail by Stepanov (1989).

As a rule the Stokes V profile observed consists of an antisymmetric Zeeman component and symmetric component due to an instrumental polarization, the Hanle effect (real crystal has small sensitivity to the linear polarization) and eventually the above-mentioned atomic polarization. During the observations we compensate this symmetric component: the linear polarization by analyzer rotation and the circular one by electrical subtraction "in phase" the simultaneously recorded I-profile multiplied by a factor (about 10^{-3}), taking advantage of our method of scanning the whole profile. Up to now we do not discriminate these "parasitic" effects. The distance of the prominence "point" observed from the optical axis of the "coronagraph + magnetograph" system does not exceed 10″ corresponding to 0.4 mm in the main focal plane.

Our magnetograph allows us to obtain the magnetic data from h>5″: the I and V Stokes profiles, doppler velocity, and prominence picture simultaneously with the spatial resolution of 3-6″, the spectral resolution of 0.12-0.33 A, the time constant of V-profile of 1-4 s, the integration time for the Hα-line of 20-40 s, the spatial resolution of prominence pictures of 0.6-1.5″ (Nikolsky et al., 1984).

2.5. Future polarimeter

To carry out magnetic observations a set of instruments consisting of a telescope, a dispersing element, a circular or linear polarization analyzer and a data registration system is needed. So present day prominence magnetic field observations are aimed at obtaining Stokes profiles. Of exceptional importance is the reduction to the minimum "parasitic" polarization. That is achieved by using coronagraphs as the feeding optics. Polarization is measured, as a rule, in "point" localized on the optical axis of the "coronagraph + magnetograph" system.

Modern polarimeters spatial resolution (>5″) is at least ten times greater than the prominence fine structure (Engvold, 1976). Considerable averaging within the large "pin-hole" may strongly distort real magnetic situation. Moreover, the integration time to get a Stokes profile is 1-2 min, which is comparable with the life time of the quiescent prominence knots. Thus, the subsecond structure magnetic field diagnostics requires that a coming polarimeter effectiveness has to increase to 100 times. This makes Leroy (1988) recognize greater possibilities of the indirect methods.

Our calculation proves that a system of a 2 m coronagraph and a magnetograph consisting of a Fabry-Perot interferometer, a $LiNbO_3$ crystal and a detector of a CCD camera type provides the "signal to noise" ratio enough for the magnetic field subsecond diagnostics. Moreover, such system would make it possible to carry out coronal magnetic field observations with the spatial resolution of 10″ and the integration time of 5-10 min. The upper limit of a lens objective diameter determined by present-day technology is 50 cm. But Zeiss (Soctur and Korendjaka, 1988) and Perkin Elmer Corporations (Terrile, 1988) successes in making superpolished mirrors, as well as testing a 5 cm diameter mirror coronagraph (Koutchmy and Smartt, 1988) give us hopes that in future a mirror coronagraph project will become feasible (Table).

Table

Polarimeter elements	Climax magnetograph	French polarimeter	American Stokes meter	Nikolsky's magnetograph	Future polarimeter
Telescope	40cm coronagraph	26cm coronagraph	40cm coronagraph	53cm coronagraph	*2m coronagraph*
Dispersing element	Grating	Lyot Filter	Grating	FPI	*FPI*
Analyzer	KDP	λ/2 plate	KD*P	$LiNbO_3$	*$LiNbO_3$*
Spectral resolution	0.1A	Several A	0.03A	(0.1-0.3)A	*(0.1-0.3)A*
Spatial resolution	15″	5″	(3.8x10)″ (6.8x10)″	(3-8)″	*0.5″*
Time constant/λ	15min/Hα	10s/D3	120s/D3	(1-2)s/Hα	*(2-4)s/Hα*
Integration time	15 min	<1 min	2 min	0.5 min	*1 min*
Height	>10″	>15″	>15″	>5″	*>2″*

3. Discussion

Let us compare the magnetographic data obtained by different groups during last decade and early. Here we can discuss only quiescent prominences and active region filaments (prominences). There are many classification of prominences based on morphology, dynamics, spectra, etc. (Tandberg-Hanssen, 1974; Zirin, 1989). Below I follow to the classification based on the relations with the solar active regions (Hirayama, 1985) and with the solar large scale magnetic fields.

Quiescent prominences (quiescents, QP) are found far from active regions **(AR)**, in the outskirts of AR, between AR. On the disk QP is observed as filament having fine structure across it. As a rule on the limb QP is observed as prominence with h>35-40″. QP have always a fine vertical structure which may be less than 1″. QP always trace the so-called neutral lines of the solar large scale magnetic field. Polar crown prominences belong to the QP class.

Active region prominences (more often called **active region filaments, ARF**) are found in active regions as fine dark threads. On the limb they are observed as prominences with h<35″. Usually there are the fine horizontal structures in ARF. ARF correspond to the young neutral lines. Analysis of our ARF shows that sometimes ARF predict the occurence of the neutral lines (Kim, 1990).

3.1. Field vector orientation in regards to photosphere (θ)

Prominence modeling requires knowledge of the field vector orientation in regards to the photosphere. Both in theoretical studies and in interpretation of the first observations by Leroy et al. (1977) it was generally accepted that the prominence magnetic field vector was localized in the horizontal plane. The first determination of the deviation of the field vector from the horizontal plane have been made by Athay et al.(1983). The mean value of θ is 3°. Here we have the so-called vertical structure paradox (Leroy, 1988): an evident fine vertical structure of QP always observed with high spatial resolution and the field vector localization in the horizontal plane. Later similar conclusions were made by Leroy et al. (1984). However they did not exclude the deviation of 30°for V-type field lines from the horizontal plane.

3.2. Angle between field direction and filament long axis (α)

Determination of α is an important problem of magnetic studies. According to Tandberg-Hanssen and Anzer (1970), the α absolute value is 15°. A brief review of α determination methods using the Hanle diagnostics is given by Hirayama (1985). In accordance with the French astronomers α is 25° (Leroy et al., 1983). For our QP we found maximum value of the longitudinal magnetic field ($B_{\parallel max}$) in each prominence, averaged these $B_{\parallel max}$ for the β angle intervals of 0-10°, 10-20°... (β is the angle between the long filament axis and the line of sight). The dependence of $B_{\parallel max}$ on β shows the peak at 15°. Supposing that the maximum value corresponds to the field vector localization along the line of sight, we conclude that α is close to 15° (Kim et al., 1990). Nevertheless, some authors (Querfeld et al, 1985) are still in favour of α closed to 90° and further determination of α are needed.

3.3. *Prominence distribution on β (the value of the angle between the long axis of filament and the line of sight)*

We suspected the dependence of the observed magnetic parameters on the filament orientation relative to the line of sight (the so-called aspect angle β). We have analyzed QP and ARF distributions on β for every year of the 1979-1988 period (Akmamedova et al.,1990). β were determined from the synoptical charts published in the SGD and Solnechnye Dannye journals. Time intervals of 3-5 months centered at our observation periods were choosen. All filaments presented in these charts were subdivided into ARF and QP families. Histograms "Occurrence-β" were compared with each other. The correlations of 0.85-0.95 between the QP distributions allow us to analyze such histograms for different epochs of the solar spot cycle and then for the whole period studied (Fig. 1a).

We note great difference in appearances of QP and ARF distributions. The QP distribution may be described by the exponent. The ARF dependence has multimodal nature. According to the Student criterion the probability of the maximum at 20-50° exceeds 98% (Fig. 1b). Further we analyzed our magnetic data taking into account the orientation of the filament long axis relative to the line of sight.

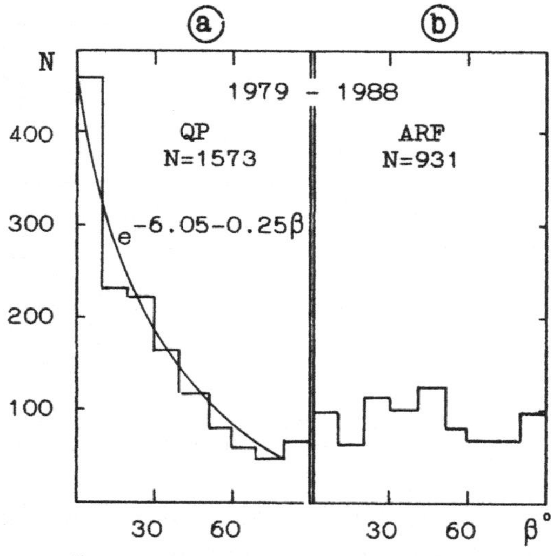

Fig.1. Filaments distribution on β
N - the total number of filaments

3.4. *Field strength*

Below we use the symbols B for the strength of the full field vector and $B_{\parallel}$ for the longitudinal one. The spatial resolution of modern polarimeters is 3-10″ while the size of the fine structure of prominences is less than 1″. However the successive measurements of $B_{\parallel}$ with the "pin-hole" of 3, 6 and 12″ and insignificant variations of $B_{\parallel}$ within QP allow us to suggest the uniform magnetic field structure in QP.

Nowadays the question about QP field strength is not so debatable as it was 15 years ago. The Hanle and the Zeeman diagnostics agreed that the QP field strength varies from 3 to 15 G, and rarely exceeds 30 G. Today a more interesting question is the prominence distribution on B(or $B_{\parallel}$): are 5 or 15 G more typical for QP and ARF? Based on their observations in 1972-1982 Nikolsky et al. (1984) noted probable peaks in the prominences distribution on $B_{\parallel}$ and compared with the analogycal distribution on B for the same epoch corresponding to tables published by Athay et al.(1983). The number of prominences both in our observations (about 30) and in those by Athay et al.(about 15) was rather small and we used the "point" measurements (Fig. 2). There are the same peaks at 8(11) G, 20(25) G and even the weak peak at 44 G in both histograms. This close correlation between the B and $B_{\parallel}$ distributions makes it possible to compare the results of the Zeeman and the Hanle diagnostics.

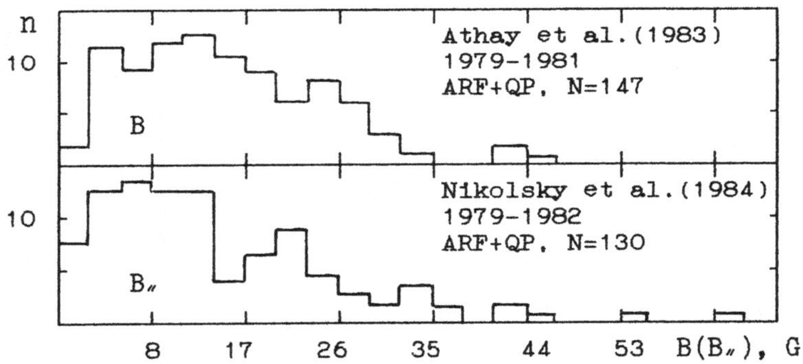

Fig. 2. Prominences "points" distribution on B ($B_{\parallel}$). N - the total number of "points" measured

We have carried out a statistical analysis of the QP and ARF data obtained by Nikolsky's magnetograph in 1979-1985. $B_{\parallel}$ was measured in 155 prominences. Each of them underwent measuring in 3-15 "points". Each point was measured at least 3 times allowing for determining the average $B_{\parallel}$ for each "point" used further as a basis for calculating the mean value for each prominence. The $B_{\parallel}$ determination error does not exceed 3 G. To do away with the artefacts it would have been correct to use a step exceeding 3 G, but for the convinence of comparing our data with those obtained by others we used the step of 3 G. In doubtful cases we checked the validity of our conclusions using the 6 G step.

Fig.3a shows the histograms "Occurrence-$B_\parallel$" for QP corresponding to figures, tables or original histograms published by different authors. We use only "statistical" data (>30 prominences observed). Analysis of this Fig. makes it possible to come to the following conclusions:

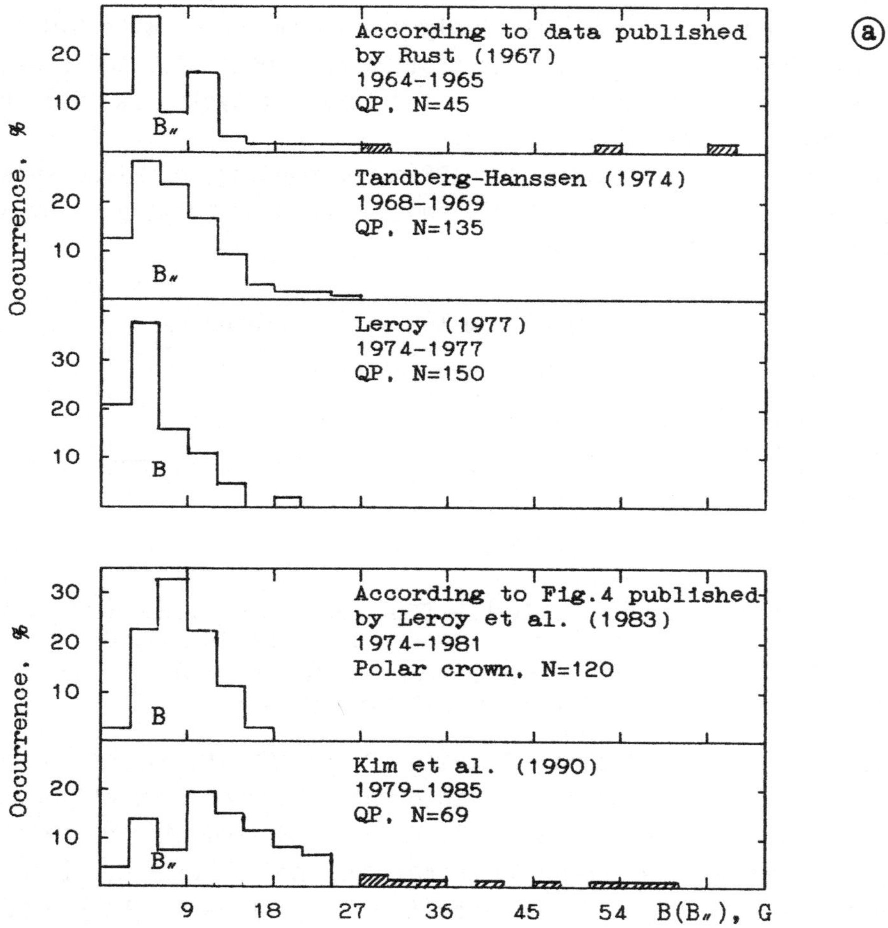

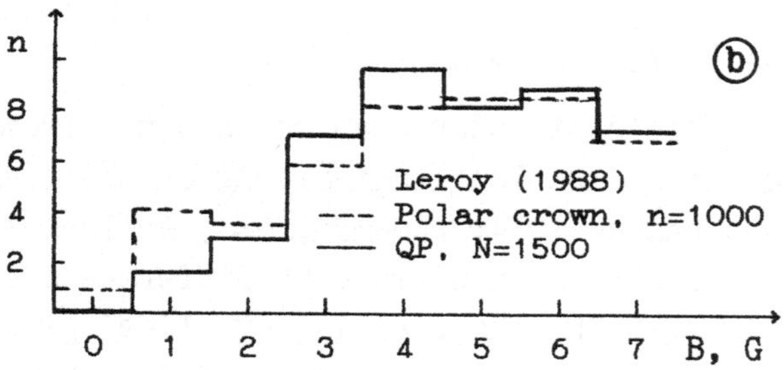

Fig. 3. a) QP distribution on B ($B_\parallel$). N - the total number of prominences; b) a detailed hystogram of the Pic du Midi measurements. N - the total number of "points"

There is an evident asymmetry of all histograms which may be caused by the multimodal character of QP distribution on $B(B_{\parallel})$. All the distributions have a small number of prominences with $B(B_{\parallel})<2$-3 G. Fig. 3b given by Leroy (1988) confirms this conclusion.

Considerable variations of $B(B_{\parallel})$ for QP have been noted. According to Rust (1966) and Kim et al. (1990) data there are QP with $B_{\parallel}>30$ G during the pre-minimum epochs of the solar activity.

In Leroy observations most QP have $B<5$ G. At the same time the Polar crown prominences have relatively many prominences with $B>9$ G. Whether this should be caused by the prominence subclasses, the observation selection, or the phase of the solar activity cycle is not known. I am in favour of the first factor.

In Fig. 4 the histograms for ARF corresponding to Nikolsky's magnetograph observations (Kim, 1990) and to the data published by Harvey (1969) and Bashkirtsev and Mashnich (1987) are given.

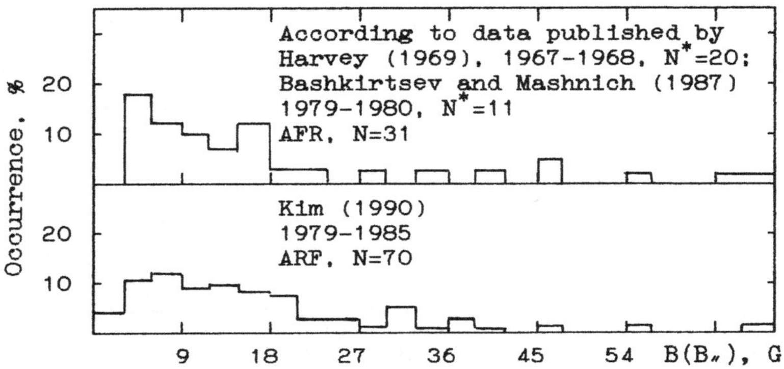

Fig.4. ARF distribution on $B_{\parallel}$. N - the total number of prominences

We can see the difference between QP and ARF histograms. $B_{\parallel}$ of ARF varies a little within the 6-21 G interval. There are relatively many ARF with $B_{\parallel}>27$ G, sometimes up to 70 G. Koutchmy and Zirker (1990) photographic observations appear to be reliable and testify $B_{\parallel}$ of several hundreds G for the fine threads localized very close to the photosphere.

3.5. The height dependence of the field strength

Indirect data of the magnetic field structure can be obtained from the studies of the vertical gradient of the longitudinal magnetic field - $dB_{\parallel}/dh$ (Anzer, 1969; Kuperus and Raadu, 1974; Pneuman, 1983; Ballester and Priest, 1987). Determination of $dB_{\parallel}/dh$ in QP made by different authors. According to Rust (1967) in most QP the positive vertical gradient of $B_{\parallel}$ averaged to 1×10^{-4} G/km. Leroy et al. (1983) found $dB_{\parallel}/dh$ of

0.5×10^{-4} G/km for 50 out of 120 Polar crown prominences. Nevertheless, from Fig. 4 (Leroy et al.,1983) it follows that not less than in 10 of 50 prominences there is a negative vertical gradient. Bashkirtsev and Mashnich (1987) found $dB_{\parallel}/dh$ of 4×10^{-4} G/km in 10 out of 18 prominences under study. And I wonder to know the height dependence of the field strength for the rest of prominence sample of the above-mentioned authors. The "narrow band" analysis of the Hanle effect (Athay et al.,1983) for 13 QP and ARF and the Zeeman diagnostics (Nikolsky et al.,1984) have not revealed an unambiguous dependence of B ($B_{\parallel}$) on h. It has been noted, that the vertical dependence of B(or $B_{\parallel}$) is a small effect which may be masked by the prominence orientation relative to the line of sight and internal variations of field strength within the prominence.

In the statistical analysis of our QP data (Klepikov, 1989a) the similarity of the magnetic structure of QP was supposed. Observed heights of "points" measured were normalized for the maximum height of every prominence and expressed in % ($\hat{h}$). The average $B_{\parallel}$ modulus was determined for every 5% height intervals of 1-5 %, 6-10 %, etc. QP were subdivided into three groups according to β intervals of 0-30°, 31-60° and 61-90°. Fig. 5a shows the regular variations of $B_{\parallel}$ with $\hat{h}$ only for β intervals of 0-30° and 61-90°. If $B(\hat{h}) = a\hat{h} + b$, then

$a = 0.20 \pm 0.03$, $b = 7.4 \pm 1.6$ for $\beta=(0\text{-}30)^\circ$,
$a = 0.04 \pm 0.04$, $b = 10.9 \pm 2.3$ for $\beta=(31\text{-}60)^\circ$,
$a = -0.11 \pm 0.02$, $b = 20.5 \pm 1.0$ for $\beta=(61\text{-}90)^\circ$.

For QP with h of 40 000 km $dB_{\parallel}/dh$ for the β interval 0-30° will be 5×10^{-4} G/km and for the β interval of 61-90° $dB_{\parallel}/dh \approx (3.3\times10^{-4}$G/km).

In Fig. 5b the $B_{\parallel}(\hat{h})$ dependence for ARF observed by Nikolsky's magnetograph in 1979-1985 is presented (Kim, 1990). No evident dependence

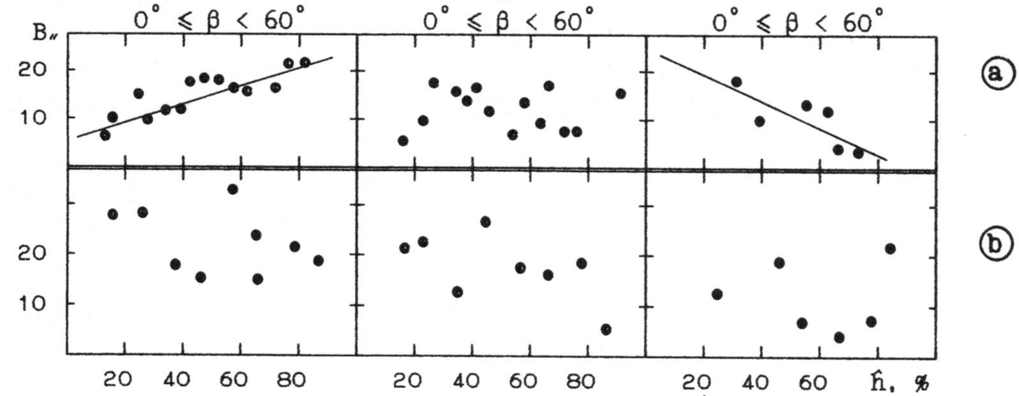

Fig. 5. $B_{\parallel}$ variations with height in 1979-1985. $\hat{h}$ - observed height normalized for the maximum one of every prominence: a) for 69 QP (Klepikov, 1989a); b) for 45 ARF (Kim, 1990)

of $B_{\parallel}$ on $\hat{h}$ is found. That may be caused by the two factors. Firstly, ARF are not the stable solar structures. Secondly, in accordance with the ARF distribution on β for this period the most part of ARF correspond to β interval of 31-60°.

3.6. *On polarity (sign) of prominence magnetic field*

The observed polarity of prominence magnetic field is a good criterion to distinguish the two main models (Anzer, 1979). For the Kippenhaan-Schluter potential-like model the observed polarity is the same as the polarity of the underlying photospheric field (UPF). For prominence model proposed by Kuperus and Tandberg-Hanssen (1967) and developed by Kuperus and Raadu (1975) the observed polarity of prominence magnetic field is inverse to the polarity of UPF (non-potential-like model).

Below we discuss the observational aspects of this problem. Following Leroy (1988) terminology we use the notations N (the normal polarity) and I (the inverse polarity) to indicate the potential-like and non- potential-like models respectively. The results of comparisons between the observed polarities of prominence magnetic field and UPF are controversial. The Zeeman diagnostics by Rust (1967) are consistent with the normal polarity. According to the "wide band" Hanle diagnostics (Leroy et al., 1984; Bommier et al., 1985;, Leroy, 1988) I prominences are presented in large number among QP. Usually the height of I prominences is more than 30 000 km. Nevertheless, there are some cases of N prominences. As a rule, these prominences are lower than 30000 km. "The narrow band" analysis of the Hanle effect (Athay et al., 1983) shows that the polarity of prominences magnetic field either coincides with the polarity of the UPF or opposite to it. But the analysis applied to the Sacramento Peak observations by Bommier et al.(1985) confirms the conclusions of the French astronomers.

We analyzed our 69 QP observed in 1979-1985. Synoptical charts published in the SGD and Solnechnye Dannye journals used for the compari-

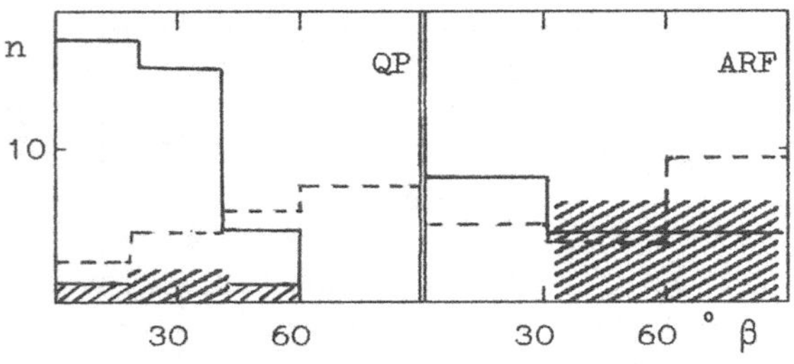

Fig. 6. The polarity comparisons for QP and ARF. n - number of prominences, —— I polarity, ---- N polarity, mixed polarity

son of the signs of the prominence field and the UPF. For 61 QP the aspect angle β have been determined with confidence (Kim et al., 1990).

The same analysis was carried out for our 70 ARF (Kim, 1990). Only for 45 ARF corresponding neutral lines of the UPF have been found. In Fig. 6 the results of these comparative analyses are presented in graphical form. The analysis of Fig. 6 shows the dependence of the sign of $B_{\parallel}$ on the orientation of the long axis of filament with respect to the line of sight. Most edge-on prominences (both QP and less evident ARF) have the inverse polarity and $B_{\parallel}$ mean value of 15 G. Side-on prominences have the normal polarity and $B_{\parallel}$ of 5 G. In prominences localized at the β interval of 31-60°a mixed polarity can be detected.

3.7. *Cyclic variations*

Prominence activity (total number, height, etc.) has cyclic character. The first identification of the $B_{\parallel}$ cyclic variations made by American astronomers has showed increasing of the $B_{\parallel}$ value for QP from 5 to 7.3 G to the maximum of the solar spot cycle in 1965-1969 (Tandberg-Hanssen, 1974). Leroy et al. (1983) noted increasing B for the Polar crown prominences from 6 to 12 G in 1974-1980. Is this increase real or the result of the observation selection? One has to keep in mind several factors. Firstly, the duration of observation periods is 2-3 months at best, which can be compared with the duration of the solar activity fluctuations influencing prominences activity. Secondly, there are considerable B (or $B_{\parallel}$) variations even for the QP subclasses whose contributions into determining B (or $B_{\parallel}$) may vary with time. Fig. 7 shows $B_{\parallel}$ (the mean value of $B_{\parallel}$ for every observation period) temporal variations for QP and ARF on the whole based on our data. We draw your attention to the maximum occurred at the pre-minimum epoch of the spot cycle. However, to back up such a conclusion both further observations and analysis taking into account the QP and ARF families and orientation relative to the line of sight are needed.

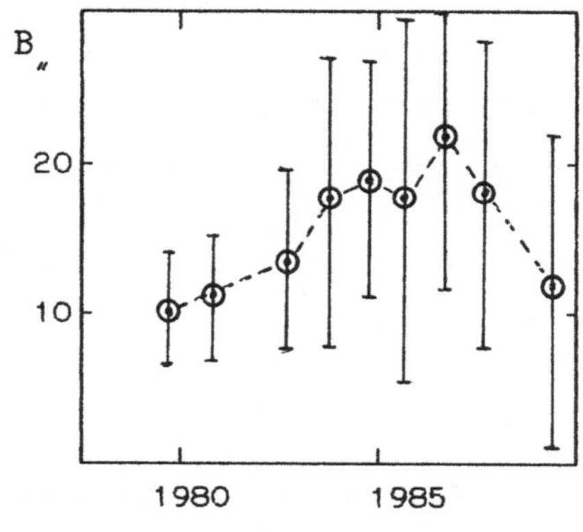

Fig. 7. Cyclic variations of $B_{\parallel}$ for QP and ARF on the whole

3.8. *Oscillations*

Observational programs for searching of prominence magnetic field oscillations are important for solving of problem of prominence magnetic configuration (Jensen, 1983). Up to now the observations of prominence magnetic field oscillations are unique. The first searches of $B_{\parallel}$

oscillations (Klepikov, 1989b) have led us to the necessity of taking into account the atmospheric effects comparable with the Zeeman signal.

The first simultaneous observations by means of Nikolsky's magnetograph and the registrator of quality of seeing conditions are presented in this issue (Geondjan et al., 1990) in detail. Here I note that within the limits of error (±3 G) no oscillations of $B_{\parallel}$ with period of 3-20 min are found. The correlation factor of 0.86 between $B_{\parallel}$ and the derivative of the image tremble amplitude indicates on the possibility of the variations of the atmospheric polarization with period of 5 min.

3.9. Some statistical properties of prominence heights

Very few observatories can carry out prominence magnetic field observations requiring complicated polarimeters. Lack of long duration magnetic data makes it impossible to use magnetic field data as an indeces of the solar activity.

Statistical analyses of magnetic data carried out recently by Leroy et al. (1984) and Stellmacher et al.(1986) suspect the prominence maximum height as a parameter characterizing these objects magnetic field structure. The magnetic field polarity of prominences with h>35″ is mainly inverse. This group includes QP and as a subclass the Polar crown prominences. Whether ARF and QP are different classes or represent two phases of evolution of the same class may depend to a great extent on the analysis of the prominence height distribution and the height variations with the phase of the solar activity cycle.

A brief review of researches devoted to the **prominence height distribution** is given by Kim et al. (1988a). All authors noted the sharp decreasing of prominences number with h>30 000 km (≈40″). In Fig. 8 our histograms "Number of prominences (n) - maximum height observed (h″)" are given. We have used only high resolution filtergrams (0.6-1.5″) obtained in 1979-1985. The error of height determination did not exceed 3″. According to the Student criterion the maximum probabilities at 20-25″ and 30-50″ are 75 and 85% respectively. The same dependences for QP (dotted line) and ARF (full line) illustra-

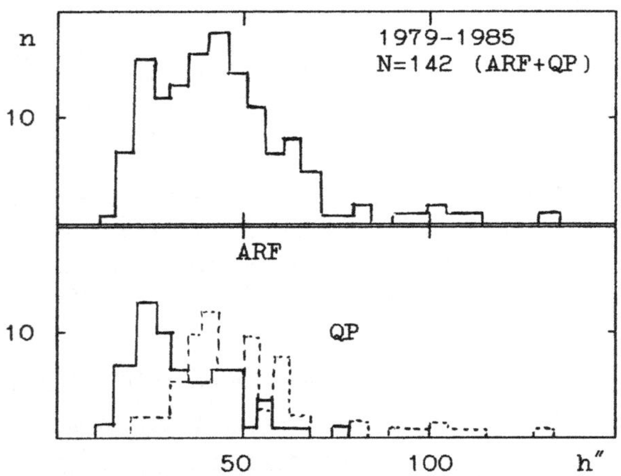

Fig.8. Prominence distribution on the height observed

te the bimodal nature of prominences height distribution. The first narrow peak corresponds to ARF, the second one relates to QP.

Cyclic variations of prominence height were noted previously (Cantu et al., 1968; Dermendjiev, 1977; Makarov, 1983; Kim et al., 1988b). The h cyclic curve maxima either forestall or coincide with the corresponding peaks of the flocculai areas cyclic curves, the h cyclic curve minima always occur during the pre-minimum epoch (Makarov, 1988; Kim et al.,1988c). Based on the bimodal prominence height distribution we noted that h cyclic variations may be caused by the variations of the relative contents of QP and ARF (Kim et al., 1988a)

Identification of prominence height as a "magnetic" parameter, occurrences of the h cyclic curve minima during the pre-minimum epoch of the solar activity, explaining this minimum by the variations of the relative contents of ARF and QP, as well as a set of researches considering the prominence and the solar wind formations in interdependence (An et al., 1985; An, 1988) set us thinking to use **ARF and QP relative contents as indexes of the solar activity.** Having employed uniform series of the Kodaikanal observatory data for the No.16 cycle Kim and Uvakina (1989) note the high correlation factor (90%) between the relative content of the prominences with $h<20''$ and Legrand and Simon (1981) recurrent geomagnetic activity.

4. Conclusions

The comparative analysis of QP and ARF magnetic data obtained by different authors results us in the conclusion that the observed magnetic data depend on the orientation of the filament long axis relative to the line of sight (the so-called aspect angle β). The most part of the edge-on QP has the inverse polarity (the sign) of the longitudinal magnetic field, the positive vertical gradient of $B_{\parallel}$ of 5×10^{-4} G/km and the mean value of the field strength of 15 G. The side-on QP have the normal polarity, the negative vertical gradient of $B_{\parallel}$ of (-3×10^{-4} G/km) and the mean value of the field strength of 5 G. For the β interval of 30-60° a mixed polarity can be detected.

The polarity of ARF has the same tendency to depend on β, but not so distinctly as in QP. There is no evident dependence of the field strength on the height. That may be caused both non-stability of ARF and the ARF distribution on β, which differs greatly from the QP one.

Our observational data set us thinking on the existence two magnetic field systems in QP and ARF. The first system oriented oppositely to the underlying photospheric field. The polarity of the second system coincides with the polarity of the underlying photospheric field. Also we can suppose the three-dimensional magnetic structure which could explain the above-mentioned observational data.

The usage of the maximum height of prominences as the solar activity index illustrates the importance of the real magnetic data for the prediction of the solar activity.

Here I did not concern with the comparative analysis of the magnetic and spectral data which was partly given by Kim et al. (1982), Leroy et al. (1983), Nikolsky et al. (1984). In my respect, taking into account β will give us more unambiguous results. I hope that useful cooperation with several institutes in the solar prominences magnetic field observations will be continued in future and together with theoretical researches will throw light upon the nature of these structures.

5. Acknowledgements

I am very thankful to Prof. E.Tandberg-Hanssen for his invariable interest to the magnetic researches of the solar prominences. I express my gratitude to all my colleagues for putting at my disposal the papers on their researches and I am extremely grateful to Prof.J.L.Leroy whose concurrent information on the French polarimetric successes helped us to draw our attention to some observational problems. Invaluable discussion with Dr S.Koutchmy helped highly the preparation of this review. I would like to thank the staff of the High Altitude Station near Kislovodsk for help in observations. Special thanks are due to Dr S.S.Asadullaev and Mrs E.V.Lavrentieva for their assistance in preparing of this manuscript.

References

Akmamedova A.Sh., Goshdzhanov M. and Kim I.S.: 1990, to be published.
An C.H., Suess S.T. and Tandberg-Hanssen E.:1985, Solar Phys. **102**, 766.
An C.H., Wu S.T., Bao J.J. and Suess S.T.: 1988, Space Science Lab. Preprint Series No. 88-107.
Anzer U.: 1969, Solar Phys. **8**, 37.
Anzer U.: 1979, in E.Jensen, P.Maltby and G.Q.Orral(eds.), *Physics of Solar Prominences*, Blindern, Oslo, p.322.
Athay G., Querfeld C.W. Smartt R.N., Landi Degl'Innocenti E. and Bommiier V.: 1983, Solar Phys. **89**, 3.
Babcock H.W.: 1953, Astrophys. J. **118**, 387.
Babcock H.W. and Babcock H.D.: 1955, Astrophys. J. **121**, 349.
Backman G. and Pflug K.: 1983, Publ. Debrecen Helia Phys. Observ. Hung. Acad. Sci. **5**, No. 2, 589.
Ballester J.L. and Kleczek J.: 1984, Solar Phys. **94**, 151.
Ballester J.L.: 1984, Solar Phys. **94**, 151.
Ballester J.L. and Priest E.R.: 1987, Solar Phys. **109**, 335.
Bashkirtsev V.S.: 1976, Solnechnye Dannye No. 7, 82.
Bashkirtsev V.S. and Mashnich G.P.: 1987 SibIZMIR preprint No. 7.
Baur T.G., Elmor D.E., Lee R.H., Querfeld C.W., and Rogers S.R.: 1981, Solar Phys. **70**, 395.

Beckers J.M.: 1968, Solar Phys. **5**, 15.
Bommier V. and Sahal-Brechot S.: 1978, Astron. Astrophys. **69**, 57.
Bommier V., Leroy J.L. and Sahal-Brechot S.: 1981, Astron. Astrophys. **100**, 231.
Bommier V., Leroy J.L. and Sahal-Brechot S.: 1985, in M.J.Hagyard(ed.), NASA Conf. Publ. No. 2374, 375.
Bommier V., Leroy J.L. and Sahal-Brechot S.: 1986, Astron. Astrophys. **156**, 79.
Cantu A.M., Godoli G. and Poleto G.: 1968, Mem. Soc. Astr. Ital. (NS), **38**, 367.
Dermendjiev V.:1977, Astrophys.Studies Bulgar. Academy of Sci. No. 2, 8.
Engvold O.: 1976, Solar Phys. **49**, 283.
Engvold O.: 1988, in E.R.Priest(ed.), *Dynamics and Structure of Quiescent Solar Prominences*, Kluver Academic Publ., p.47.
Gnevyshev M.N., Nikolsky G.M. and Sazanov A.A.:1967, Solar Phys.**2**, 223.
Grigoriev V.M.: 1977, Izv. Krymsk. Astrof. Obs. **56**, 166.
Harvey J.W.:1969, NCAR Co.Thesis No.17, University of Colorado, Boulder.
Hirayama T.: 1985, Solar Phys. **100**, 414.
House L.L.: 1970, in R.D.Howard(ed.), *Solar Magnetic Fields*, D.Reidel Publ.Co., Dordrecht, Holland.
House L.L. and Smartt R.N.; 1982, Solar Phys. **80**, 53.
Hyder C.L.: 1965, Astrophys. J. **141**, **1374**.
Idlis G.N., Karimov M.T., Delone A.B., Obashev C.O.: 1956, Izv. Astrof. Inst. Kaz. AN SSR, **2**, 71.
Ioshpa B.A.: 1962, Geomagnetism i aeronomiya **2**, 149.
Ioshpa B.A.: 1968, *Solnechnaya aktivnost'*, Nauka, Moskva, p. 44.
Jensen E.: 1983, Solar Phys. **89**, 275.
Kim I.S.: 1985, Abastumani Astrophys. Obs. Bull. No. 60, 141.
Kim I.S.: 1990, to be published.
Kim I.S., Kim Gun-der, Klepikov V.Yu. and Stepanov A.I.: 1988a, Astron. tsirkular No. 1532, 17.
Kim I.S., Klepikov V.Yu., Koutchmy S., Stepanov A.I. and Stellmacher G.: 1988b, Solnechnye Dannye No. 1, 75.
Kim I.S., Klepikov V.Yu., Koutchmy S., Stepanov A.I. and Stellmacher G.: 1988c, Solnechnye Dannye No. 5, 77.
Kim I.S., Klepikov V.Yu., Koutchmy S., Stepanov A.I. and Stellmacher G.: 1989, Solnechnye Dannye No. 6, 98.
Kim I.S., Klepikov V.Yu., Koutchmy S., Stepanov A.I. and Stellmacher G.: 1990, Pis'ma v Astron. Zh. No. 3.
Kim I.S., Koutchmy S., Nikolsky G.M. and Stellmacher G.: 1982, Asrton. Astrophys. **114**, 347.
Kim I.S., O En Den, Stepanov A.I.: 1984, Solnechnye Dannye No. 7, 90.
Kim I.S. and Uvakina V.F.:1989, in R.A.Gulyaev(ed.), *Atmosfera Solntsa, mezhplanetnaya sreda, atmosfery planet*, IZMIRAN AN SSSR, Mockva, p. 125.
Klepikov V.Yu.: 1989a, Preprint IZMIRAN No. 47 (873).
Klepikov V.Yu.: 1989b, in R.A.Gulyaev(ed.), *Atmoosfera Solntsa, mezhplanetnaja sreda, atmosfery planet*, IZMIRAN AN SSSR, Moskva, p.12.
Kippenhahn R.and Schlüter A.Z.: 1957; Z. Astrophys. **43**, 36.
Kulikova G.N., Molodensky M.M., Starkova L.I. and Filippov B.P.: 1986, Solnechnye dannye No. 10, 60.
Koutchmy S. and Smartt R.N.: 1988, in Oskar Von der Luc(ed.), *Proceedings of the 10th Sacramento Peak Summer Workshop "High Spatial Resolution Solar Observations", August 22-26, 1988.*
Koutchmy S. and Zirker J.B.: 1990, this issue.
Kuperus M. and Raadu M: 1974, Astron. Astrophys. **31**, 189.
Kuperus M. and Tandberg-Hanssen E.: 1967, Solar Phys. **2**, 39.
Lamb F.K.: 1970, Solar Phys. **12**, 186.
Landi Degl'Innocenti E.: 1982, Solar Phys. **79**, 291.
Landi Degl'Innocenti E.: 1986, in A.I.Poland(ed.), NASA Conf. Publ., No. 2442, p. 203.
Lee R.H., Rust D. and Zirin H.: 1965, Appl.Opt. **4**, 1081.

Lee R.H., Harvey J.W. and Tandberg-Hanssen E.: 1969, Appl. Opt. **8**, 2370.
Legrand J.P. and Simon P.A.: 1981, Solar Phys. **70**, 173.
Leighton R.B.: 1957, Astrophys. J. **130**, 366.
Leroy J.L.: 1977, Astron. Astrophys. **60**, 79.
Leroy J.L.: 1978, Astron. Astrophys. **64**, 247.
Leroy J.L.: 1979, in E.Jensen, P.Maltby and F.A.Orral(eds.), *Physics of Solar Prominences*", Blindern, Oslo, p.56.
Leroy J.L.:1985, in Hagyard M.J.(ed.), NASA Conf. Publ. No. 2374, p.121.
Leroy J.L.: 1988, in Priest E.R.(ed.), *Dynamics and Structure of Quiescent Solar Prominences*, Kluver Academic Publ.
Leroy J.L., Bommier V. and Sahal-Brechot S.: 1983, Solar Phys. **83**, 135.
Leroy J.L., Bommier V. and Sahal-Brechot S.: 1984, Astron. Astrophys. **131**, 33.
Makarov V.I.: 1983, Solnechnye Dannye, No. 4, 100.
Martin S.: 1986, in A.J.Poland(ed.), NASA Conf. Publ. No. 2442, 73.
Mc Intosh P.S.: 1972, in Golburn D.S., Sonett C.D. and Wilcox J.M. (ed.),"*Solar Wind*" NASA SP-308, p. 136.
Milovanov V.N. : 1974, Trudy AN Kaz. SSR, **26**, 164.
Molodensky M.M. and Filippov B.P.: 1988, Astron. Zh. **65**, 1047.
Nikolsky G.M., Kim I.S. and Koutchmy S.: 1982, Solar Phys. **81**, 81.
Nikolsky G.M., Kim I.S., Koutchmy S., Klepikov V.Yu. and Stepanov A.I.: 1984a, Solnechnye Dannye No. 9, 88.
Nikolsky G.M. Kim I.S., Koutchmy S. and Stellmacher G.: 1984b, Astron. Astrophys. **140**, 112.
Nikolsky G.M., Kim I.S., Koutchmy S., Stepanov A.I. and Stellmacher G.: 1985, Astron. Zh. **29**, 669.
O En Den: 1978, Thesis, IZMIRAN, Moscow.
O En Den, Kim I.S. and Nikolsky G.M.: 1977, Solar Phys. **52**, 35.
Ohman Y.: 1929, M.N.R.A.S. **89**, 479.
Pneuman G.W.: 1983, Solar Phys. **88**, 219.
Querfeld C.W., Smartt R.N.: 1984, Solar Phys. **91**, 299.
Querfeld C.W., Smartt R.N., Bommier V., Landi Degl'Innocenti E. and House L.L.: 1985, Solar Phys. **96**, 277.
Ratier G.: 1975, Nouv. Rev. Optique **6**, 149.
Rompolt B.:1988, in E.R.Priest (ed.), *Dynamics and Structure of Quiescent Solar Prominences*, Kluver Academic Publ.
Rust D.: 1966, NCAR Co. Thesis, Univ. of Colorado.
Rust D.: 1967, Astrophys. J. **150**, 313.
Rust D.:1972, Air Force Cambridge Research Laboratory Preprint No. 237.
Sahal-Brechot S.: 1981, Space Sci. Rev. **30**, 99.
Sahal-Brechot S., Bommier V. and Leroy J.L.: 1977, Astron. Astrophys. **59**, 223.
Schmieder B., Malherbe J.M., Mein P. and Tandberg-Hanssen E.: 1984, Astron. Astrophys. **136**, 81.
Smolkov G.Ya. and Bashkirtsev V.S.: 1973, in Sykora J.(ed.), "*7th Regional Consultation on Solar Physics*", Slovak Academy Sci. p.175.
Soctur D.G. and Korendjaka C.W.: 1988, B.A.A.S. **20**, No. 4, 990.
Shpitalnaja A.A. and Vjalshin G.F.: 1970, Solnechnye Dannye, No. 4, 10.
Stellmacher G., Kim I.S., Klepikov V.Yu., Koutchmy S. and Stepanov A.I.: 1986, Pre-publ. No. 131, Institut d'Astrophys. du CNRS, France.
Stenflo J.O.: 1976, Astron Astrophys.**46**, 61.
Stenflo J.O.: 1985, Solar Phys. **100**, 189.
Stepanov A.I.: 1989, Thesis, IZMIRAN, Moscow.
Tandberg-Hanssen E.: 1970, Solar Phys. **15**, 359.
Tandberg-Hanssen E.; 1974, "*Solar Prominences*", D. Reid. Publ. Co., Dordrecht Holland.
Tandberg-Hanssen E. and Anzer U.: 1970, Solar Phys., **15**, 158.
Tandberg-Hanssen E., Malville J. McKim: 1974, Solar Phys. **39**, 107.
Terrile R.J.: 1988, A.I.A.A. Pap., No. 555, 1.
Thiessen G.: 1951, Z. Astrophys. **30**, 8.
Zirin H. and Severny A.B.: 1961, Observatory **81**, 155.
Zirin H.: 1988, *Astrophysics of the Sun*, Cambridge University Press.

DISCUSSION

PRIEST: For your active-region filaments, what is the error in determining the magnetic polarity? In view of the observations of flows and structure along such filaments, is it possible that the field is also exactly along them?

KIM: The error depends on the synoptical charts accuracy. Our observations do not exclude the field direction along the filaments. The field strength (horizontal field $B_{\parallel}$) corresponding to β = 30° - 40° angle between the long axis of the filament and the line of sight is 9-15 G.

RADIO OBSERVATIONS OF PROMINENCES

F. CHIUDERI DRAGO
Department of Astronomy and Space Science, University of Florence
Largo Enrico Fermi, 5
50125 Florence, Italy

1. Introduction

Radio signatures of the presence of prominences have been observed both in their quiescent and eruptive phase: millimetric and centimetric radio emission is mostly associated with quiescent prominences, metric events with the eruptive ones. There are a few exception of this generale rule: Axisa et al. (1971) and Dulk and Sheridan (1974) observed at m-λ enhanced radiation on the solar disk in regions overlying quiescent filaments, while Kundu and Lantos (1977) observed an eruptive prominence at mm-λ.

Radio events associated with eruptive prominences and Coronal Mass Ejections (CME) are mostly type II and moving, as well as stationary, type IV radio bursts. It is well known that this type of events are typically coronal ones and they can be associated either with eruptive prominences or with flares. The eruptive prominence acts as a gun-ball that, crossing the corona, triggers longitudinal plasma oscillations at the local plasma frequency which are then converted into electromagnetic waves. All the physical parameters that can be derived from the observations of these bursts refer therefore to the corona and do not give any information on the exciting prominence, apart from its speed, if a coronal model is assumed. For this reason, radio observations associated with eruptive prominences and CME, although an extremely interesting subject in itself, will not be treated in the present revue.

Let us see what radio observations can tell us about quiescent prominences.

It is very well known that, given a plasma whose electron density is N, there exists a critical frequency: $\nu_p = 9\ 10^3\ N$ such that no electromagnetic wave of frequency $\nu \leq \nu_p$ can propagate. Since in the upper solar atmosphere the electron density decreases with height, we can define for each frequency a critical level which represents the deepest level from which information can be acquired at that frequency.

Typical values of the electron density in the upper part of the solar atmosphere are listed in Table I together with the corresponding critical frequency and wavelength.

TABLE I

N (cm^{-3})	ν_p	λ (cm)
10^{11}	$3\ 10^9$	10
10^{10}	$9\ 10^8$	30
10^9	$3\ 10^8$	100
10^8	$9\ 10^7$	300

It would appear from the above table that a plasma whose electron density lies in the typical prominence range (10^{10}-10^{11} cm^{-3}), could be investigated by observing at radiowavelengths lower than 30 cm.

However the crucial limiting factor for cm- λ comes from considerations of the optical depth. In fact, while in the upper corona radio observations may reach levels very close to the critical one, for thicker and colder plasmas, a high optical depth is reached much above the critical level. This is easily seen by computing the geometrical thickness l inside a prominence, corresponding to an optical depth $\tau = 1$ at different radiowavelengths, by using the expression of the free-free absorption coefficient ($k = \xi N^2 T^{-3/2} \lambda^2/c^2$ with $\xi \cong 0.15$) and values of electron density and temperature typical for prominences: $N = 5\ 10^{10}$ cm^{-3}, $T = 7000$ K (Schmieder, 1989). Results are presented in Table II which shows that radio observations at wavelengths $\lambda > 1$ cm cannot give any information about the prominences parameters.

TABLE II

λ(cm)	l (km)
0.3	120
1.0	10
3.0	1.2
10.0	0.1

Moreover, if we take into account the contribution of the Prominence-Corona Transition Region (PCTR), we find that at $\lambda > 2$ cm an optical depth $\tau > 1$ is already reached in the PCTR itself. We then conclude that radio observations at $\lambda > 1$ cm can only give information on the prominence environment and not on the prominence itself.

2. Observations and interpretation of radio data

Filaments and prominences appear on radio maps of the Sun taken in the mm and cm range of wavelengths as they appear in H_α : namely as dark structures on the disk and bright ones at the limb. Radio observation of prominences at the limb are however fewer than those on the disk. For a complete list of observations of radio filaments see Table I of Hiei et al. (1986).

The first observations of "radio filaments" at mm-λ were performed in 1959 by Khangil'din (1964) using a 22 m mirror which gives, at $\lambda = 8$ mm, a resolution of 1'.7: "dark regions", observed on the disk above H_α filaments, were explained, in analogy with these latter, assuming that the filament is optically thick, with a temperature $T_f \cong (5\text{-}6)\ 10^3$ K and absorbs the radiation coming from the underlying quiet chromosphere, which has a temperature $T_{ch} > T_f$.

A prominence model, the only one based on radio observation at millimetric wavelengths: 4 mm $\leq \lambda \leq 8.6$ mm, was deduced by Apushkinskii and Topchilo (1976). They selected a sample of 70 radio filaments out of a total number of 370 observed by three different radiotelescopes in Soviet Union and derived a model in which the temperature increases monotonically from 6300 to 8300 K.

Radio depression are often observed on the solar disk even without any optical counterpart: in one of the earlier observation of mm-filaments, Buhl and Tlamicha (1970) reported that in some dark radio regions with no corresponding optical feature, very often an H_α filament appeared within a few solar rotations. Hiei et al. (1986) observed a radio depression at 36 GHz which darkened the next day when an H_α filament appeared in the same position.

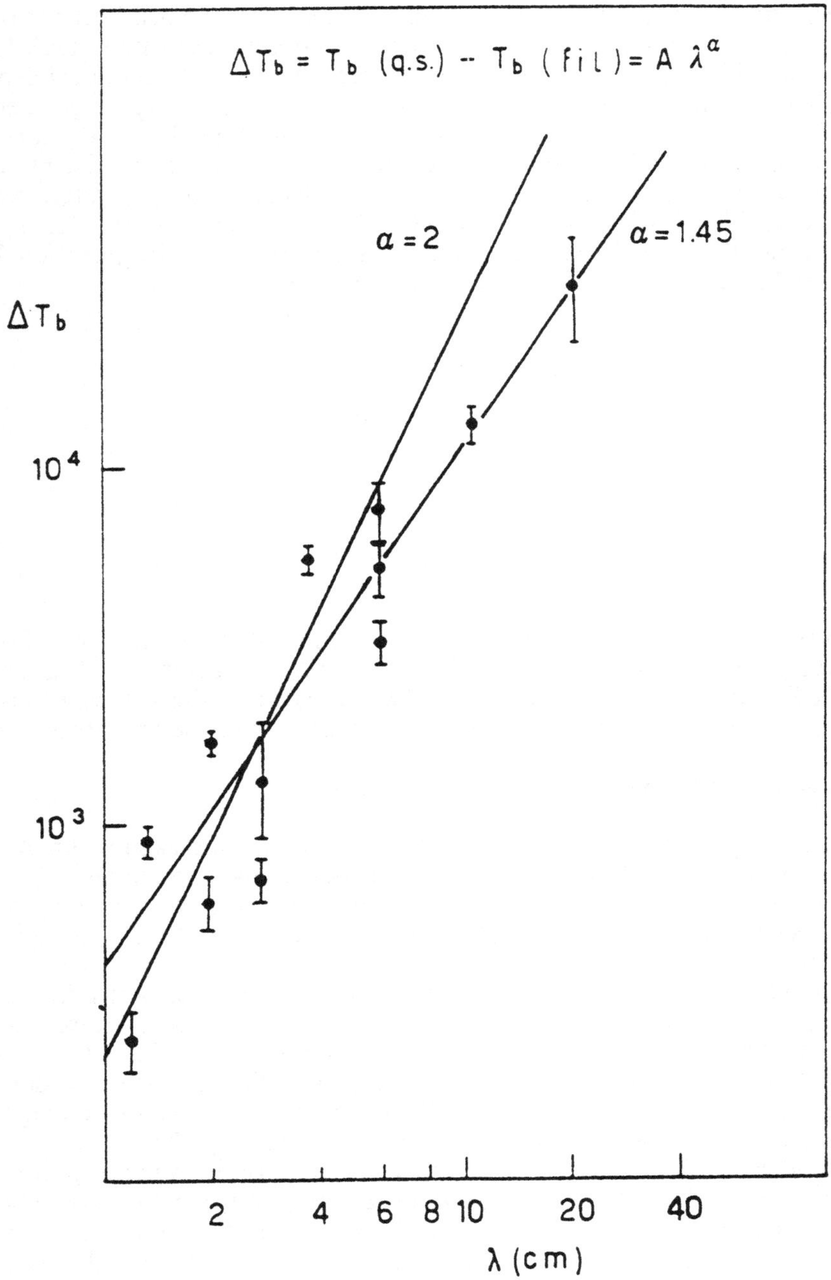

Fig. 1. - Radio spectrum of the brightness temperature depression observed above filaments. Notice that the spectrum does not scale as λ^2 as expected if the cause of depression is totally due to the lock of coronal emission.

The association rate between filaments and radio depressions has been statistically analyzed by Schmahl et al. (1981) who found that only 2/3 of the depressions observed at 22 GHz are associated with visible filaments (the association rate depends on the "darkness" of the filament), but all of them, with a very few exceptions, lie on magnetic neutral lines or on their natural extensions. The above authors conclude that "microwave observations can, therefore, supplement optical observations in identifying neutral lines".

Due to the lower angular resolution, the first solar radio maps showing "radio filaments" at cm-λ were obtained, about 10 years later than the corresponding ones at mm-λ by Chiuderi Drago and Felli (1970), using the NRAO 140 ft dish which has an angular resolution of 2'.1 at $\lambda = 1.95$ cm. At this wavelength the explanation that the radio depression is due to the filament absorption of the underlying radiation cannot be applied, since an optical depth $\tau > 1$ is reached above the top of the prominence. The above authors assumed therefore that the observed depression was due to the lack of coronal emission in the region actually occupied by the prominence, finding reasonable values for the coronal density and for the prominence height.

During the following years several maps showing radio depression in filament regions were obtained at different wavelengths up to $\lambda = 11$ cm, mostly using the 100 m radio telescope in Effelberg (Furst et al., 1973; Chiuderi Drago et al., 1975; Butz et al., 1975; Kundu et al., 1978) and the 40 m dish in Haystack (Straka et al., 1975; Pramesh Rao and Kundu, 1977; Schmahl et al., 1981).

Due to the low angular resolution of the above instruments at cm-λ it has been widely discussed if radio filaments have a larger size than the optical ones or not. Most of the authors found the former result, but Raoult et al. (1979), statistically analyzing their results, have shown that, within the errors, radio and optical filaments have the same size for 1.2 mm $\leq \lambda \leq$ 6 cm. This uncertainty on the radio filament dimension strongly affects the determination of the corresponding brightness temperature.

When the radio spectrum of the filaments became available, it was clear that the depression could not be ascribed to the coronal lack of emission alone, since it does not scale as λ^2 as required by an optically thin plasma radio emission (fig. 1).

Straka et al. (1975) assume that the origin of the depression observed at $\lambda = 3.8$ cm is due to the coronal cavity, a region of lower electron density than the quiet corona, surrounding the filament (Saito and Hyder, 1968; Saito and Tandberg Hanssen, 1973).

The presence of a cavity of size much larger than the filament would give a larger size of the filament at radio wavelengths. Moreover it reduces the observed brightness temperature in two ways: *a)* by reducing further the coronal contribution, and *b)* by decreasing the total optical depth in such a way that the layer corresponding to $\tau = 1$ (from which most of the radiation comes) shifts at lower heights in the TR. Considering the high temperature gradient in the TR, this fact can appreciably decrease the corresponding brightness temperature. From the observed depression, Straka et al. derived a ratio for the electron density in the cavity and in the quiet corona $N_c/N_q = 0.5$.

This interpretation for the radio depression is in conflict with Raoult et al. (1979) statistical results and with the observations made by Schmahl et al. (1981) of a radio filament during several days as it approached the limb, without showing the brightening predicted by Straka model. It is instead supported by an eclipse observation at 10.7 GHz performed by Bracewell and Graf (1981) who found that a lower brightness temperature is present on both sides of a filament, while the filament itself shows a brightness temperature $T_b(f) \cong T_b(q.s)$. Following these authors the attribution of the radio brightness temperature depression to the filament is due to a lack of resolution.

In more recent years radio filaments on the disk have been observed at $\lambda =$ 3, 6 and 20 cm using large synthesis instruments such as VLA (Pramesh Rao and Kundu, 1980; Dulk and Gary, 1983; Kundu et al., 1986; Gary, 1986), and Westerbork (Chiuderi Drago et al., 1977). The high resolution observations in principle would allow a perfect determination of the radio filaments size previously deduced by deconvolution, which is one of the most serious sources of uncertainty in deriving the filament brightness temperature $T_b(f)$.

The VLA radio map at $\lambda = 3$ cm, observed by Gary (1986), shows a rather strange feature, namely the presence of two regions of enhanced brightness on both sides of the radio depression coinciding with the optical filament. If this is not an instrumental effect, it is exactly the opposite of what found by Bracewell and Graf (1981). No quantitative data on the filament size and brightness temperature was given by the above author.

At $\lambda = 6$ cm Pramesh Rao and Kundu (1980) measured a radio filament size slightly larger than the optical one while Kundu et al. (1986) found a radio size $\Delta_r \sim 2 \div 3\,\Delta_{opt}$. Also the measured brightness temperature in the former observed filament was lower by about 30 % than in the latter one.

At $\lambda = 20$ cm Chiuderi Drago et al. (1977) did not observe any depression on Westerbork I map above a very long filament, while, on the V map, two large regions of opposite polarity separated by a narrow line following the optical filament along all its length were present. The sense of polarization was in agreement with the sense of the underlying photospheric field indicating that there was no change of field polarity between the photosphere and the level where the 21 cm radiation is formed. At the same wavelength Dulk and Gary (1983) noticed radio depressions above filaments, but no measurement of the corresponding T_b was given. Finally, Kundu et al. (1986) observed a clear absorption at $\lambda = 20$ cm above a long filament and measured the corresponding brightness temperature during two consecutive days. Using the VLA observations at $\lambda = 6$ and 20 cm together with previous observations at several wavelengths, they deduced a PCTR model that will be extensively discussed in the next section.

3. PCTR models

The necessity to postulate the existence of a PCTR from radio observation was first put forth by Chiuderi Drago et al. (1975). In the following years attempts to deduce the PCTR parameters from radio observations were done by Butz et al. (1975) and Kundu et al. (1978). These latter authors observed the same filament at 5 wavelengths between 0.4 and 11 cm and found a perfect fit of the observed T_b scaling, by a factor 0.8 the electron density of a quiet sun model deduced from radio data by Furst et al. (1973).

The first physical model able to account for radio observation was proposed by Pramesh Rao and Kundu (1977). Their calculation was based on the following assumptions:

a) constant pressure $p = p_o$

b) energy balance between conduction and radiation:

$$\frac{d}{dz}\left(c\,T^{5/2}\,\frac{dT}{dz}\right) = E_R \qquad (1)$$

where for E_R a composite power law was used.

The two basic parameters of the model p_o and F_o, the conductive flux entering the prominence, were found by fitting the theoretical radio spectrum to the observations at 8 different frequencies: the five ones previously used by Kundu et al. (1978) plus three new observations of another filament made by the authors using the Haystack RT.

It appears that all data points are included between the curves corresponding to $0.03 \leq p_0 \leq 0.2$ dyn/cm^2 (assuming $F_0 = 0$) but none of them gives a nice fit of the points in the whole spectral range.

The same problem was considered again by Kundu et al. (1986) who added, at the previously mentioned set of data points, new VLA observation at $\lambda = 6$ and 20 cm. They repeated the calculation of Pramesh Rao and Kundu by replacing the assumption $p = p_0$ with $p = a\,T^n$. Increasing the number of free parameters, the fit became extremely good. It appears that, in order to fit radio observations, the pressure in the PCTR must vary from 0.33 dyn/cm^2 at $T = 10^4$ K to $2\,10^{-2}$ dyn/cm^2 at $T = 10^6$ K.

The above model, in spite of the extremely good agreement with observations, presents 3 weak points:

1) First of all it contains an inconsistency between the temperature gradient derived from equation 1 and the hydrostatic equilibrium which for a pressure varying as $p = a\,T^n$ gives: dT/dh = const.

2) The balance between conduction and radiation holds only in the lower part of the TR where the conduction is an input of energy. Above the point where $T^{5/2}\,dT/dh$ presents its maximum ($T > 10^5$), its derivative changes sign and the conduction becomes also a loss of energy. Therefore it cannot balance the radiation losses. This point however is not very important for the cm-λ emission since it is mostly originated in the lower part of the TR.

3) The expression of the conductive flux used by the above authors holds either for zero magnetic fields or for magnetic fields parallel to the temperature gradient. If, as it happens in prominences, the magnetic field forms an angle $\theta \sim 90°$ with the temperature gradient (Leroy, 1989) then the conducted energy strongly depends on θ.

A more consistent model of the PCTR, should take into account points 2 and 3 above. The energy balance in a stationary state is therefore given by

$$E_H = E_R + \nabla \cdot F_c \tag{2}$$

where E_H represents the heating function. If the magnetic field $\underline{B}$ forms an angle $\theta \neq 0$ with the temperature gradient ∇T, the second term on the right hand side of equation 2 becomes

$$- \nabla \cdot F_c = \frac{d}{dz}\left(k_{//}\,T^{5/2}\,\frac{dT}{dz} + k_{\perp}\,T^{-5/2}\,\frac{dT}{dz}\right) \tag{3}$$

where, if $\dfrac{BT^{3/2}}{N} >> 10^{-4}$ (Priest, 1982) :

$$k_{//} = c\,\cos^2\theta\,, \quad c \sim 1 \cdot 10^{-6} \quad \text{and}$$

$$k_{\perp} = 2 \cdot 10^{-11}\,c\,\mathrm{sen}^2\,\theta\,\frac{p^2}{4k^2\,B^2} \quad \text{(in c.g.s. units).}$$

Assuming in equation (3) $\theta \neq 90^o$ and $d\theta/dz = 0$, eq. (2) becomes:

$$k_{//} \frac{d}{dz} \left(T^{5/2} \left(1 + \frac{\varepsilon}{T^5}\right) \frac{dT}{dz} \right) = E_R - E_H \qquad (4)$$

with

$$\varepsilon = \frac{k_\perp}{k_{//}} = 2.6\ 10^{20}\ tg^2\,\theta\ p^2/B^2$$

It appears that, for reasonable values of p and B in a prominence, the term ε/T^5 can exceed 1 only for $\theta > 85^o$ and for $T < 10^5$ k.

Eq. (4) can be analytically integrated (Chiuderi and Chiuderi Drago, in preparation), in the following assumptions:

a) $p = const$;

b) $E_R = \frac{p^2}{4k^2} A_i\ T^{\alpha_i - 2}$ (Rosner, 1979);

c) the heating function E_H is also parametrized in the form:

$$E_H = \frac{p^2}{4k^2} A_x\ T^{\alpha x} .$$

In fact, defining

$$\eta(T) = T^{5/2}\,(1 + \varepsilon/T^5)\ \frac{dT}{dz}$$

and using $\frac{d}{dz} = \frac{d}{dT}\frac{dT}{dz}$, we get:

$$\eta(T) = \eta(T_o) + \frac{p^2}{4k^2 k_{//}} \int_{T_o}^{T} (A_i\ T^{\alpha_i - 2} - A_x\ T^{\alpha x})\ dT \qquad (5)$$

In equation (5) $\eta(T_o)$ is related to the conductive flux F_o entering the prominence by the relations

$$\eta(T_o) = 1 \cdot 10^6\ F_o / \cos^2\theta$$

and the parameters A_x and α_x must be such that

$$\eta(T_c) = 0 \quad .$$

If we assume, for instance, $\eta(T_o) = 0$, then we have to assume also

$$\left(\frac{d\eta}{dT}\right)_{T=T_o} = 0$$

in order to have an inflection point in the temperature profile at the prominence top. In this case A_x and α_x are fully determined by the radiative losses at $T = T_o$ and by the coronal temperature T_c. If $\eta(T_o) \neq 0$, then one of the two parameters can be given arbitrarily.

Once the function $\eta(T)$ is known, we may integrate the radio transfer equation to get the filament brightness temperature. In fact the radio optical depth is given by:

$$d\tau = -\frac{\xi}{\nu^2}\frac{N^2 dh}{T^{3/2}} = -\frac{\xi}{4k^2\nu^2}\frac{p^2}{\cos\phi}\frac{dz/dT}{T^{7/2}}\,dT \qquad (6)$$

where ϕ is the angle between the line of sight and the local vertical z and dz/dT is related to the function $\eta(T)$ by the relation:

$$\frac{dz}{dT} = \frac{T^{5/2}(1+\varepsilon/T^5)}{\eta(T)} \quad .$$

By integrating equation (6) we get

$$\tau(T) = \frac{1}{\cos\phi}\frac{a}{\nu^2}p^2\int_T^{T_c}\frac{1+\varepsilon/T^5}{T\cdot\eta(T)}\,dT + \tau_c \quad .$$

Assuming the corona in hydrostatic equilibrium at a constant temperature T_c,

$$\tau_c = \frac{1}{\cos\phi}\frac{\xi}{\nu^2}\frac{N_c^2 H_c}{2\,T^{3/2}} = \frac{b\,p^2}{\cos\phi\, T_c^{5/2}}$$

where $H_c = RT/\mu g_o$ is the scale height and N_c is the density at the top of the PCTR (at $T = T_c$).

The constants a and b are given by:

$$a = \frac{\xi}{4k^2} \cong 2.6\;10^{30} \qquad \text{(c.g.s. units)}$$

and

$$b = \frac{a\cdot R}{2\mu g_o} \cong 6.5\;10^{33} \qquad \text{(c.g.s. units)}$$

We want to stress that, in the case of $\eta(T_o) = 0$, $\eta \propto p/\cos\theta$ and therefore $d\tau \propto p^2\, dz/dT \propto p\cos\theta$. This means that an increase of the angle θ is equivalent, as far as the TR optical depth is concerned to a decrease of the gas pressure.

Finally, knowing τ and $d\tau$ as a function of T, we can compute the filament brightness temperatures

$$T_b(f) = \int_{T_o}^{T_c} T e^{-\tau} \, d\tau + T_o \, e^{-\tau(T_o)} + T_c \, (1-e^{-\tau_c}) \qquad (6)$$

using different values of the parameters, and compare them with the observations. The set of observations used for this comparison, all those available in the literature at $\lambda >$ 1 cm, are listed in table III.

TABLE III

λ (cm)	Tb (fil)	Instruments		Observers
1.2	9270 ± 50	100 m	Effelsberg	Kundu, Furst, Hirth and Butz (1978)
1.35	8900 ± 100	40 m	Haystack	Rao and Kundu (1977)
1.95	10400 ± 150	43 m	NRAO	Chiuderi Drago and Felli (1970)
2.	9330 ± 150	40 m	Haystack	Rao and Kundu (1977)
2.	7200 ± 500	100 m	Effelsberg	Butz, Furst, Hirth and Kundu (1975)
2.8	11700 ± 700	100 m	Effelsberg	Butz, Furst, Hirth and Kundu (1975)
2.8	12300 ± 100	100 m	Effelsberg	Kundu, Furst, Hirth and Butz (1978)
3.8	12540 ± 400	40 m	Haystack	Rao and Kundu (1977)
6.	14500 ± 1900	100 m	Effelsberg	Chiuderi Drago, Furst, Hirth and Lantos (1975)
6.	17800 ± 500	100 m	Effelsberg	Kundu, Furst, Hirth and Butz (1978)
6.	12500 ± 1500		VLA	Rao and Kundu (1980)
6.	15000 ± 1000		VLA	Kundu, Melozzi and Shevgaoukar (1986)
11.	32000 ± 1500	100 m	Effelsberg	Kundu, Furst, Hirth and Butz (1978)
21.	50000 ± 10000		VLA	Kundu, Melozzi and Shevgaoukar (1986)

Most of the authors subtracts from the observed $T_b(f)$ the coronal contribution $T_c(1-e^{-\tau_c}) \approx T_c \, \tau_c$ and then compare them with the computed T_b taking into account only the first two terms on the right hand side of eq. (6). If the theoretical parameters used in the T.R model are changed, the correction to the observations should be changed accordingly. This is not always done and in fact it turns out that the coronal pressure used by some authors to compute $T_c \, \tau_c$ is larger than the pressure at the top of the TR.

During this meeting Lang has reported a VLA observation of a filament a $\lambda =$ 91.6 cm (Lang and Wilson 1989) which is not included in our set of observation. The above authors claim that their observation of a $T_b = 3.5 \; 10^5$ K agrees with the variable pressure model by Kundu et. al (1986) better than with the constant pressure model of Pramesh Rao and Kundu (1977). Actually Lang and Wilson did not realize that in Kundu et al. fig 6a, that they report in their own paper, the plotted T_b are those

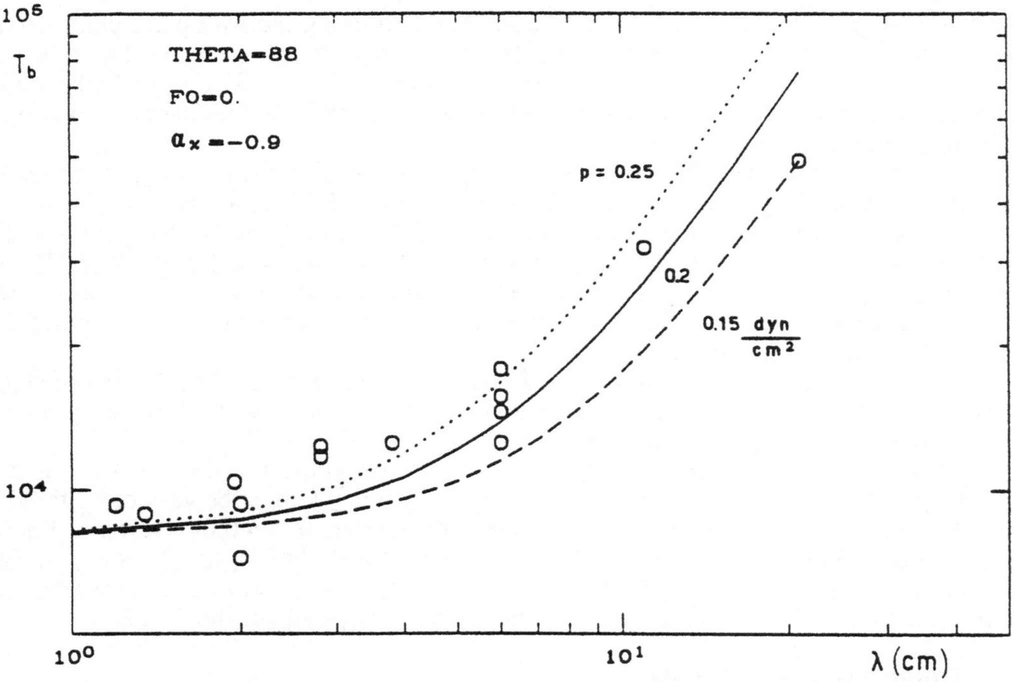

Fig. 2. - Observed and computed filament radio spectrum at a fixed angle θ between the magnetic field and the temperature gradient and for different values of the pressure.

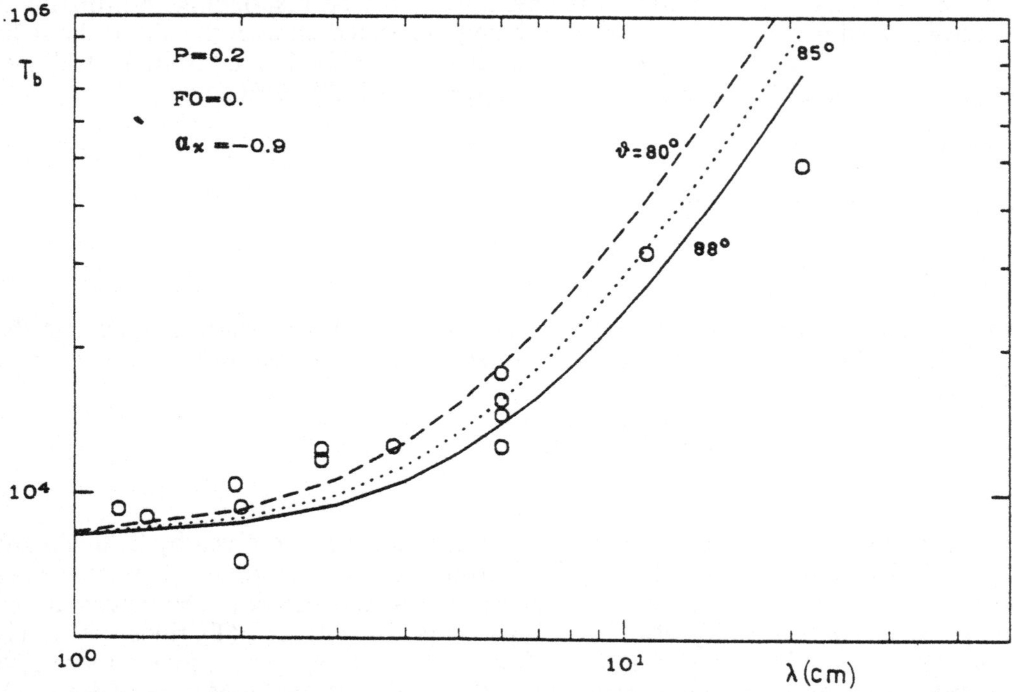

Fig. 3. - Same as fig. 2 for a fixed value of the pressure and for different values of the angle θ.

obtained after subtraction of the coronal contribution using the following parameters: $T_c = 1.6\ 10^6$ K, p = 0.1 dyn/cm^2. Therefore in their figure they compare the corrected T_b given by Kundu et al. with their direct observation at $\lambda = 91.6$ cm. If the same subtraction is applied to their observation, the resulting brightness temperature becomes negative and looses any significance.

In the following calculation the term $T_c\ \tau_c$ is included in the routine in such a way that it is always changed consistently with the changes of PCTR parameters, p and T_c.

The coronal temperature is assumed $T_c = 1.5\ 10^6$ and the lower temperature T_0 has been put equal to 8000 K since, following Apushkinskii and Topechilo (1976) model, the optical depth of the prominence at this temperature is $\tau > 1$ at all $\lambda > 1$ cm.

In fig. 2) and 3) the fit obtained with different values of the pressure p and of the angle θ between $\underline{\nabla T}$ and $\underline{B}$ are shown.

In all calculation we have assumed the angle between z and the line of sight $\phi = 0$. Since it changes from case to case, this is probably the largest cause of scattering observed among data points.

Figures (2) and (3) show that, using acceptable value for the pressure in the PCTR, p = 0.1, 0.2 dyn/cm^2, the angle θ between $\underline{B}$ and $\underline{\nabla T}$ must be very close to 90°. If $\theta \cong 0°$ the pressure needed to give a good fit of the data is p = 0.04 dyn/cm^2. Radio observation could therefore supply a good estimate of the angle θ between the magnetic field and the temperature gradient, if the pressure is known or viceversa. It appears therefore that radio data alone are not a good diagnostic of the PCTR.

4. Comparison with UV data

It is very well known that the best diagnostic of the transition region both in the quiet sun and in active regions is provided by UV line intensity. In fact, since they are formed in a very narrow range of temperatures, the corresponding information is directly related to that portion instead of being integrated all over the TR as the radio emission does. The physical parameter that can be directly derived from the UV lines intensity is the so called differential emission measure (DEM) defined as:

$$Q(T) = N^2 \frac{dz}{dT} \qquad (7)$$

or, following Engvold (1989),

$$F(T) = S_{eff}\, p^2 \frac{dz}{dT} \qquad (8)$$

where $S_{eff} \leq 1$ is the effective emitting area which will be assumed equal 1 in the following. The DEM's F(T) or Q(T) are related to the function $\eta(T)$ defined above by the relation:

$$F(T) = 4k^2T^2\, Q(T) = \frac{p^2\, T^{5/2}(1+\varepsilon\, T^{-5})}{\eta(T)}$$

Observed values of Q(T) and F(T) for prominences are given by Schmahl and Orral (1986) and by Engvold (1989). Both sets of data refer to prominences observations at the limb, performed during the Skylab mission. However, if one compares the corresponding DEM in the same units, for instance F(T), finds a relatively good agreement at low temperature ($T \leq 8.10^4$) and a strong disagreement, more than one order of magnitude, in the upper part of the PCTR. If the DEM derived from UV lines is used to compute the radio brightness temperature, the curves shown in fig. 4 are obtained.

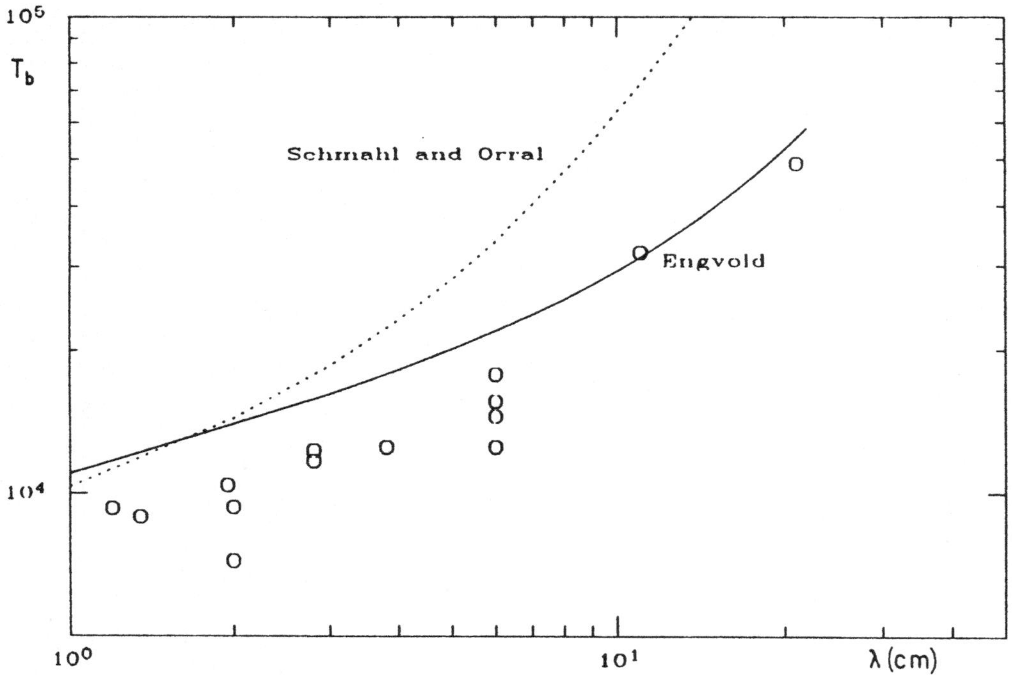

Fig. 4. - Computed radio brightness temperature using the DEM's derived from UV line intensities.

Fig. 4 shows that Engvold's DEM gives the corrected T_b only at low frequencies, while the brightness temperatures derived from Schmahl and Orral DEM are systematically larger than the observed ones all over the observed radio spectrum.

It will be now shown that the discrepancy between Engvold (1989) and Schmahl and Orral (1986) data, as well as the disagreement between UV and radio data can be explained in terms of a different angle θ between the magnetic field $\underline{B}$ and the temperature gradient $\underline{\nabla T}$. We want first to recall that radio observations are performed on the disk and therefore they refer to the top PCTR, while UV line intensities are measured at the limb and hence they refer to the side PCTR.

According to Leroy (1989) the magnetic field in prominences lies in a plane parallel to the solar surface and in this plane forms an angle α with the prominence axis.

A histogram of the α values obtained by Leroy et al. (1983), on a large sample of polar prominences, shows that α can vary from 0° to ~ 60° with the most probable value given by $\alpha = 25° \pm 5°$. Therefore, when we observe the prominences from the top (filaments on the disk) at radio frequencies, the angle θ between B and ∇T is $\theta \cong 90°$ while when the prominence is observed at the limb (UV lines) the angle can be $30° < \theta < 90°$ with the most probable value given by: $\theta = 65° \pm 5°$. The relationship between the angles α and θ is in fact $\theta = 90° - \alpha$. In Fig. 5 the function $F(T) = p^2\, dz/dT$ is derived from the previously defined function η (T) and it is compared with Engvold (1989) and Schmahl and Orral (1986) data. The parameters used for this computation are the same as those used to compute the fit of radio brightness temperature shown in fig. 3 and the angle θ is varied from 60° to 88°. We see that the upper part ($T > 8\ 10^4$) of the computed DEM can reproduce both sets of data by assuming different angles θ between $\underline{B}$ and $\underline{\nabla T}$ all within in the range observed by Leroy et al (1983).

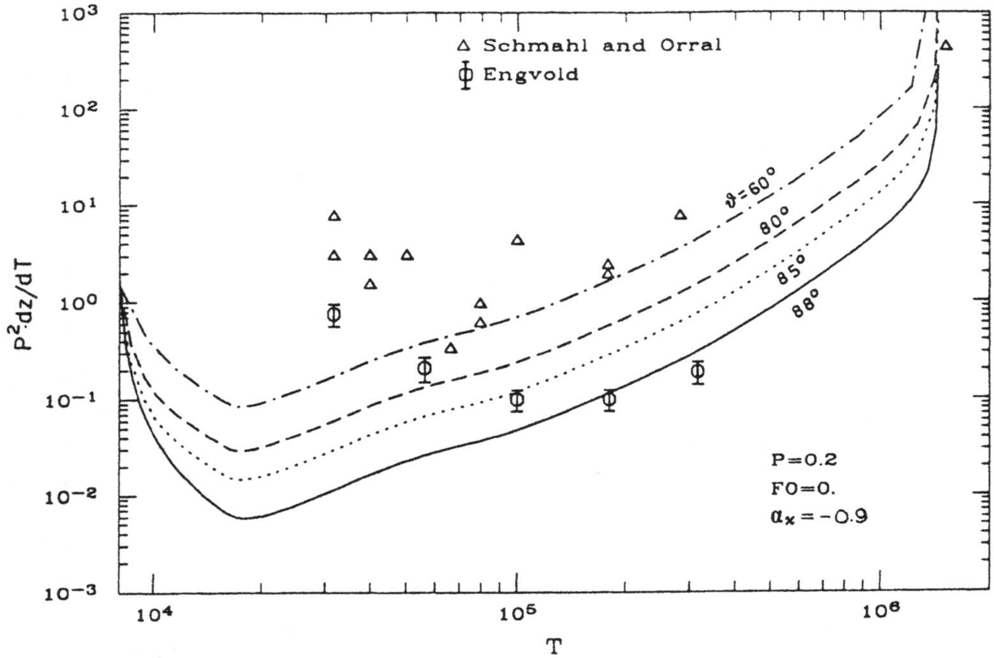

Fig. 5. - Comparison between the observed and computed DEM's for different values of the angle θ between the magnetic field and the temperature gradient.

On the contrary, in the lower part of the TR, where the observed data are in better agreement to each other, there is no agreement at all between the observations and the computed differential emission measure.

The computation has been repeated using different values of F_o and α_x. The results show that it is always possible to find reasonable values of the TR parameters which can reproduce the radio spectrum and the high temperature line intensities, but it appears impossible, in the above framework, to move the minimum of F(T) up to temperatures $T \sim (8 \div 10)\ 10^4$ as required by UV data. I think that among all the approximations done in this model the most severe ones are the assumption of a unique heating mechanism in the whole transition region between $T < 10^4$ and $T > 10^6$ and the assumption of a steady state (v = o).

5. Conclusions

Prominences can be analyzed by means of radio observations only using very short wavelengths ($\lambda \leq 1$ cm). Longer wavelengths observations, on the contrary, can supply interesting information on the PCTR. It has been shown that radio observations of filaments could give an excellent determination of the angle θ between the temperature gradient and the magnetic field, if the pressure in the TR is known. The comparison of radio data with UV line intensities strongly supports the determination of the magnetic field direction done by Leroy and coworkers. A PCTR model, derived by assuming a balance among the energy radiated and conducted and a heating function:

$$E_H = A\,T^{\alpha}$$

can reproduce the observed radio brightness temperature and the UV line intensities at $T > 8.10^4$, if the proper 3D field geometry (i.e. the angle θ) is correctly taken into account. This model however fails in reproducing the low temperature line intensities.

A calculation is now in progress (Chiuderi and Chiuderi Drago) aimed to determine a heating function, capable of reproducing the observed UV line intensity over the whole range of temperatures: $10^4 \leq T \leq 10^6$.

References

Apushkinskii, G.P. and Topchilo, N.A.: 1976, Sov. Astron. 20, 2, 393.

Axisa, F., Avignon, Y., Martres, M.J., Pick, M. and Simon, P.: 1971, Solar Phys. 19, 110.

Bracewell, R.A. and Graf, W.: 1981, Nature, 290, 758.

Buhl, D. and Tlamicha, A.: 1970, Astron. Astrophys. 5, 102.

Butz, M., Furst, E., Hirth, W. and Kundu, M.R.: 1975, Solar Phys. 45, 125.

Chiuderi Drago, F. and Felli, M.: 1970, Solar Phys. 14, 171.

Chiuderi Drago, F., Furst, E., Hirth, W. and Lantos, P.: 1975, Astron. Astrophys. 39, 429.

Chiuderi Drago, F., Felli, M. and Tofani, G.: 1977, Astron. Astrophys. 61, 79.

Dulk, G.A. and Sheridan, K.V.: 1974, Solar Phys. 36, 191.

Dulk, G.A. and Gary, D.E.: 1983, Astron. Astrophys. 124, 103.

Engvold, O.: 1989, "Dynamic and Structure of Quiescent Solar Prominences", E.R. Priest ed., Kluwer Acad. Pub., p. 47.

Furst, E., Hachenberg, O., Zinz, W. and Hirth, W.: 1973, Solar Phys. 32, 445.

Gary, D.E.: 1986, "Coronal and Prominence Plasmas", NASA Conference Publ. 2442, 121.

Hiei, E., Ishiguro, M., Kosugi, T. and Shibasaki, K.: 1986, "Coronal and Prominence Plasmas", NASA Conference Publ. 2442, 109.

Khangil'din, U.V.: 1964, Sov. Astron. 8, 234.

Kundu, M.R. and Lantos, P.: 1977, Solar Phys. 52, 393.

Kundu, M.R., Furst, E., Hirth, W. and Butz, M.: 1978, Astron. Astrophys. 62, 431.

Kundu, M.R., Melozzi, M. and Shevgaonkar, R.K.: 1986, Astron. Astrophys. 167, 166.

Lang, K.R. and Willson, R.F.: 1989, Ap.J 344, L73.

Leroy, J.L.: 1989, "Dynamic and Structure of Quiescent Solar Prominences", E.R. Priest ed., Kluwer Acad. Pub., p. 77.

Leroy, J.L., Bommier, V. and Sahal Brechot, S.: 1983, Solar Phys. 83, 135.

Pramesh Rao, A. and Kundu, M.R.: 1977, Solar Phys. 55, 161.

Pramesh Rao, A. and Kundu, M.R.: 1980, Astron. Astrophys. 86, 373.

Priest, E.R.: 1982, "Solar Magneto-Hydrodynamic", Reidel Publ. Co.

Raoult, A., Lantos, P. and Furst, M.R.: 1981, Solar Phys. 71, 311.

Saito, K. and Hyder, C.L.: 1968, Solar Phys. 5, 61.

Saito, K. and Tandberg-Hanssen, E.: 1973, Solar Phys. 31, 105.

Schmahl, E.J., Bobrowsky, M. and Kundu, M.R.: 1981, Solar Phys. 71, 311.

Schmahl, E.J. and Orral, F.: 1986, "Coronal and Prominences Plasmas", NASA Conference Publ. 2442, 127.

Schmieder, B.: 1989, "Dynamic and Structure of Quiescent Solar Prominences", E.R. Priest ed., Kluwer Acad. Pub., p. 15.

Straka, R.M., Papagiannis, M.D. and Kogut, J.A.: 1975, Solar Phys. 45, 131.

DISCUSSION

VIAL: Ambipolar diffusion and the induced ionization increase could provide the extra energy input you look for.

CHIUDERI-DRAGO: I have not taken into account the ambipolar diffusion so far. This term as well as the enthalpy flux in the energy balance will be taken into account in the final version of this model.

VIAL: You mentioned that the radio emission (cm) increased before formation, and stayed after the disappereance. Why? Is it a thermal effect?

CHIUDERI-DRAGO: Absorption at μ-waves is larger than in Hα, it is therefore possible that the prominence plasma can be thick enough to be seen at radio waves, but not enough to be seen in Hα.

PRIEST: I am glad to see you constructing an energy balance model, since the earlier model $p = a\ T^n$ did not seem to contain much physics. Have you thought of trying a turbulent thermal conductivity or to include an enthalpy term, since Engvold found the latter to be important in his models?

ENGVOLD: Since there are substantial flow velocities ($\gtrsim 5$ km s^{-1}) in the P-C transition region I agree that you should possibly include the enthalpy flux in your energy equation.

MASS MOTION IN AND AROUND PROMINENCES

B. Schmieder
Observatoire de Paris, Section de Meudon, Dasop
F 92195 Meudon Principal Cedex, France

ABSTRACT

The mass of a quiescent prominence is equivalent to one-tenth of the all coronal plasma. It is obvious that this crucial problem is resolved, now if we consider the dynamical nature of prominences. Observations of motions of filaments will be reviewed in regard to their time scales:

. solar cycle (slow migration of filaments, pivot points, convection)

. days or hours (stationary motions, oscillations)

. hours or minutes (appearance or disparition brusque, eruption)

These motions will be discussed in view of a better understanding of the formation of filaments (chromospheric injection or coronal plasma condensation), stability of the fine structures, existence of the feet, relationship of the DB, and the coronal mass ejections.

1 Introduction

Mass motion in and around prominences is directly connected to the question of the formation of the prominence and its stability. The origin of the material and the transport of the energy are puzzling problems.

Several reviews are now available on this topic (Tandberg-Hanssen, 1974; Hirayama,1985; Poland,1986; Schmieder,1989; Zirker,1989). Until recently, only static models were proposed with normal magnetic configuration (Kippenhahn and Schlüter,1957-KS) or inverse magnetic configuration (Raadu and Kuperus,1973-KR). But now it seems essential to take into account the dynamics of filaments and their environment, the convective motions in the sub-filament layers, the upward motions of filaments as an overall structure (0.5 km s^{-1} in Hα and 5 km s^{-1} in C IV) and larger ones in the horizontal direction, and the eruptions of filaments, as consequences of the evolution of the magnetic field (large and small scales) or of thermal instabilities. New models tentatively integrating some of these parameters will be presented in the relevant sections.

2 Cyclic Motions

A study of long time-scale motions shows the importance of filaments as tracers of the general solar magnetic field. Hα filaments are located between large regions of opposite polarity appearing at

20 degrees latitude. During their lifetime, most of them become aligned in an east-west direction due to differential rotation.

2.1 Large-scale Magnetic Pattern

Makarov and Sivaraman (1989) show a latitude zonal structure of the large-scale magnetic field during solar cycles, using Hα charts of 1905 to 1982 (CSCS Meudon and MacIntosh,1979). The filaments trace the trajectory of the poleward boundaries of magnetic regions of one dominant polarity as they migrate to high latitude during the solar cycle with a variable velocity from 5 to 30 ms^{-1} depending on the phase of the cycle . Ribes (1986) uses a somewhat similar method with the Meudon synoptic maps. The magnetic pattern consists of several torii during the active phase of the cycle. New torii occur at intervals of 2 to 3 years. The large-scale organization of the magnetic field seems to disappear near sunspot minimum, with each torus moving poleward.

2.2 Rolls

The original work of Ribes (review 1989) was to compare this pattern with the zonal meridional circulation pattern deduced from the migration of the sunspots from 40 degrees to the equator during the solar cycle. Using the collection of spectroheliograms of Meudon and a digital analysis based upon complex image processing to correct the solar image for photometric and geometric distortions (Mein and Ribes, 1990). Ribes, Mein, and Mangeney (1985) found a complex zonal meridional circulation following the trajectory of young sunspots with an amplitude between 15 m s^{-1} and 100 m s^{-1}. The borders of unipolar regions defined by the filaments lie at the latitudes where the meridional circulation reverses. This coincidence suggests the existence of east-west rolls characterized by a magnetic polarity and a direction of rotation. The new rolls drift equatorwards, and the pattern of the rolls moves poleward with time . *Converging motions or diverging motions* may exist from one side to the other side of filaments. Such motions are very important in understanding the mass motion of prominences. A large flow pattern in the convection zone correlated with filament has been already suggested by Schmieder *et al.* (1984b).

The torsional oscillation signal (Dermendjiev *et al.* 1989) is well explained by the roll pattern (Ribes, 1989). Latitude bands of faster and slower than average rotation will form and move equatorwards, following the appearance and disappearance of rolls. The boundary between the latitude bands moves in a latitude range of $\pm$ 20 degrees (Duchlev *et al.* 1988). The amplitude of the drift is small (3m s^{-1}), however may be important in the stability of prominences in view that they are deeply anchored in the photosphere with the footpoints.

The convection of the rolls with rigidly rotating layers favors a solar dynamo located below the convection zone possibly in the radiative interior which rotates like a solid body. The rolls can be theoretically interpreted as the convective response of the toroidal field in the framework of the dynamo theory . It is not clear, however, that the rolls represent an unique system of convective motions in the generation of magnetic activity as suggested Snodgrass and Wilson (1987). Gilman and Miller (1986) also proposed theoretical models with meridional cells. But which is dominant, the Laplace forces or the Coriolis forces? If the solar cycle is a balance between two large-scale convective modes, one would expect to detect meridional cells near sunspot minimum (Ribes, 1986).

2.3 Singularities of Surface Differential Rotation

The surface (Ribes, 1989) of the Sun may exhibit some singularities in differential rotation with a time scale of the month. Recently, following the behavior of long lived filaments during 7-10 rotations, some points rotating with the Carrington rate have been detected (Soru-Escaut, Martres, and Mouradian,1984). They can be located in a large latitude range (±35 degrees). It is tempting to associate these points to the boundary between latitude bands found by Duchlev *et al.* 1988. In the vicinity of pivot points, new active centers (Martres *et al.* 1987), new emerging fluxes (Mouradian *et al.* ,1987) and flaring activity (Soru-Escaut, Martres, and Mouradian,1985; Bumba and Gesztelyi 1987) have been observed. The pivot point plays an important role with the reappeareance of filaments after Disparition Brusque (see Section 4).

3 Stationary Motions Over Day

Stationary motions in prominences (time scale: few hours or day) may be the keys to understanding the formation, the stability of prominences. The problem of the stability is connected to the knowledge of the photospheric motions: footpoints, diverging or converging motions of the subphotospheric layers. Models of injection of chromospheric material into the corona or condensation of coronal plasma itself may be tested directly by the dynamics of the chromosphere and prominence-corona transition zone plasma.

During the last 10 years, many observations have been made on the disk or at the limb, using coronagraphs, spectrographs, and filtergram techniques. The observations of filaments were made in a wide wavelength-range: UV lines with space telescopes (UVSP on SMM, HRTS) or Ca II, He I and Hα , and Fe I with ground-based instruments. On the disk, dopplershifts are interpretable directly as velocities, when the filament is thick enough. If this is not the case, different techniques may be used, based on the "cloud model" method (Beckers 1964, Grossmann-Doerth and Von Uexküll,1971). If the velocity values are large ($\geq$5 km s^{-1} in Hα), the method developed by Mein and Mein (1989) gives good accuracy (Schmieder *et al.* 1988b). Concerning the observations at the limb, the difficulty is to appreciate the diplacements as velocities of material. Apparent motions may be due to waves going through the plasma increasing the density, to radiative transfer, or to dynamics changing the ionization degree.

What have we learned since the new observations? We can summarize the results around four points: the filament as a global structure, the footpoints, the fine structure, and the oscillations.

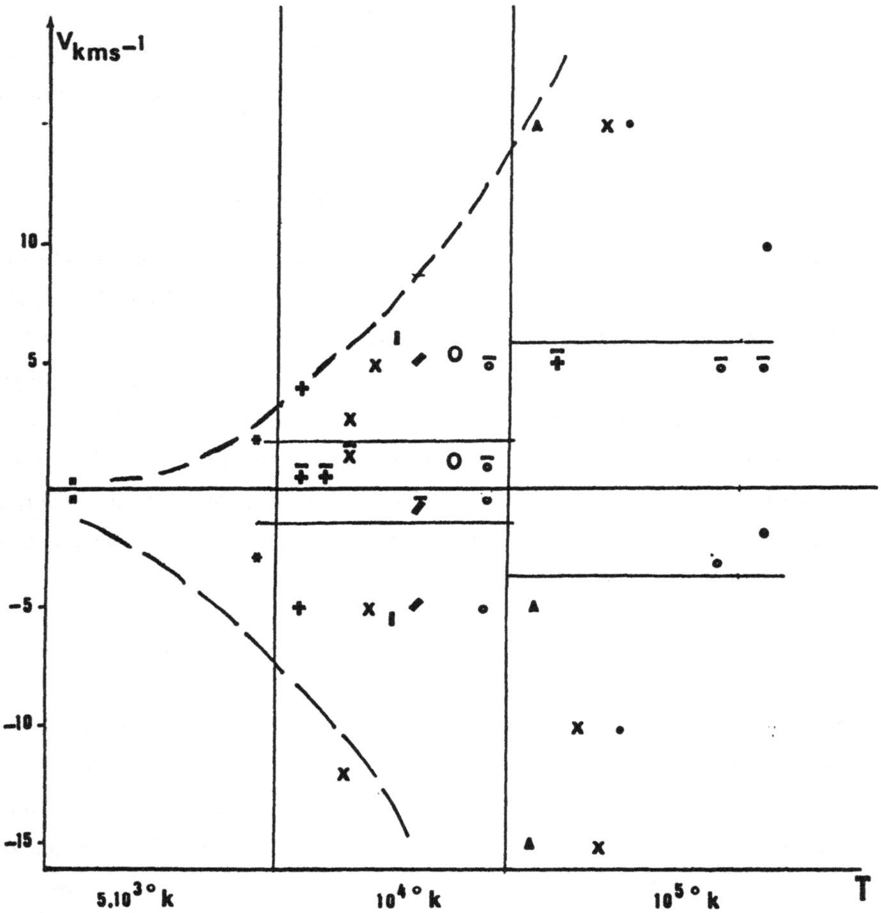

Figure 1: Vertical velocities in filaments on the disk according to the formation line temperatures. The dispersion of the values increase with the temperature. In spite of the dispersion the mean velocity values (overlined signs) are upward.

■ Martres, Rayrole, and Soru-Escaut,(1976) ✱ Engvold and Keil (1986) ■ Martres, *et al.* (1981)

✚ Schmieder *et al.* (1984a) ✖ Schmieder *et al.* (1985) ✖ Simon *et al.* (1986)

⁄ Kubota and Uesugi (1986,1986) ✚ Malherbe *et al.* (1983) O Kiryukhina (1988)

● Dere, Bartoe, and Brueckner(1986a,b) ● Klimchuk (1986) ❙ Kubota (1980)

○ Schmieder *et al.* (1988b) ▲ Engvold *et al.* (1985)

3.1 Global Structure

Most of the recent filament *dopplershift* measurements are presented in the Figure 1. The Dopplershifts correspond to up and downward motions with the largest values for the highest formation line temperatures. Nevertheless the average values are always blueshifted (upward motions):

Fe I lines ($5x10^3$ K)	$\bar{V} \leq 0.3$	km s^{-1}
Hα, He I lines (10^4 K)	$\bar{V} \sim 0.5$	km s^{-1}
C IV line (10^5 K)	$\bar{V} \sim 5$	km s^{-1}

An explanation for such an upward motion has been suggested by Malherbe and Priest (1983). They proposed a qualitative model with magnetic configuration, either normal (N) or inverse (I) magnetic type (respectively KS and KR types). The new coronal plasma enters symmetrically on both sides of a current sheet where it condenses and cools. The upward motions in filaments could be due to converging (or diverging) photospheric motions causing a steady reconnection of magnetic field lines below the prominence in an I configuration (or N). Sakai, Colin and Priest (1987) investigated quantitatively such a model . Filaments lie along the inversion line between two regions of opposite polarity in a bipolar region (N) or between two bipolar regions (I) according to Tang (1987). The formation of a current sheet is produced by the approach of such two regions. With temporal compression, condensation occurs in the current sheet. The dynamics of magnetic collapse exhibits nonlinear oscillation of the current sheet and upflow motions.

Using the observed ratio between the upward motions in chromospheric and transition lines, some physical parameters may be deduced: density and thickness of the transition zone filament-corona. If the flow at the base of prominences is negligible, as argued Malherbe and Priest (1983), mass conservation holds between input of material on both sides of the sheet and material flowing out at the top of the sheet. Thus, the mass flux of cold Hα plasma should be equal to the flux crossing the transition region at the top of filaments. The relation of gas pressure conservation ($P = \rho\ T$) between the cold (Hα, $T = 10^4$ K) and the hot (C IV, $T = 10^5$ K) plasmas postulates a ratio between the densities equal to 0.1 which leads to a ratio of vertical mass fluxes per unit surface ($F = \rho\ V$) equal to 1, assuming the ratio between the velocities equals 10. Thus, mass conservation holds, which means that the observed upflow in C IV could correspond to the flow crossing the transition at the top of filaments when returning to a coronal state. Using the rate between the densities and the ratio between the C IV intensities measured in filament and outside ($I = kN^2L/T$), the thickness of the transition region L_f filament-corona can be compared with that of the quiet Sun L_c. For the classical electronic density value ($N = 10x^{11}cm^{-3}$) , $L_f \leq L_c$; for lower values, it is the contrary. The problem cannot be resolved.

It is interesting to speculate that the observed updraught of prominence is a source of solar wind. The high speed solar wind comes from the coronal region void between streamers (MacQueen,Sime, and Priest, 1983) while, through open magnetic structures , material could contribute to low velocity solar wind (Saito and Tandberg-Hanssen, 1973). The slow upward prominence could provide the elements in solar wind. Démoulin made a computation based on the separation of elements. Five prominences with a density n = 10^{16}-10^{17} m^{-3} are sufficient to produce the low velocity solar wind.

The horizontal motions are evidenced in filaments observed near the limb and well visible in the two wings of Hα line (Figure 2, top). They have larger amplitudes (10-20 km s^{-1}) than the vertical ones :

lines	V_{max}	authors
Hα	10 km s^{-1}	Mein*et al.* , 1989 (Hvar communication)
Mg II	20 km s^{-1}	Vial *et al.* , 1979 , Lemaire, Samain, and Vial, 1988
Ca II	30 km s^{-1}	Engvold, Malville, and Livingstone, 1978 (edges of filament)
C IV, Si IV	10 km s^{-1}	Athay, Jones, and Zirin, 1985; Lites *et al.* 1976; Engvold, Tandberg-Hanssen, and Reichmann, 1985

Is the horizontal flow along or perpendicular to the filament axis? Using observations of prominences at the limb, it is difficult to conclude. The geometry of prominences is needed to answer to this question with some assumptions (Figure 2,bottom). Analysis of Hα center to limb filament observations with the Meudon Multichannel Subtractive Double Pass (MSDP) spectrograph showed the existence of a large horizontal component of the velocity vector. Fast horizontal motions (∼ 5 km s^{-1}) are suggested at the edges of active region filament (Malherbe *et al.* 1983) with a direction slightly inclined toward the prominence axis (20 degrees), a similar direction found by Leroy, Bommier, and Sahal-Bréchot, (1984) for the magnetic field lines. The material circulation may be a process of continuous material supply from the chromosphere or the corona to the prominence (siphon flow model of Pikel'ner 1971).

3.2 The Footpoints

The footpoints in the prominence play an important role in the formation and the equilibrium of prominences. Most of the models concern the overall structure (2D) and do not pay attention to the anchorage of the prominence in the photosphere. Rotational or translation motions near the footpoints cause disruption of filament (Section 4). Looking at dynamical movies of prominences, it is frequent that vertical flow along tubes is visible, but what kind of model may be investigated if we realize that the vertical extension is 100 times the pressure scale height? Magnetic fields play a fundamental role. Observations with the Pic du Midi coronagraph always indicate horizontal magnetic fields but they were made 10,000 km over the limb and the feet could not be seen. The observations of the footpoint dynamics could supply if we realize that material is frozen in magnetic field lines. Schmieder *et al.* (1985) have detected strong downflows using the MSDP and the UVSP spectrographs in active filaments. Their magnitudes were comparable in Hα and C IV ($\leq$ 10 km s^{-1}). They were located at the end of filaments. Their lifetime can be around 1 to 10

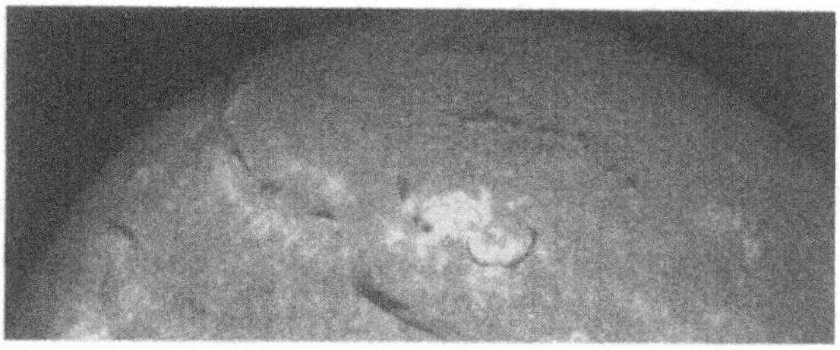

Hα

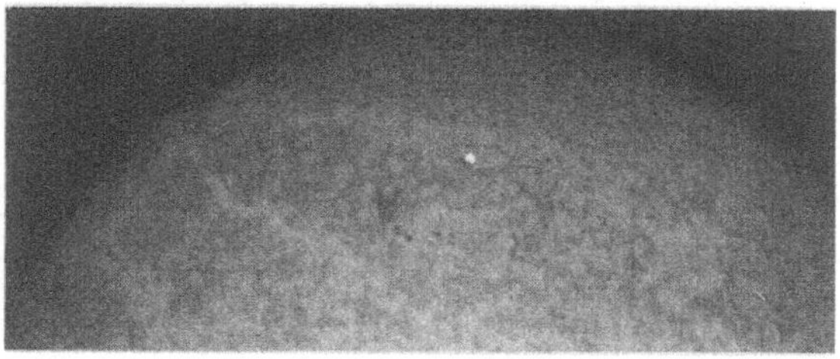

Hα - 0.5 Å

Hα + 0.5 Å

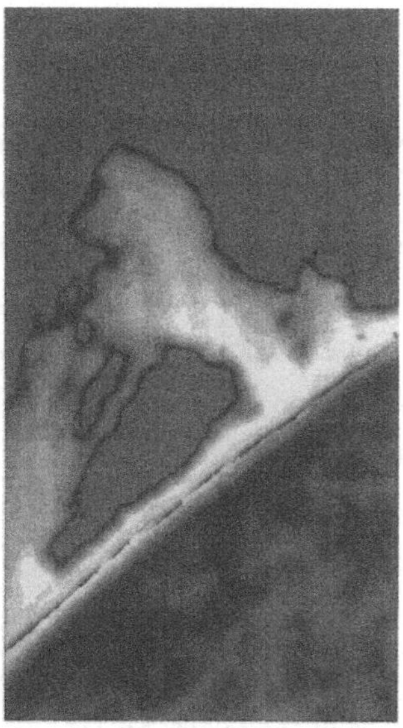

I

V

Figure 2: Horizontal velocities, Top panel: Meudon Heliograms obtained on May 5, 1989, at 17:40 UT. North is up. Note the long filament at the east visible in the both wings which indicates strong horizontal flow and a surge in the active region. Bottom panel: Prominence observed at Pic du Midi on June 7, 1988, with the MSDP. Note the white regions corresponding to blueshifts up to 10 km s^{-1} and the dark ones to redshifts in the structure parallel to the limb illustrating shear velocities (Mein, IAU Colloq.n^0 117,1989).

hours. Up and down motions ($\sim$ 6 km s^{-1} in Hα) were also observed at footpoints by Kubota and Uesugi (1986) and Kubota *et al.* (1989-Hvar communication).

The interaction of the footpoints with convection cells in the photosphere is obvious now. Their location is still controversial. Some arguments are in favor of the centers of supergranules, such as the distances between feet of prominences or "suspended legs," comparable with supergranule sizes. But observations frequently show footpoint anchorages at the boundaries of several supergranules when the contours of the supergranules are not vanished in filament channels (Plocieniak, and Rompolt, 1973; Hermans and Martin, 1986). Schmieder and Mein (1989-Hvar communication) observed such a configuration with the high resolution telescope of Pic du Midi (Figure 3). Using a cloud model method, upward and downward motions reaching 10 to 20 km s^{-1} have been detected at the footpoints which correspond to spicule-like speed. Such upflow could provide material that injection models need (An, Wu, and Bar, 1988).

A three dimensional model has been investigated by Démoulin, Priest, and Anzer (1989), modeling the external field with a linear force free field and the filament by a current line, which exhibits the two possibilities of anchorage of the filament footpoints. With a (I) configuration the feet are at the center of the convective cells, while with a (N) configuration, they are located at the edges. The magnetic field is less sheared and stronger at the feet. This model is quasi-stationary. Convection motions have to be taken into account to understand how they could deform the magnetic field in order to induce condensation and eruption.

3.3 Fine Structure

The situation becomes more confused for quiescent filaments observed with high resolution at the Pic du Midi, even if the mean velocity is upward on the disk (Landman, 1985; Simon *et al.* , 1986; Démoulin *et al.* , 1987) where the velocity cell size is small and limited by the spatial resolution. The filament consists of vertical threads or fine loops imbedded in a large configuration and the velocity becomes enigmatic with no general behavior; it is average over many fine structures (Engvold and Keil, 1986). The Dopplershift cells that we observe do not coincide spatially with intensity contours, so it appears that there is a problem due to the radiative transfer in the Hα line. The measured velocity and intensity probably do not come from the same region. Models with Alfv̀en waves as support have been proposed by Jensen (1986,1989-Hvar review).

3.4 Oscillations

The problems of the existence of oscillations are well summarized in the review by Tsubaki (1988). We did not present "winking" phenomena occurring in prominences during activity or flare. The recent paper of Vrs̃nak (1984) on an oscillating prominence is more relevant to this phenomenon than to stationary oscillations in quiescent prominences (Section 4).

We have reported in Table 1 most of the relevant observations of the oscillation detection in

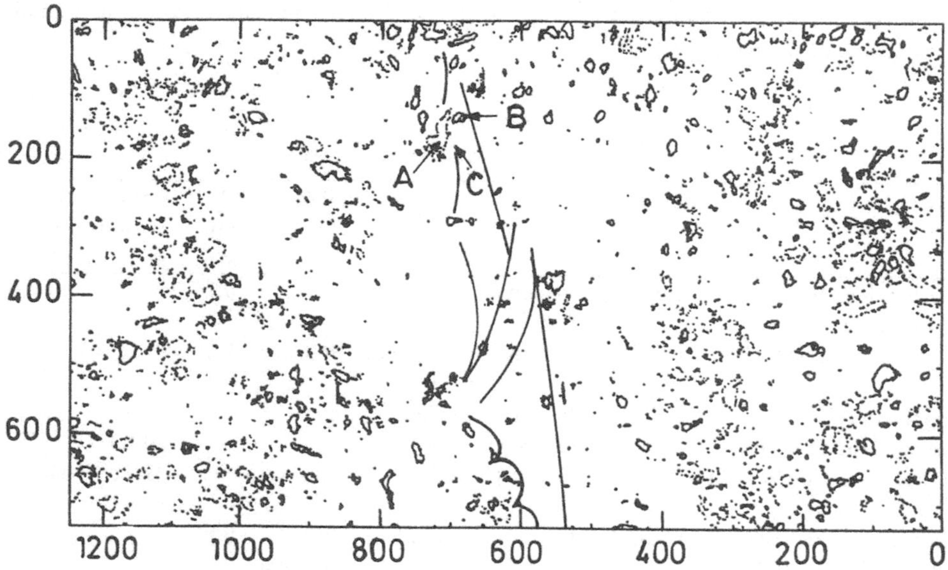

Figure 3: Velocity contour map ($\pm$ 5 km s^{-1}), exhibiting small cells of high values (continuous light lines/dashed correspond to upward /downward motions), at the boundaries of supergranules and at the footpoints of the filament (heavy lines) on June 17, 1986, (Schmieder and Mein IAU Colloq.n^0 117,1989).

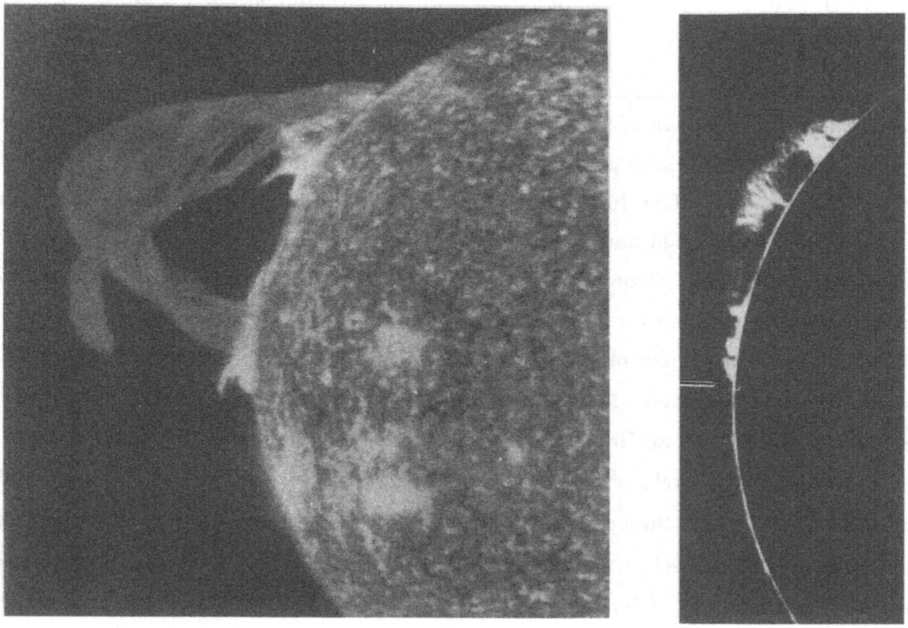

Figure 4: *Large prominence observed in* He_{II} *line (304 Å , 50000 K) with Skylab on Dec.18, 1973 rising with a velocity of 100 km* s^{-1} *(NRL). Six hours later, the prominence footpoints were visible in* $H\alpha$ *at the same position than in He II with material between (Meudon, Courtesy of Dr.Martres).*

prominences. Two kinds of oscillations have been detected in the velocity field : short period ($\leq$ 400 s), long period (40-80 min).

Table 1. Oscillations in Solar Prominence.

Authors	Line	Method	Region	Short Period s	*A* km s^{-1}	Long Period min	*A* m s^{-1}
Malherbe, Schmieder,	Hα	Velocity	whole	no	0.15		
and Mein,1981;		F F T					
Malherbe *et al.* ,1987	C IV	Velocity	filament	200-250	5-10		
Bashkirtsev,Kobanov,	Hβ	Velocity	5"x4"	no		76-82	
and Mashnich,1983,1987		difference	and			40-80	
Bashkirtsev,Mashnich,1984		(Sayan)	9"x9"	no		42-82	200
Tsubaki *et al.* 1987;1988	Ca II K	F.F.T.	points	160	0.8		
Tsubaki, Takeuchi,1986		Spectro	along	240-400	2		
Suematsu *et al.* 1989	Hβ	(Hida)	the slit	240-830		60	
Wiehr, Stellmacher,	Hα	Zeeman		180-300		50-64	
Balthasar, 1984							
Balthasar *et al.* 1986		Polari-		210-400	2	48	
Balthasar, Stellmacher,		meter	80"x80"	180-300		14	
Wiehr, 1988		(Tenerife)	48"x48"			70	

The problem of the detection of the oscillations is questionable. The small periods correspond , more or less, to chromospheric or photospheric frequencies. Suematsu *et al.* (1989) indicate that they are inconsistent with Alfv̌en like waves. Are these oscillations due to resonant phenomena as suggested Koutchmy, Zugžda and Ločans (1983)? Did they really exist, in the filament and even in the solar atmosphere? Geonjian (Hvar communication, 1989) explain the 3 and 5 min by reconnections and reflections in the terrestrian atmosphere layers. Wiehr (Hvar communication, 1989) is now doubtful with his new observations concerning the same prominence obtained with two different instruments at Canaries. The two sets of data do not show reliable periods. Thompson and Schmieder have analyzed an Hα filament near the limb in order to detect oscillations in the horizontal component of the velocity. Only chromospheric periods were detected. Tsubaki *et al.* (1988) found for each vertical thread forming a prominence its own period. Such motions may be generated from below by convection actions. This idea has been invoked to interpreted the increase of the power spectrum of a C IV line near the feet of an active filament (Malherbe *et al.* , 1987).

Long-period waves are related more or less to activity like the observation of Malville and Schindler (1981) or transient phenomena (Suematsu *et al.* , 1989). Oscillations could precede the onset of a flare. The russian group (see table 1) detected in 15 observations oscillations with periods from 42 to 82 min. The oscillation amplitudes are not stationary but undergo periodical decreases or increases in magnitude, and the chromosphere shows the same long period oscillations, not the photosphere. Again it could be indicated that the oscillations in prominences are due to chromospheric ones. A process of resonance may be investigated.

4 Evolution of Prominences in Some Hours or Minutes

4.1 Description of Disparition Brusque and Eruption

Different words are commonly used to define evolution such as: Disparition Brusque and eruption. "Disparition Brusque" (DB) of filament was observed at first in 1889 by Deslandres using a spectroheliograph and well studied by D'Azambuja and D'Azambuja (1948). Moreover DBs correspond to disappearances of filaments between two consecutive daily observations. A large quiescent filament has an expected life of the order of three rotations. Some are nevertheless visible during a year, as shown in the statistical analysis of D'Azambuja concerning 206 filaments. DB is a common event in the life of the filament in two third of the cases. It is also obvious that the same filament can have two or more successive DBs occurring during its life. Before DB, the filament becomes darker in Hα (brighter at the limb), indicating an increase of either density or microturbulence (Doppler broadening) or scatter light. Unresolved velocities (or microturbulence) of the order of 20 to 40 km s^{-1} in a filament have been measured in Hα and C IV lines by Mein and Schmieder (1988).

Eruptions of prominences are observed on the limb and correspond to fast motions ($\sim 100\,\mathrm{km\,s^{-1}}$). They are well visible in Hα during few hours ($\sim$ 4 hours) (Figure 5 and Rompolt, Hvar review, 1989). Nevertheless ejections of material during flares look like eruptions of prominences at the limb (Engvold, Jensen, and Andersen, 1979, Mein and Mein, 1982). Only the continuity of the observations is a good test of the existence of filament before a DB. As the loop expands, the electron density becomes low: Ne $\sim 10^8$ cm^{-3} (Athay, Low and Rompolt,1987; Illing and Athay, 1986) and could explain the difficulty of completly following an eruption on the disk .

These phenomena could be due to heating of plasma , principally if the DB is temporary (Mouradian, Martres, and Soru-Escaut 1981; Mouradian, Soru-Escaut, 1989) or to acceleration of cool material (Raadu *et al.* , 1987).

In fact, DB and eruption represent similar physical phenomena with more or less heating and dynamics of plasma (Figures 4 and 6). Coordinated observations using space data in UV lines and measuring velocities are absolutly necessary to determine the principal cause of DBs. Fontenla and Poland (1989) observe an eruption of prominence in lines formed in a wide temperature range (10^4 to 10^5 K) and show effectively that a dynamical process can occur within a heating process.

Mouradian, Soru-Escaut, (1989) point out that DBs are final if no pivot point exists. The DBs are often temporary; they last a few minutes to some hours but no more than a few days (5 to 6); progressively dark points and then dark absorbing matter reappear between the feet in order

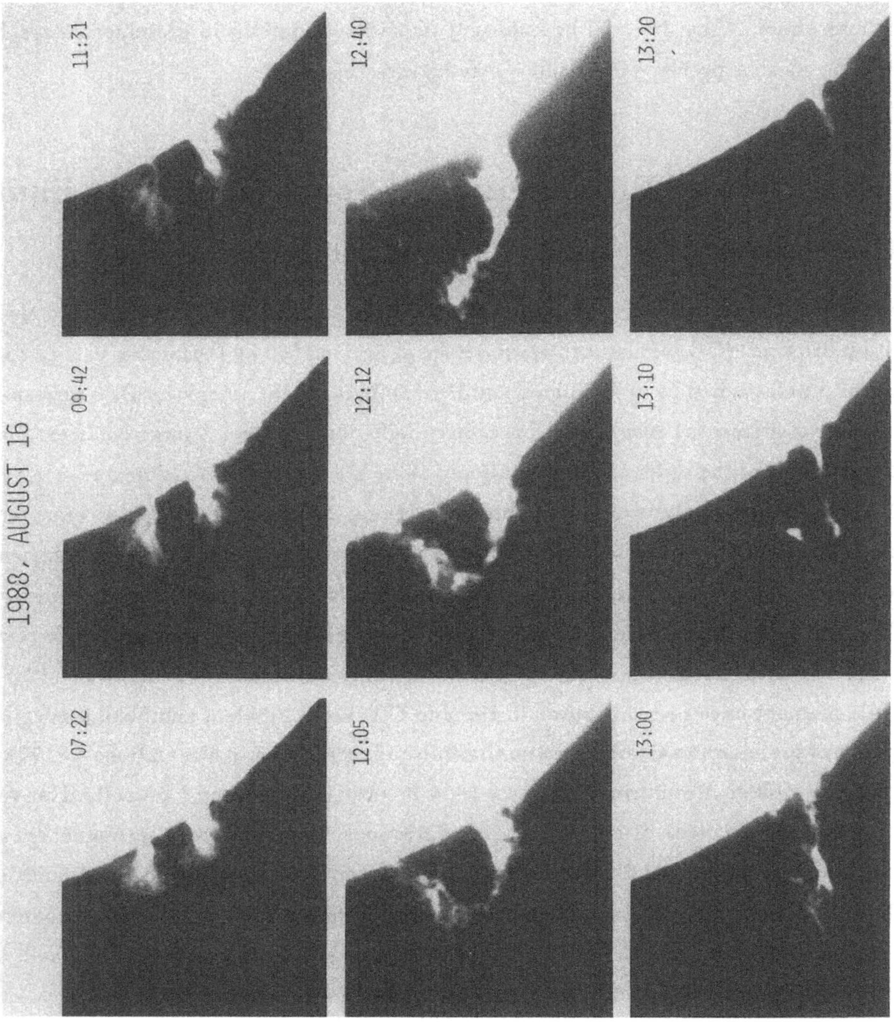

Figure 5: Dynamical DB of a quiescent prominence observed with the Meudon heliograph (Mouradian, and Soru-Escaut, IAU Colloq.n^0 117,1989)

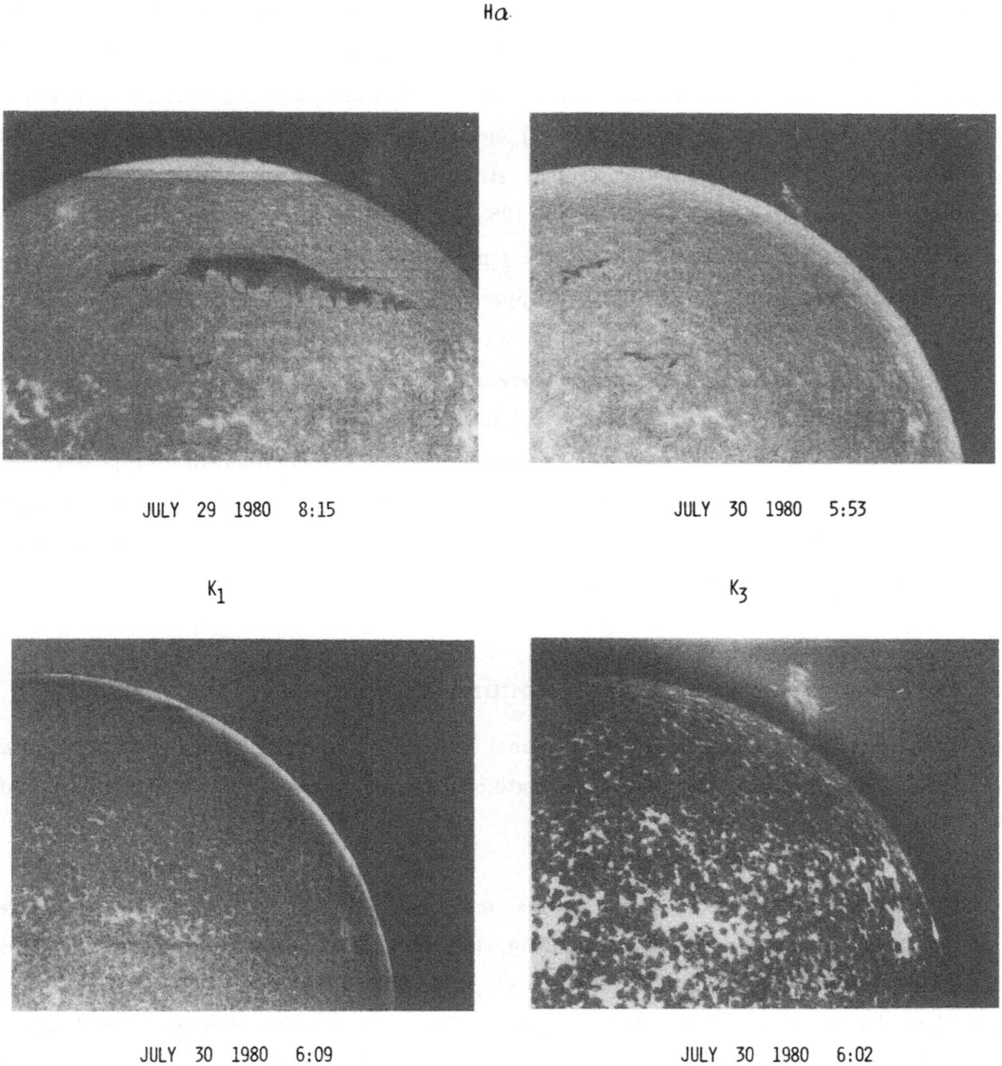

Figure 6: Large polar crown filament observed on July 29, 1980 at Meudon in Hα as a very dark structure. On July 30, at 06:02 UT, the filament is no more visible, only the corridor of the filament and its two extreme ends. Material of the filament is visible over the limb as a large loop reaching an altitude more than 150,000 km at 07:00 UT and higher than 300 000 km in K_3 (0.5 R). During the DB, the footpoints between the two extreme ends of the filaments were disconnected from the photosphere. The prominence is observed in K_{1V} indicating the presence of blueshifted plasma ($\sim$ 100 km s^{-1}); the expansion of the filament loop is probably not radial. The material was ejected and heated like in the example of Figure 4 (courtesy of M.J. Martres).

to reconstruct a filament. It is not well defined if this matter corresponds to material cooling while it is going down or if the conditions of condensation to form the filament are satisfied again. During DB, some observations of active filaments suggest that the filament material is always at the same place but heated (no more visible in Hα); ejected cool material is provided by condensation in a new current sheet overlying the filament, structured into threads by kink instabilities and accelerated by magnetic forces (Raadu *et al.* , 1988). In these observations the filament reappears completely suddenly (few minutes), when the temperature is decreasing. Disparition Brusque, due principally to a heating mechanism and Apparition Brusque, will be reversible phenomena, as suggested Malherbe (1989). This process is more commonly observed in active regions.

During filament disappearances, energy is released by mechanical and radiative ways as in flare or in surge events. The causes of DB could be similar to flare processes (Rust, 1984). But DBs are not really related to flare events, except in the two-ribbon flares where filament disappearance is a precursor phenomenon (Tang, 1985,1987, Kalher *et al.* , 1986, Kaastra 1985). Harrison, Rompolt and Garczynka (1988) determined an energy loss for eruptive prominences as 10^{29} ergs in 90 min which is relatively low compared with flare energy.

4.2 Causes of Instabilities in Prominences

The instabilities in prominences are of a thermal or magnetic nature (Priest, 1982). What can induce the increase of the heating rate, the length of the loops, or the shear of the magnetic field? Many causes may be invoked:

1. The **local photospheric activation** may induce destabilization: vortex motions (Martres *et al.* , 1982) and convection motions : shearing, stretching (Martin *et al.* , 1983), new emerging flux (Martres and Soru-Escaut, 1977; Simon *et al.* , 1984,1986; Raadu *et al.* , 1988, Apushkinskij, 1988; Merlenko, Palamanchuk and Polyakov, 1983). Uralov (1988) describes what could happen in case of **new emerging flux**. He imposes boundary conditions at the footpoints of loops with currents antiparallel to B. The loops rise and then bend, they connect and form knots; then it is the rupture of reconnected loops. We observe a long magnetic filament and short, flat small loops below.

Hermans and Martin (1986) studied small-scale eruptive filaments in the quiet Sun and found that the majority of these structures were related to **cancelling magnetic features** in video-magnetograms.

Priest (1987) points out the important role of reconnection processes in the cancellation of photospheric magnetic features. He proposes the reconnection submergence; the reconnection takes place above the photosphere, the curvature force balances the magnetic buoyancy so that the lower field line moves down as the separate poles approach. The upper line goes up. Van Ballegooijen

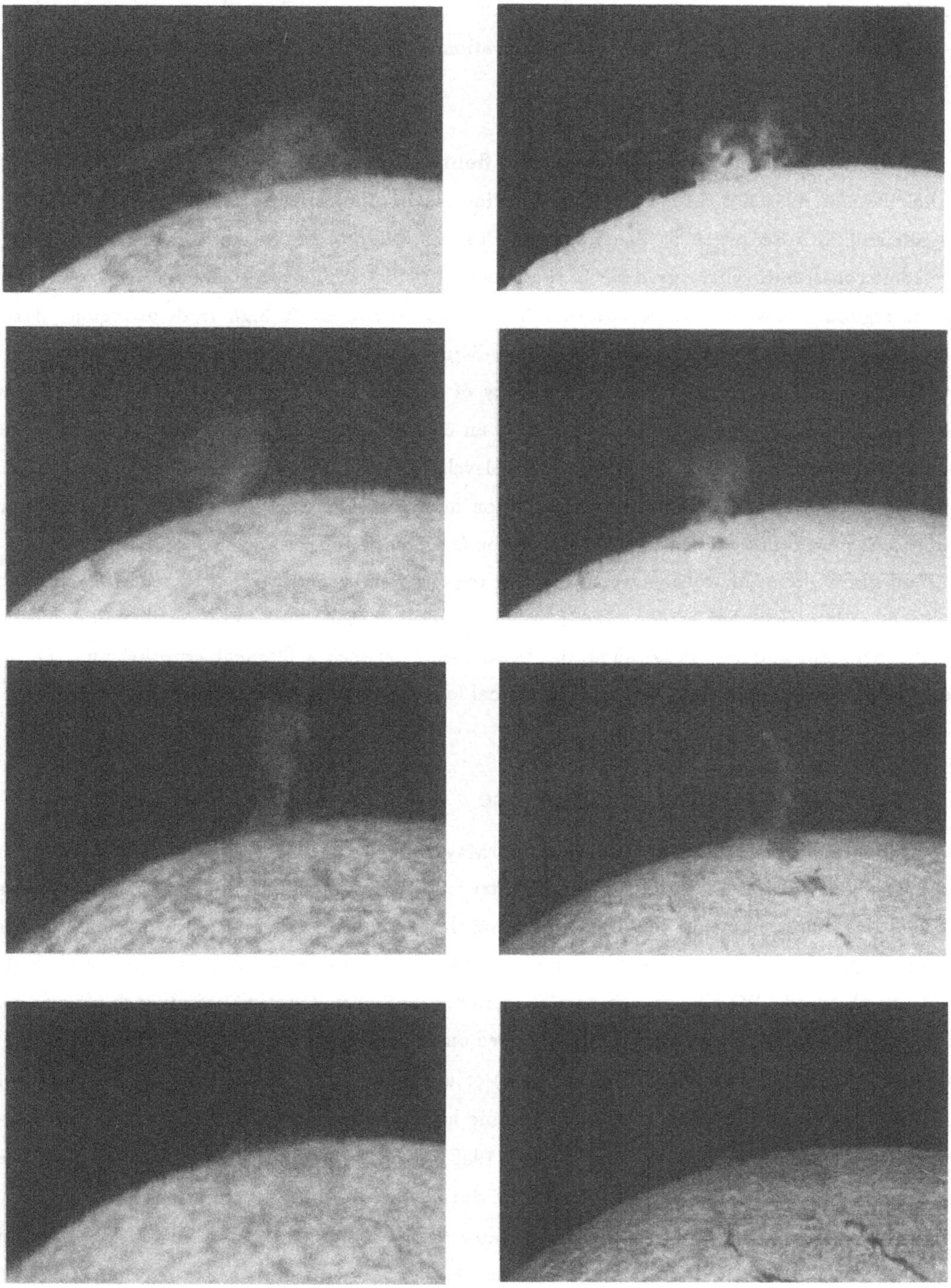

Figure 7: Torsional motions in a prominence during its diruption (top to bottom May 31, June 2, 4, 5, 1989). Note the different structures visible in Hα, indicator of density (right panel),and in K_3, indicator of temperature (left panel) (Meudon spectroheliograph).

and Martens (1989) propose a model where flux cancellation in a sheared magnetic field drive a reconnection process which produces a configuration of the Kuperus-Raadu type, capable of supporting prominence plasma.

2. The topology of **large-scale magnetic field** is primordial.The instability is well forecast if the filament does not follow the inversion line of the photospheric magnetic field (Martres, Rayrole,and Soru-Escaut,1976; Maksimov and Prokopiev, 1988). Maksimov and Ermakova (1986) give some constraints on the gradient of the magnetic field ∇B. Quiescent filament can be formed only in regions of low ∇B values corresponding to large corridors. In high gradient regions, flares may occur. Gaizauskas (1989) explains filament disruption by the slow evolution of bipolar regions (expansion and contraction). At the boundary of a spiral spot, Schmieder *et al.* (1989) show the formation and the partial disappearance of an active filament due to large-scale anti-parallel converging mass motions at the photospheric levels; it is a good example of the sketch given by Rompolt and Bogdan (1986). Destabilization may occur by changes of filament equilibrium conditions: increase of current in a bipolar region (Démoulin and Priest, 1988).

Hot giant coronal loops overlying active regions visible in X rays (Schmieder, Raadu and Malherbe, 1985; Schmahl *et al.* , 1982) or in microwaves (6-20 cm) have destabilizing effects. Kundu, Schmahl and Fu (1989) show the relationship between a filament eruption and the increase of microwave emission (20 cm) in a coronal loop using VLA observations.

4.3 Helical Structure Prominence

The prominence observed by Vrŝnak (1984) on May 26, 1982, is a typical case of destabilization without effective eruption. At first, flow of matter is going out of the prominence; as the pressure becomes low, the prominence could rise, torsional threads are observed in the prominence always tied in the photosphere, then, one footpoint is disconnected, untwisting threads at the other footpoint are observed. We give an example of torsional motion in eruptive prominence (Figure 8). This kind of destabilization is frequently observed on the disk (Schmieder, Raadu, Malherbe,1985; Gaizauskas 1985; Malville and Schindler, 1981) or at the limb (Rompolt, 1975, Athay, Low and Rompolt, 1987; Vrŝnak *et al.* , 1988). Only some helical prominence structures are eruptive according to their pitch angle ϕ (Vrŝnak *et al.* , 1988). Recently Vrŝnak (1990) has measured the variation of ϕ in a prominence with the altitude during 150 min until its eruption.

A twisted flux tube with free ends is unstable to the helical kink instability. The effect of line-tying at the ends of a flux tube is stabilizing. A uniform twist-force-free tube requires a twist larger than 2.6 π before it becomes unstable (Hood and Priest,1980). Vrŝnak *et al.* (1988) shows that if we take into account the curvature of the tube and the "mirror current" effect, the critical twist is reduced to 2.38 π. The eruptive prominences that they have observed are in the unstable region of its diagram ϕ/D (D = d/r_0) / Z (the altitude), the other helical prominences observed are in the stable part. Priest, Anzer, and Hood (1989) have developed a model of torsional flux tube. The torsion is created by subphotospheric motions (differential rotation or Coriolis force) The prominence is formed if $\phi \geq \phi_{cri}$; the instability increases with magnetic shear (Démoulin, Priest and Anzer, 1989). The currents in the solar atmosphere cannot be tested. Making the assumption that the matter is frozen along magnetic field lines, the observation of twisting tubes could be very instructive to connect the atmosphere to the lower layers and understand convection and dynamo theory.

4.4 Post-Flare Loops

During the gradual phase of two-ribbon flares, dense and cold features called "post-flare loops" are formed. Fine observations were made at Yunnan Observatory using the Hα- SSHG spectroheliograph and the Hα telescope (Gu and Li 1988) that allowed them to the reconstruct the geometry and to find the forces interacting (Hanaoka, Kurokawa, and Saito,1986). Postflare loops have been studied extensively by Schmieder *et al.* (1987,1988a). Heinzel and Karlicky (1987) computed theoretical redshifted profiles of Hα lines to explain the brightening appearance of Hα off-band observations of dense loops (Loughead, Wang, and Blows,1983). Reconnection modes are consistent with the observations. They explain the formation of the post-flare loop by the compression of the plasma due to a fast shock occurring at the X-reconnection point. Hot material is condensed and high density post-flare loops are formed (Forbes and Malherbe, 1986; Forbes *et al.* , 1989).

CONCLUSION

In conclusion, it is very interesting to notice that quiescent filaments are located between large-scale magnetic structures of opposite polarity, the rolls, with converging or diverging motions, one toward the other, with a non-uniform amplitude of velocity 15 to 100 m s^{-1}. Zonal structures are accelerated or decelerated compared with the normal differential rotation; such behavior leads to shear of the magnetic field and torsional effects. As the filaments are anchored deep in the photosphere, these motions favor conditions of formation (creation of current sheet) and disruption of filaments (increase of current) . Theoretical 3D models are required to take into account such motions in order to explain the instabilities in filaments. The problem of the anchorage of filaments in the photosphere is in progress but not resolved. The stability of quiescent prominences composed of a lot of fine threads is difficult to understand. The direction of horizontal flow compared to the filament axis would lead to a better understanding of the filament formation (condensation process or chromospheric injection). Coordinated observations of magnetic field and velocity field in and around filaments are necessary to make progress in this topic. Soho and Themis are very promising for the future.

REFERENCES

- An, C.H., Wu, S.T., and Bao, J.J.: 1988, in J.L. Ballester and E.R. Priest (eds.), Dynamics and structure of Solar Prominences, in Proceeding of Mallorca workshop, p.51.
- Apushkinskij, G.P.: 1988, *Astronomichesky Journal*, **65**, 1319.
- Athay, R.G., Jones, H., and Zirin, H.: 1985, *Astrophys. J.*, **288**, 363.
- Athay, R.G., Low, B.C., and Rompolt, B.: 1987, *Solar Phys.*, **110**, 359.
- Balthasar, H., Knolker, M., Stellmacher, G., and Wiehr, E.: 1986, *Astron. Astrophys.*, **163**, 343.
- Balthasar, H., Stellmacher, G., Wiehr, E.: 1988, *Astron. Astrophys.*, **204**, 286.
- Bashkirtsev, V.S. and Mashnich, G.P.: 1984, *Solar Phys.*, *91*, 93.
- Bashkirtsev, V.S., Kobanov, N.I., and Mashnich, G.P.: 1983, *Solar Phys.*, **109**, 399.
- Bashkirtsev, V.S., Kobanov, N.I., and Mashnich, G.P.: 1987, *Solar Phys.*, **109**, 399.
- Beckers, J.M.: 1964, Thesis, Utrecht University.
- Bumba, V. and Gesztelyi, L.: 1987, *Bull. Astron. Inst. Czechosl.*, **38**, 351.
- D'Azambuja, L. and D'Azambuja, M.: 1948, *Ann. Obs. Paris-Meudon*, **6**, 7.
- Démoulin, P., Raadu, M.A., Malherbe, J.M. and Schmieder, B.: 1987, *Astron. Astrophys.*, **183**, 142.
- Démoulin, P. and Priest, E.R.: 1988, *Astron. Astrophys.*, **206**, 336.
- Démoulin, P., Priest, E.R. and Anzer, U.: 1989, *Astron. Astrophys.*, in press.
- Dere, K.P., Bartoe, J-D.F., and Brueckner, G.E.: 1986a, *Astrophys. J.*, **310**, 456.
- Dere, K.P., Bartoe, J-D.F, and Brueckner, G.E.: 1986b, *Astrophys. J.*, **305**, 947.
- Dermendjiev, V.N., Duchlev, P.I., Velkov, K.P., and Zlateva, E.B.: 1989, *Astrophys.Investig.*, Sophia, 7.
-Duchlev,P.I., Dermendjiev,V.N.,Zlateva,E.B.:1988, Proceedings of the XII Regional Consultation on Solar Physics, Odessa.
-Engvold, O., Malville, J.M., and Livingstone, W.: 1978, *Solar Phys.*, **60**, 57.
- Engvold, O., Jensen, E., and Andersen, B.W: 1979, *Solar Phys.*, **62**, 331.
- Engvold, O., Keil,S.: 1986, in A.I. Poland (ed.), Coronal and Prominence Plasmas, NASA CP-2442,169.
- Engvold, O., Tandberg-Hanssen, E., and Reichmann, E.: 1985, *Solar Phys.*, **96**, 35.
- Fontenla, J.M., and Poland, A.: 1989, *Solar Phys.*, in press.
- Forbes, T.G., and Malherbe, J.M.: 1986, *Astrophys. J. Lett.*, **302**, L67.
- Forbes, T.G., Malherbe, J.M., and Priest, E.R.: 1989, *Solar Phys.*, **302**, L67.
- Gaizauskas, V.: 1985, in C. de Jager and C. Biao (eds.) Proceedings Workshop on Solar Physics and Interplanetary Travelling Phenomena, p. 710.
- Gaizauskas, V.: 1989, Proceedings of the Chapman Conference on the Physics of Magnetic Flux Ropes, Bermuda.
- Gilman, P.A., and Miller, J.: 1986, *Astrophys. J. Suppl.*, **61**, 585.
- Grossmann-Doerth, U. and Von Uexküll, M.: 1971, *Solar Phys.*, **20**, 31.
- Gu, Xiamo and Li, Qiusha: 1988, *Chinese Astronomy and Astrophys.*, **12**, 1, 19, ed., Kiung.
- Hanaoka, Y., Kurokawa, H., and Saito, S.: 1986, *Solar Phys.*, **105**, 133.
- Harrison, R.A., Rompolt, B., and Garczynka, I.: 1988, *Solar Phys.*, **116**, 61.
- Heinzel, P., and Karlicky, M.: 1987, *Solar Phys.*, **110**, 343.
- Hermans, L., and Martin, S.F.: 1986, in A.I. Poland (ed.), Coronal and Prominence Plasmas, NASA CP-2442, p. 369.
- Hirayama, T.: 1985, *Solar Phys.*, **100**, 415.
- Hood, A.N. and Priest, E.R.: 1980, *Solar Phys.*, **66**, 113.
- Illing, R.M.E., and Athay, G.: 1986, *Solar Phys.*, **105**, 173.

- Kaastra, J.S.: 1985, Ph.D. Thesis, Rijkuniversiteit-Utrecht-Netherlands.
- Kalher, S., Cliver, E.W., Cane, H.V., McGuire, R.E., Stone, R.G., and Sheeley, N.R.: 1986, *Astrophys. J.*, **302**, 504.
- Kippenhahn, R. and Schlüter, A.: 1957, *Astrophys. J.*,**43**, 36.
- Kiryukhina, A.I.: 1988, *Proceedings of XIII Consultative meeting of KAPG*, Odessa.
- Klimchuk,J.A.: 1986, in A.I.Poland (ed.), Coronal and Prominence Plasmas, NASA CP-2442,183.
- Koutchmy, S., Zugzda, Y.D., and Ločans, V.: 1983, *Astron. Astrophys.*, 120, 185.
- Kubota, J.: 1981, in F. Moriyama and J.C. Henoux (eds.), Proceedings of the Japan-France Seminar on Solar Physics, p. 178.
- Kubota, J. and Uesugi, A.: 1986, Publ. Astron. Soc. Japan, 38, 903.
- Kubota, J. and Uesugi, A.: 1989, Hvar Publ.
- Kundu, M.R., Schmahl, E.J., and Fu, Q.J.: 1989, *Astrophys. J.*, **336**, 1098.
- Landman, D.A.: 1985, Astrophys. J., 295, 220.
- Lemaire, P., Samain, D., and Vial, J.C.,: 1988, *Adv. Space Res.*, COSPAR meeting.
- Leroy, J.L., Bommier, V., and Sahal-Bréchot, S.: 1984, *Astron. Astrophys.*, **131**, 33.
- Lites, B.W., Bruner, E.C.,Chipman, E.G., Shine, R.A., Rottman, G.J., White, O.R., and Athay, R.G.: 1976, *Astrophys. J. Lett.*, **210**, L111.
- Loughead, R.E., Wang Jia-Long, and Blows, G.: 1983, *Astrophys. J.*, **274**, 883.
- MacIntosh, P.S. 1979, NOAA, Boulder, Colorado.
- MacQueen, R.M., Sime, D.G., and Picat, J.P.: 1983, *Solar Phys.*, **83**, 103.
- Makarov, V.I. and Sivaraman, K.R.: 1989, *Solar Phys.*, **119**, 35.
- Maksimov, V.P. and Ermakova, L.V.: 1986, *Contributions of the Astro. observ. Stainete*, **15**, 65.
- Maksimov, V.P. and Prokopiev, A.A.: 1988, *Astrophys. J. Moscow*, **79**, 90.
- Malherbe, J.M., Schmieder, B., and Mein, P.: 1981, *Astron. Astrophys.*, **55**, 103.
- Malherbe, J.M., Schmieder, B., Ribes, E., and Mein, P.: 1983, *Astron. Astrophys.*, **119**,197.
- Malherbe, and J.M., Priest, E.: 1983, *Astron. Astrophys.*, **123**, 80.
- Malherbe, J.M., Schmieder, B., Mein, P., Tandberg-Hanssen, E.: 1987, *Astron.Astrophys.*, **172**, 316.
- Malherbe, J.M.: 1989, in E. Priest (ed.), *Dynamics and Structure of Quiescent Solar Prominences*, Kluwer Academic Publishers, p.115.
- Malville, J.M. and Schindler, M.: 1981, *Solar Phys.*, **70**, 115.
- Martin, S.F., Deszo, L., Gesztelyi, L., Antalova, A., Kucra, A., and Harvey, K.L.: 1983, *Adv. Space Res.*, **2**, (11), 39.
- Martres, M.J., and Rayrole, J., and Soru-Escaut, I.: 1976, *Solar Phys.*, **46**, 137.
- Martres, M.J., and Soru-Escaut, I.: 1977, *Solar Phys.*, **53**, 225.
- Martres, M.J., Mein, P., Schmieder, B., and Soru-Escaut, I.: 1981, *Solar Phys.*, **69**, 301.
- Martres, M.J., Rayrole, J.,Semel, M., Soru-Escaut, I., Tanaka, K., Makita, M., Moriyama, F., and Unno, W.: 1982, *Publ. Astr. Soc. Japan*, **34**, 299.
- Martres, M.J., Mouradian Z., Ribes E., and Soru-Escaut I.: 1988, in J.L. Ballester and E.R. Priest (eds.), "Dynamics and Structure of Solar Prominences", in Proceeding of Mallorca workhop, p.25.
- Mein, P. and Mein, N.: 1982, *Solar Phys.*, **80**, 161.
- Mein, P. and Schmieder, B.: 1988, in J.L. Ballester and E.R. Priest (eds.), Dynamics and Structure of Solar Prominences, in Proceeding of Mallorca workshop, p.17.
- Mein, P. and Ribes, E.: 1990, *Astron. Astrophys.*, in press.
- Mein, P. and Mein, N.: 1989, *Astron. Astrophys.*, **203**, 162.

Heliophysical Observatory, **5**, 293.
- Mouradian, Z., Martres, M.J., and Soru-Escaut, I.: 1981, in F. Moriyama and J.C. Hénoux (eds.), Proceedings of the Japan-France Seminar on Solar Physics, 195.
- Mouradian, Z., Martres, M.J., Soru-Escaut,I., Gesztelyi, L.: 1987, *Astron.Astrophys.*, **183**,129
- Mouradian, Z. and Soru-Escaut, I.: 1989, *Astron. and Astrophys.*, **210**, 410.
- Nakagawa, Y. and Malville, J.M.: 1969, *Solar Phys.*, **9**, 102.
- Pikel'ner, S.B.: 1971, *Solar Phys.*, **17**, 44. 11.
- Plocieniak, S. and Rompolt, B.: 1973, *Solar Phys.*, **29**, 399.
- Poland, A.I.(ed.): 1986, Coronal and Prominence Plasmas, (CPP) workshops, NASA CP-2442.
- Priest, E.R.: 1982, *Solar Magnetohydrodynamics, Geophysics and Astrophysics Monographs*, D. Reidel Publ. Co., Dordrecht, Holland.
- Priest, E.R.: 1987, in E. Schröter, Vasquez, and Wyller (eds.) *The Role of the Fine Scale Magnetic Fields on the Structure of the Solar Atmosphere*, p.297.
- Priest, E.R., Anzer, U., and Hood, A.: 1989, *Astophys. J.*, in press.
- Raadu, M.A. and Kuperus, M.: 1973, *Solar Phys.*, **28**, 77.
- Raadu, M.A., Malherbe, J.M., Schmieder, B., and Mein, P.: 1987, *Solar Phys.*, **109**, 59.
- Raadu, M.A., Schmieder, B., Mein, N., and Gesztelyi, L.: 1988, *Astron. Astrophys.*, **197**, 289.
- Ribes, E., Mein, P., and Mangeney, A.: 1985, *Nature*, **318**, 170.
- Ribes, E.: 1986, *C.R. Acad. Sc. Paris*, **302**, series II, 14.
- Ribes, E. and Laclare, F.: 1988, *Geophysical and Astrophysical Fluid Dynamics*, **41**, 171.
- Ribes, E.: 1989, in A. Cok and W. Livingston (eds.) *Solar Interior and Solar Atmosphere*, University of Arizona press, .
- Rompolt, B.: 1975, *Solar Phys.*, **41**, 329.
- Rompolt B. and Bogdan T.: 1986, in A.I. Poland (ed.), Coronal Prominence Plasmas, NASA, CPP-2442, Workshop proceedings, p.81.
- Rust, D.M.: 1984, *Solar Phys.*, **93**, 73.
- Saito, K. and Tandberg-Hanssen, E.: 1973, *Solar Phys.*, **31**, 105.
- Sakai, J., Colin, A., and Priest, E.R.: 1987, *Solar Phys.*, **114**, 293.
- Schmahl,E.J., Mouradian,Z., Martres,M.J., Soru-Escaut, I.: 1982, *Solar Phys.* 77, 121.
- Schmieder, B., Malherbe, J.M., Mein, P., Tandberg-Hanssen, E.: 1984a, *Astron.Astrophys.*, **136**, 81.
- Schmieder, B., Ribes, E., Mein, P., and Malherbe, J.M.: 1984b, *Mem. S.A. It.*, **55**, 319.
- Schmieder, B., Raadu, M.A., and Malherbe, J.M.: 1985, *Astron. Astrophys.*, **142**, 249.
- Schmieder, B., Malherbe, J.M., Poland, A.I., Simon, G.: 1985, *Astron. Astrophys.*, **153**, 64.
- Schmieder, B., Forbes, J.G., Malherbe, J.M, Machado M.E.: 1987, *Astrophys. J.*, **317**, 956.
- Schmieder, B., Mein, P., Forbes, J.G., Malherbe, J.M: 1988, *Adv.Space Res.*, **8**, 11, 45.
- Schmieder, B., Poland, A.I., Thompson, B., Démoulin, P.: 1988, *Astron. Astrophys.*, **197**, 288.
- Schmieder, B., Dere, K., Raadu, M., and Démoulin, P.: 1989, *Astron. Astrophys* **213**, 402.
- Schmieder, B.: 1989, in E. Priest (ed.), Dynamics and Structure of Quiescent Solar Prominences, Kluwer Academic Publishers, 15.
- Simon, G., Mein, N., Mein, P., and Gesztelyi, L.: 1984, *Solar Phys.*, **93**, 325.
- Simon, G., Schmieder, B., Démoulin, P., Poland, A.I.: 1986, *Astron. Astrophys.* **166**, 319.
- Snodgrass, H. and Wilson, P.: 1987, *Nature*, **328**, 696.
- Soru-Escaut, I., Martres, M.J., and Mouradian, Z.: 1984, in P. Simon (ed.), *Solar and Terrestrian Predictions*, T 299, series II, 9, 545.

- Suematsu, Y., Yoshinaga, R., Terao, N., and Tsubaki, T.: 1989, *Publ. Astron. Soc. Japan*, to be published.
- Tandberg-Hanssen, E.: 1974, *Solar Prominences*, D. Reidel Publ. Co.
- Tang, F.: 1985, *Solar Phys.*, **102**, 131.
- Tang, F.: 1987, *Solar Phys.*, **107**, 233.
- Tsubaki, T.: 1988, in R.C. Altrock (ed.), Solar and Stellar Coronal Structure and Dynamics, Sac Peak, 140.
- Tsubaki, T. and Takeuchi, A.: 1986, *Solar Phys.*, **104**, 313.
- Tsubaki, T., Ohnishi, Y., and Suematsu, Y.: 1987, *Publ. Astron. Soc. Japan*, **39**, 179.
- Tsubaki, T., Toyoda, M., Suematsu, Y., and Gamboa, G.A.R.: 1988, *Publ. Astron. Soc. Japan*, **40**, 121.
- Uralov, A.M.: 1989, *proceedings of the Chapman Conference* on the physics of Magnetic flux Ropes, Bermuda.
- Van Ballegooijen, A.A. and Martens, P.C.H.: 1989, *Astrophys. J.*, **343**, in press.
- Vial, J.C., Gouttebroze, P., Artzner, G., and Lemaire, P.: 1979, *Solar Phys.*, **61**, 39.
- Vršnak, B.: 1984, *Solar Phys.*, **94**, 289.
- Vršnak, B., Ruždjak, V, Brajša, R., and Džubur, A.: 1988, *Solar Phys.*, **116**, 45.
- Vršnak, B.: 1990, *Solar Phys.*, in press.
- Wiehr, E., Stellmacher, G., and Balthasar, H.: 1984, *Solar Phys.*, **94**, 285.
- Zirker, J.B.: 1989, *Solar Phys.*, **119**, 341.

Questions

S. Martin: Does the upflow in CIV occur uniformely along the length of a prominence or at discrete locations?

B. Schmieder: The upflow of 5 km s^{-1} in C IV is an averaged value over a large dispersion of values generally (see Figure 1).

E. Priest: You said that the magnetic field is horizontal in prominence feet. This is reasonable theoretically because otherwise the fall off of pressure with height would be larger than observed. However, the observations of Leroy refer to the region above 10, 000 km, so are there other observations from other sources below that height?

I.Kim: Nikolsky magnetograph allow us to obtain magnetic data at height > 7". For ARF, the horizontal field strengh is 7-15 G.

S. Koutchmy: Magnetographic measurements with a coronagraph are not the best method to look at the vertical component of the magnetic field, at least with the Zeeman effect. It is better to look at the center of the disk. This is what we started to do with the Sacramento Peak VTT, combining the use of the UBF, circularly polarized video frames and digital integration in the wings of Hα. We see a homogeneous vertical field significantly above the noise level, at the height of formation of the Hα line in absorption on the disk.

THE PROMINENCE-CORONA INTERFACE

J.-C. VIAL
Institut d'Astrophysique Spatiale, B. P. 10, 91371 Verrières-le-Buisson Cédex, France

ABSTRACT

The existence of cool and dense material in the hot and diluted corona implies specific mass and energy transfers between the two media. This is true for all steps in prominence lifetime : formation, quiescence and disappearance. Much theoretical work has been done recently on the formation by coronal condensation, but observational signatures are scarce, probably because of the long duration involved. On the contrary, the "Disparition Brusque" phenomenon has been observed in different wavelengths (temperatures) and shown to be either essentially magnetic or thermal. Line ratios have been used for the density diagnostics of eruptive prominences and point to a small filling factor. As for the quiet PCTR, the increase of Differential Emission Measure at lower temperatures, extensively studied with Skylab, is still a puzzle. With the help of both u-v (HRTS) and radio (VLA) new data, temperature gradients have been derived. The DEM increase could be explained by such heating process as waves or transients and also (at low temperature) by the reduction of radiative losses in optically thick lines. UVSP observations on SMM indicate upflows and downflows in the PCTR. Their positions with respect to the magnetic field lines are unknown simply because no magnetic measurement exists in the PCTR. There is much activity in modeling prominences as a superposition of fine structures (threads, loops,..) in thermal equilibrium and in comparing with the uv emission. Obviously, we now have some information on pressure and temperature gradient in the PCTR but we do not know the geometry, the magnetic field nor the heating process. Further decisive progress will be made with the spectrometers and coronagraphs on SOHO.

Introduction

Many recent general reviews on prominences have been published during the two last years which discuss the Prominence Corona Transition Region (PCTR). Let us mention Hirayama (1985), the Proceedings of the C.P.P. workshop (Poland, 1986), Zirker (1989). The latest reviews about the PCTR are from Engvold in Priest's book (1989) and in the 9thSacramento Peak Symposium Proceedings (1988). As we shall see, some significant work has been made recently and is going on. We discuss below the PCTR during the different stages of prominences : formation, disappearance and quiet life. We pay more attention to the determination of thermodynamic quantities such as temperature, densities, pressures in order to evaluate the energy exchanges through the PCTR. We also discuss the results obtained in modeling prominences in mhd and in radiative transfer. The importance of a complete temperature diagnostic with the best spatial resolution is emphasized.

I. Formation

The origin of material in prominence formation (photosphere-chromosphere or corona) is still debated and much work is being done on photospheric magnetic signatures of formation process (see Sarah Martin's review in this issue). Syphon models have been improved (see e. g. Poland and Mariska, 1986). We are concerned here with the mechanism known as the **coronal condensation** i.e. mass flows that originate from the corona. One will find in Malherbe's thesis a resume of the progress made in modeling the mhd of coronal condensation. Basically, the converging motions at the footpoints of a X configuration lead to a rise of material and a slow convergence of coronal gas along the nearly horizontal field lines (Figure 1, from Malherbe and Priest, 1983).

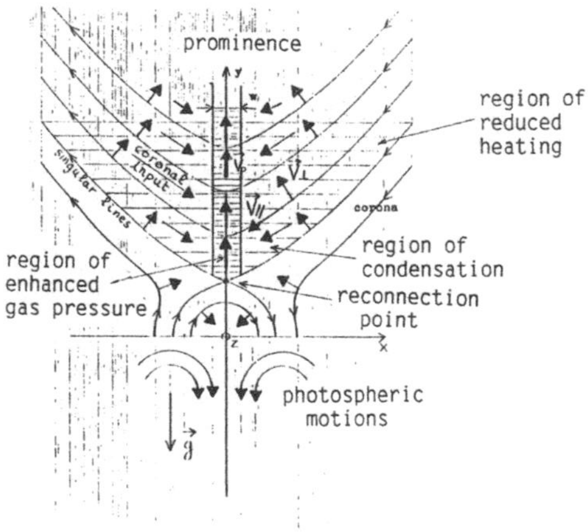

Figure 1 (from Malherbe and Priest 1983):
Schematic representation of the prominence formation through coronal condensation. Solid headed arrows represent the velocity field, i.e. the coronal mass input.

Both upward velocities (see Schmieder et al 1988) and a Kuperus-Raadu type magnetic configuration (Leroy, Bommier and Sahal 1984) exist but no coronal mass input has been directly observed up to now. An indirect signature has been suggested by Saito and Tandberg-Hanssen as early as 1973, namely the existence of a **coronal cavity** around Quiescent Prominences. Skylab and radio data have not provided a decisive answer (see e.g. Kundu 1986). Could the "missing" density around prominences (by factors 1/2 or 1/4) be enough for the formation of the cool prominence material ? Is there any Doppler signature of convergent velocities such as an anomalous line broadening (Fig. 1) ? Evidence of such a process going on during the lifetime of quiescent prominences has been provided by Toot and Malville (1987, see in § III.2). The necessary conditions for such a decisive observation of prominence formation include the appropriate timing for the prominence formation preferably at the East limb, the full wavelength (temperature) coverage (especially in uv lines), the best spatial resolution to investigate a possibly very thin transition region and of course a powerful instrumentation in order to record very faint emissions.

II. Disparition

Prominences seem to follow two different paths when disppearing (see Rompolt's review, this issue).

Magnetic disappearance : it can take the form of field lines untwisting, expanding (erupting prominences), etc.. Gaizauskas (1989) noticed flows and untwisting before an eruption along with major changes in the magnetic flux cells far from the filament. I recall here the observation of an erupting prominence of the polar crown during the 1981 eclipse (!) by Stellmacher, Koutchmy and Lebecq (1986). In spite of a short time sequence, these authors could measure some apparent motion of material from the top of the faint prominence towards the coronal helmet with a velocity of 160 km/s (Figure 2).

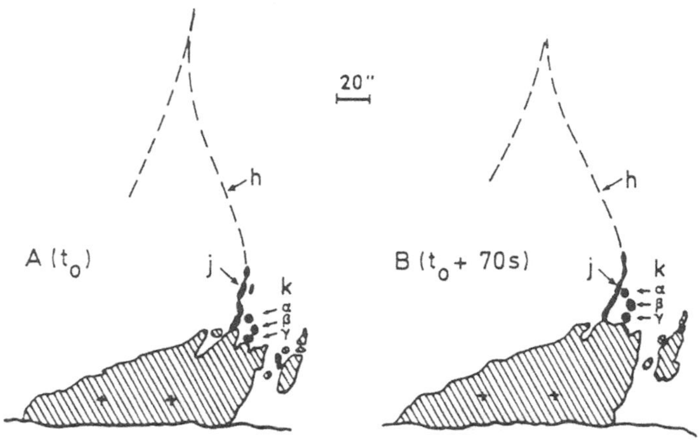

Fig. 2 : Sketch of the evolution of apparent motions of pieces of material above a faint (polar crown) prominence observed by Stellmacher, Koutchmy and Lebecq during the 1981 eclipse.

They concluded that a field as low as 3.5 G was enough to build the (magnetic) driving force. However, the way prominence and coronal magnetic field lines are connected is not clear.

Thermal Disappearance : It had been noticed in radio that prominences did not completely disappear after a Disparition Brusque (see e.g. Lantos and Raoult, 1980).
Enhancements have been recently observed in microwave before the disappearance (Kundu, Schmahl and Fu 1988). Analyzing Skylab data, Mouradian, Martres and Soru-Escaut (1980) also found enhancements in uv lines (Fig 3) and more recently, Mouradian et al (1989) working on simultaneous observations in euv, X and Hα explained the temporary nature of some DB by the heating of the prominence.

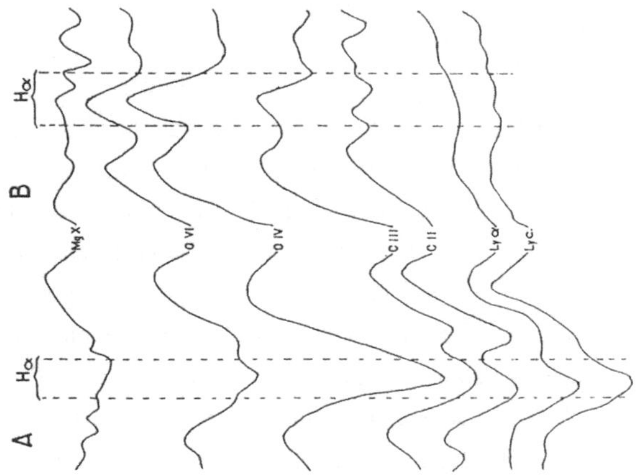

Fig. 3 : Cuts across a filament in different uv lines corresponding to higher temperatures (from bottom to top) obtained by Mouradian, Martres and Soru-Escaut (1980).

Erupting Prominences :
Their travel in the corona is a special case of interface with the corona. Recent work shows some problems in the determination of the density. Athay and Illing (1986) derive about $10^8 cm^{-3}$ for the electron density (at $2\ 10^4$ K) but Widing, Feldman and Bathia (1986) find about $8\ 10^9$-$3\ 10^{11} cm^{-3}$ from line ratio measurements. Such a discrepancy points at a very small filling factor in eruptive prominences.
Above the erupting structure observed with UVSP/SMM, Fontenla and Poland (1989) found a reduced electron density : would there be material still drained from the corona into the prominence even when the whole structure is lifted ?
An other interesting observation by Harrison, Rompolt and Garczynska (1988) shows an X ray emission localized at the foot of the prominence to erupt (Figure 4). This seems to be a disparition of the "third kind" since the activation of the prominence may be due to intense X ray radiation.

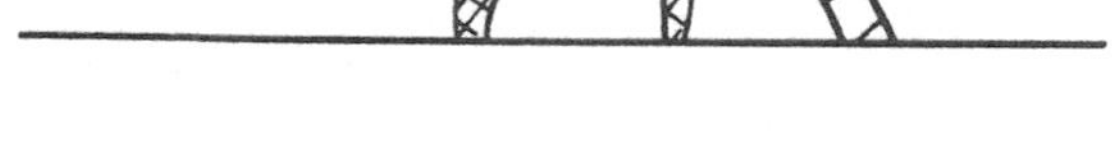

Fig. 4 (from Harrison et al 1988) : Sketch of the interaction between the prominence foot (seen in Hα) and a small X ray loop which activates the prominence.

III. Quiet Prominence-Corona (PC) interface

1/ Basic quantities (temperature, densities, pressure, thickness)

We recall briefly the early results, especially the ones gathered with the access to uv radiation. Orrall and Zirker (1961) had derived a strong temperature gradient in the PC interface. From cuts accross prominences in uv lines recorded with Skylab, Schmahl et al (1974) concluded that the PCTR was very thin (40 km ?). From eclipse observations, rather different results were obtained : Yang et al (1975) found an electron density of $10^{9.75}$ at $10^{5.5}$ K, implying a high gaseous pressure; on the contrary, Orrall and Speer (1974) derived a pressure of 0.04 dyn cm^{-2}. From line ratios measured again with Skylab, Moe et al (1979) determined a range of pressure of 0.01-0.2 dyn cm^{-2}. Let us also note that Orrall and Schmahl (1976, 1979) discovered around prominences the existence of Hydrogen absorption, a feature to be found probably also in the CCTR. These early results can be summarized as follows :

- ° a very thin transition region of a few ten km, as the CCTR is
- ° a pressure lower than in the CCTR
- ° and a temperature gradient less steep than in the CCTR, (Engvold et al 1987)

2/ The problem of the Differential Emission Measure (DEM)

The quantity $n_e{}^2 dl/dT$ or $A_{eff}\, P_e{}^2/(dT/dh)$ characterizes the variation of the emitting power of the layer with temperature. Here Aeff is the efficient area, that is the actual area of emitting material as seen along the line of sight. As shown in Fig. 5 (from Engvold, 1988), the DEM of the PCTR looks like the DEM of the PCTR.

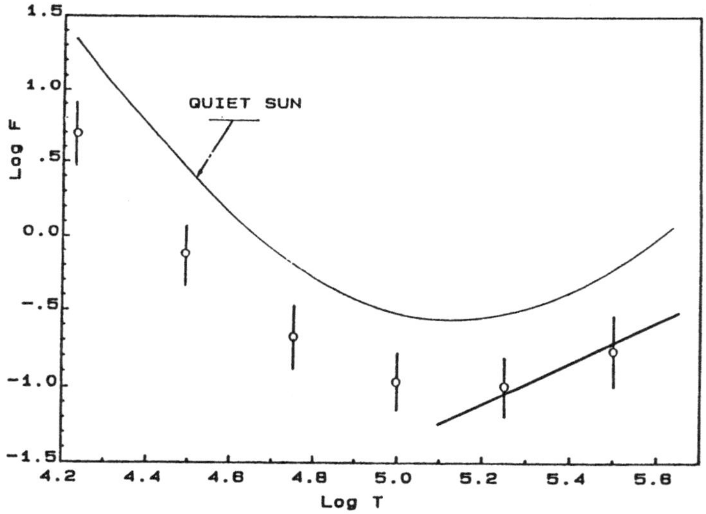

Fig. 5 : (from Engvold 1988)
Emission measures for 3 quiescent prominences observed with ATM/Skylab. The upper curve represents the quiet Sun (Engvold et al 1987). The straight line for $T>10^5$ K shows the relation expected when thermal conduction is the only energy input.

The increase of DEM at temperature lower than 10^5 K is as difficult to explain for the PCTR as for the CCTR. The heat conduction alone cannot be efficient at lower temperature. Since one sheath between cool and hot regions does not work, one can think to a superposition of sheaths. Orrall and Schmahl (1980) showed that about 4 to 10 such elementary transition regions were necessary in order to explain the observed (hydrogen) absorption. However, these authors (Schmahl and Orrall (1986) proved that multiple sheaths do not provide the increase of DEM, a not so surprising result in the optically thin approximation. These authors showed also that a mixture of isothermal threads at <u>different</u> temperatures observed with UVSP/SMM (Poland and Tandberg-Hanssen, 1983) does not work. They finally tried with some success threads with <u>longitudinal</u> (// B) temperature gradients (grad B). However, such a model faces some difficulties : the evidence of vertical threads <u>and</u> horizontal field lines (at least in cool regions) for instance, or the quasi impossibility to add up many transition regions along the field. Another model suggested by Rabin (1986) takes into account the angle between grad T and B and includes both parallel and perpendicular conductivity. The ratio between the dimensions of structures in parallel and perpendicular directions is very high in his model (thicknesses would be as low as a few km) and although it is below the observing capabilities and cannot be excluded, it is difficult to make the model work for both the PCTR and the CCTR.

<u>Recent Results :</u>

We already mentioned HRTS uv data. The analysis of Engvold et al (1987) starts from the expression of the DEM where they take p_e and A_{eff} constant and derive the temperature gradient dT/dh. The main results (see Figure 6) are the following: assuming a low filling factor, they obtain a low dT/dh. The gradient is not very sensitive to the angle of the structure with the vertical. An empirical relation between the length of the tube (L in km), the filling factor and the gas pressure (cgs) can be built : $L.A_{eff}.p_g=2.6$.

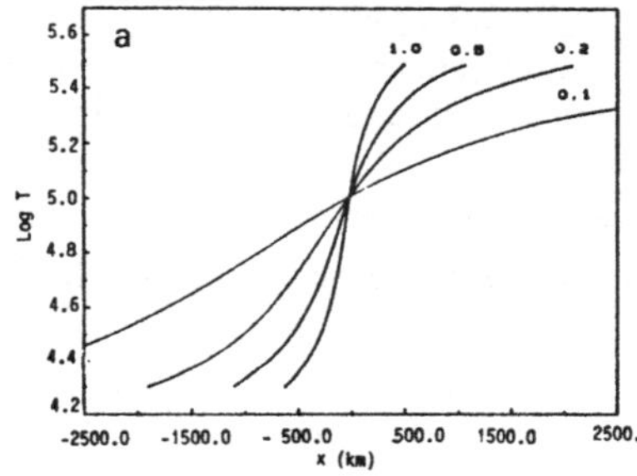

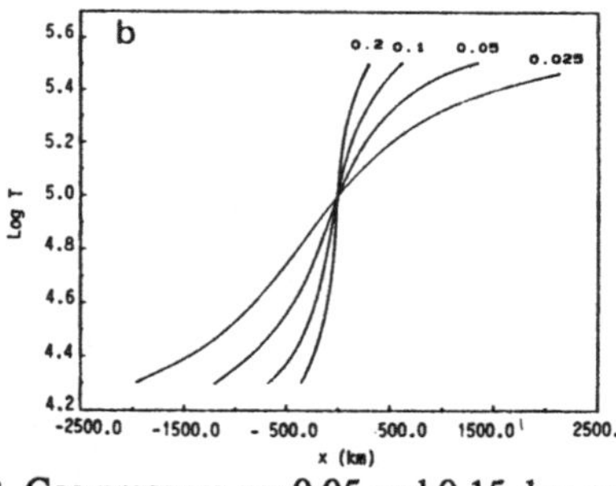

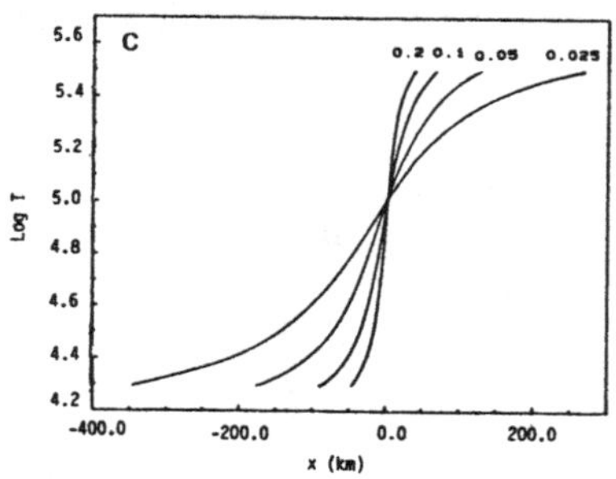

<u>Fig. 6 : (from Engvold 1988)</u>

Computed temperature structures of thin tubes in the PCTR. Gas pressure are 0.05 and 0.15 dyn cm^{-2} (a and b respectively) and surface filling factor ranges from 0.025 to 1. Figure 6c shows the temperature structure for the quiet Sun with a gas pressure of 0.45 dyn cm^{-2}.

For instance, with p_g=0.1 dyn cm^{-2}, L is the order of 1300 km. It should be mentioned that the emission filling factor (taken here as 0.02) is a surface factor, which implies that the actual volume filling factor is still lower

With the VLA, the diagnostic of such faint structures as filaments has been made possible, when the solar activity is minimum. Kundu et al (1986) noticed a depression at 20 cm, that could be interpreted as a coronal cavity around the filament. Having measured the brightness temperatures T_b at 20 cm and 6 cm (not an easy task because of the quiet Sun background), and assuming that the conductive flux balances the radiative losses at all temperatures, they could show that a constant pressure in one slab could not fit the observed T_b vs λ. On the contrary, a temperature variation of the pressure as $p=a\ T^{-b}$ gave an agreement (see Figures 7 for the values of a and b).

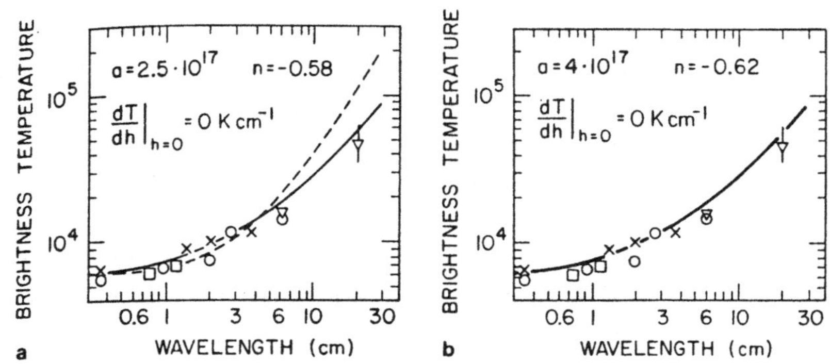

Fig. 7 : (from Kundu et al 1986)

Computed brightness temperatures versus wavelength for models with $p=aT^n$ (see text). The dotted line represents the spectrum with p=3 10^{14} cm^{-3}.

The derived run of temperature across the slab (Figure 8) displays a temperature gradient around 10^5 K similar to the one found by Engvold (10^{-2}-10^{-3} K/cm).

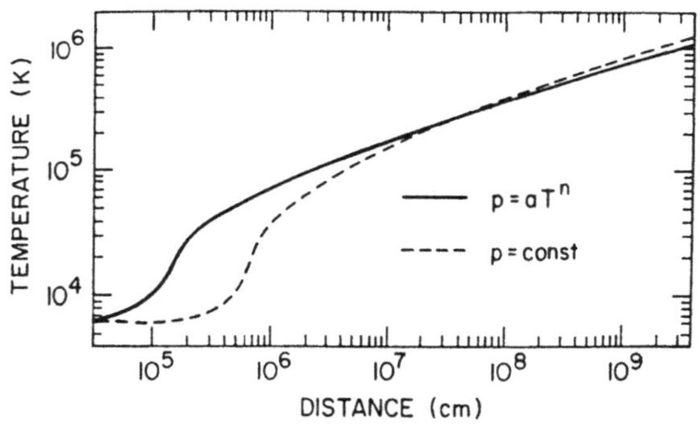

Fig. 8 : (from Kundu et al 1986)

Variation of temperature with distance in the filament for the two type of pressure models

Latest data with the VLA (Lang and Willson 1989) confirm the above run of temperature, since the observed brightness temperature at 92 cm lands on the computed one (Figure 9).

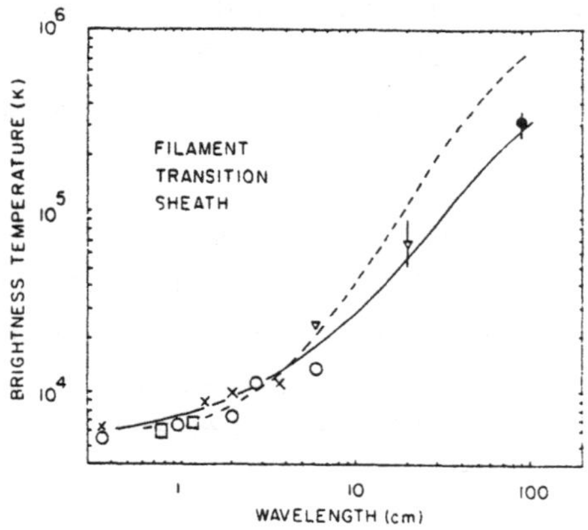

Fig. 9 (from Lang and Willson 1989)
Computed brightness temperatures versus wavelength for a models with p=2.5 10^{17} $T^{-0.58}$ (see text). The dotted line corresponds to the model with p=3 10^{14} cm^{-3}. The observed temperature at 91.6 cm is represented by the filled circle.

However, F. Chiuderi-Drago (Hvar meeting, this issue) questioned the validity of analyses where radiative losses are balanced by the conductive flux considered as an energy input everywhere. She also criticized the hypothesis of a temperature gradient parallel to the field, since magnetic field measurements (see e.g. Leroy et al 1977) indicate a horizontal field at an angle of about 20° with the structure and consequently at an angle of 80° with the temperature gradient. Perpendicular conduction must then be taken into account (see her Hvar contribution).
However, we are left with an excess of emission below 6 10^4 K that remains to be explained. We have two ways to solve the issue : a/ introducing some extra heating process or b/ decreasing the radiative losses at low temperatures.

A/ Energy input :
Enthalpy flux : as mentioned by Engvold (1988) quoting O.K. Moe, "enthalpy flux could contribue 90% of the total energy flow in a tube". It makes especially important the mass flow budget at all temperatures in filaments or prominences (see below).
Wave energy : velocity oscillations have been reported to take place in the cool part of prominences (Malherbe et al 1987, Balthasar et al 1986, Tsubaki and Takeuchi 1986, Tsubaki et al 1987, 1988, and also Wiehr 's talk in Hvar). Periods range from 3 minutes to one hour. Loop prominence oscillations at 8 minutes have been observed by Vrsnak (1984). Oscillations at shorter periods in the corona have been noticed by Koutchmy (1981) and Tsubaki (1977). But no measurement exists in the PCTR. The only proof of some mechanical dissipation is the existence of a non-null turbulence (see below).
Impulsive brightenings and velocity transients : impulsive events have been detected in Hα near the edges of prominences by Toot and Malville (1987). The large profile reversal, the magnitude of the Doppler shift and the rather long lifetime of such events (up to 60 minutes) show that there is an

important mass (and energy) transfer. Their location may be the indication of continuous condensation of coronal material.

Ambipolar Diffusion (see Fontenla, Avrett and Loeser, 1989): because of the steep temperature gradient in the PCTR, a transfer of ionization energy is possible towards the cool layers; the increase of electron density could be reflected in the DEM. This is discussed in § III.4.

B/ Exact radiative losses:

At, say 20 000 K, the optically thin approximation is no longer valid. Athay (1986) has shown that for the Lα line, the formula $Q = h\nu C_{12} n_1$ for the radiative loss is still valid but n_1 must be computed properly. For this line which is the main contributor around 2 10^4 K, the deviation from the coronal approximation increases with increasing opacity and decreasing electron density. Such conditions exist wherever the structure is large enough to prevent the ionization by the chromospheric L_c radiation.

3/ Flows and microturbulence

A complete review is given by B. Schmieder in this issue. In the PCTR, the measurement of flows and turbulence lies on the spectral capability of uv instruments. Skylab recorded lines intensities only. With UVSP/SMM and HRTS, profiles were obtained in many different lines but most works concentrated on the C IV line at 155 nm which is formed at 10^5 K. With UVSP/SMM, Dopplergrams evidenced upflows (Schmieder et al 1988) of about 5 km/s but also downflows : from a center-to-limb study, Simon et al (1986) showed that vertical velocities were larger than horizontal ones; and Engvold et al (1985) explained the presence of both up- and downflows by the model of Figure 10, where the ascending mass goes into the streamer above.

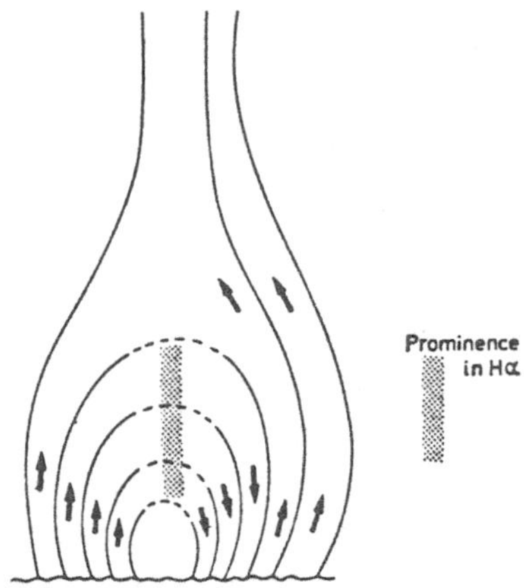

Fig. 10: (from Engvold et al 1985)
Schematic representation of up- and downflows in a prominence and in the streamer above.

With HRTS, full profiles are recorded along the slit. Schmieder et al (1989) analyzed the C IV profiles in an active region filament in connection with simultaneous MSDP observations in Hα (Figure 11). These authors notice a strong correlation between Hα and C IV; in the PCTR up- and downflows reach

50 km/s Their field computations (force free with α constant) show that the magnetic field is constant with height in the filament.

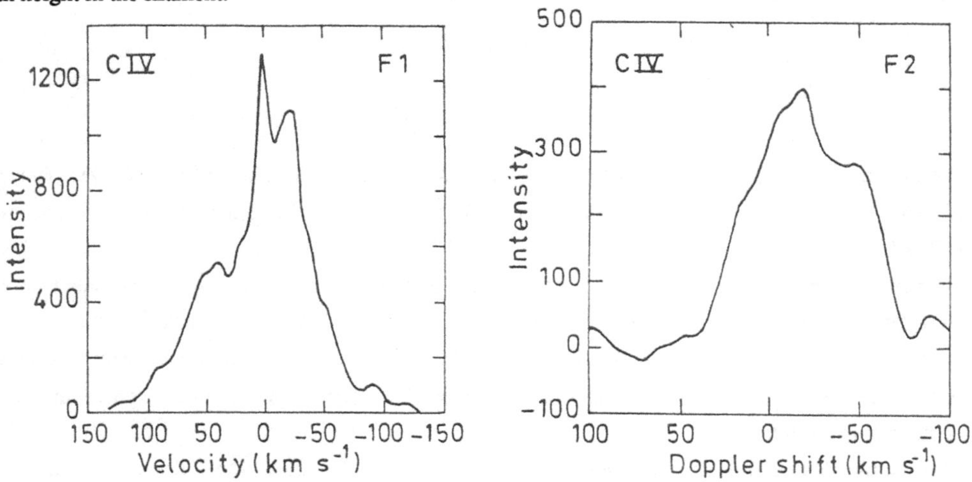

Fig. 11: (from Schmieder et al 1989)
C IV profiles in filaments obtained with HRTS.

Let us note the nongaussian shape of C IV profiles which probably indicates that different velocity components are mixed along the line of sight. Such an addition may explain the rather high values of turbulence derived from low spectral resolution observations. Let us mention however a rather unique measurement in the O VI line (at 103.2 nm formed at 3 10^5-10^6 K) obtained with OSO8 (Vial, 1988). It indicates that the turbulence is necessarily lower than 30 km/s (a similar result was obtained above an active prominence by Vial et al 1980).

4/ Magnetic Fields

The situation is very simple : NO measurement of the magnetic field in the PCTR, as in the CCTR and as in the corona. Many computations have been performed (see in this issue) with some assumption on the external field (potential, force free with α constant, ..); they all concern cool regions (Ballester and Priest, 1987).

5/ Modeling

The basic question to answer first is what transition region in what geometry ?

Geometry :

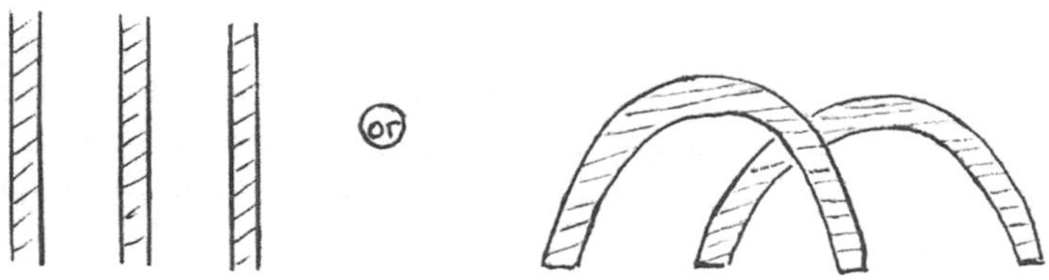

Transition Region : do we have a "warm" PCTR around a cool core or the contrary (Rabin 1986)? Could we have the transition <u>along</u> the structure (probably along the field lines)?

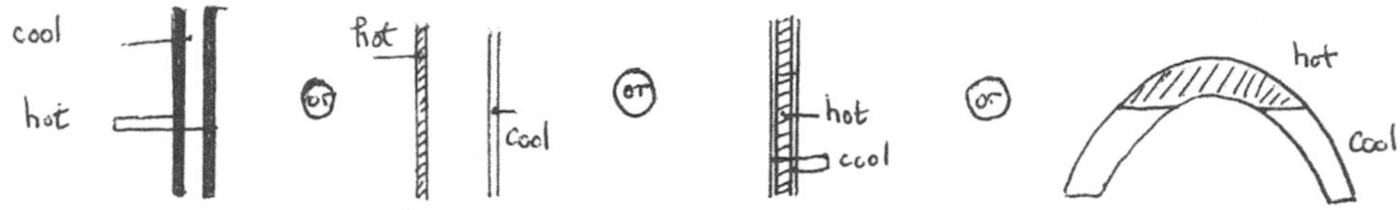

Present computations are more modest but point at the necessity to take some PCTR (or PCTRs) into account, even for the hydrogen lines system, usually supposed to be formed at rather low temperature ($<2\ 10^4$ K).

Heinzel, Gouttebroze and Vial (1988) simply drafted some adhoc transition regions (derived from the VAL chromospheric models) and showed that such an addition "improved" the $L\alpha/L\beta$ ratio (see Vial et al, this issue). But the $L\alpha$ and $L\beta$ emissions increased too, up to unobserved values. Moreover, most of these PCTRs still retained the temperature plateaux characteristic of the chromosphere and rather unrealistic in prominences.

More sophisticated models had been built by Fontenla and Rovira (1985). The prominence was considered here as a superposition of (identical) threads, the structure of which was determined from the energy equation (radiative losses=conductive flux) and a fixed central temperature. Such computations have been repeated with two basic improvements : a better treatment of the resonance scattering (Partial frequency redistribution) and the inclusion of ambipolar diffusion in the atmospheric model computation (see Fontenla, Avrett and Loeser 1990). From the comparison with observed $L\alpha$, $L\beta$ and $H\alpha$ lines, the authors (Vial et al, this issue) derive a model of about 100 threads with pressure around 0.1 dyn cm^{-2}. Because of the ambipolar diffusion and its related heat increase, the $H\alpha$ emission is too high. Moreover, contrary to Heinzel (this issue) and Zharkova (this issue), the radiative interaction has not been taken into account.

Finally, mhd computations begin to deal with some multitemperature models (the first step towards a full PCTR).

Steele and Priest (1989) consider an arcade as the addition of loops. Looking after thermal equilibrium in one loop, they find 3 possible solutions : hot, cool and hot-cool loops. Such structures would be easy to trace with a multispectral instrument working at the subarcsecond resolution.

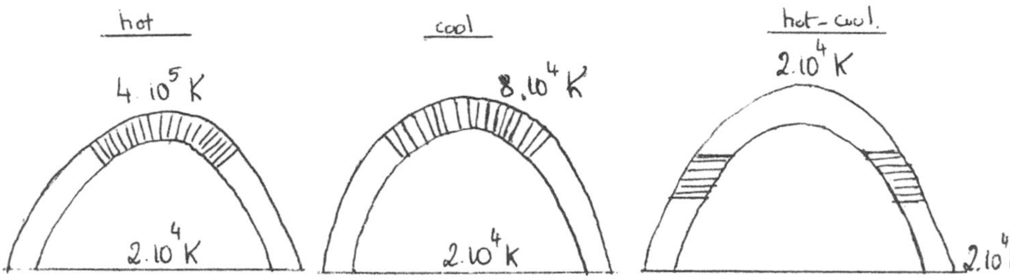

Ballester and Priest (1989) consider active region filaments as made of fibrils. They show (Figure 12) that the formation of a dip (useful for supporting the material) is possible for reasonables values of densities, external field, etc... and x_h, a parameter that is nothing else than the thickness of the PCTR (a few thousand km).

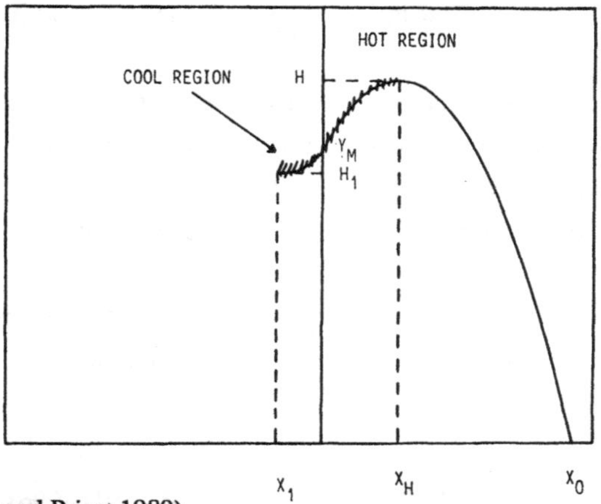

Fig. 12: (from Ballester and Priest 1989)
Schematic representation of the temperature profile along the flux tube from one end to the other. T_c and T_p are coronal and prominence temperatures.

IV. Future Observations of the PCTR

It is obvious that important progress in the observation and the understanding of the PCTR have been made with the advent of space experiments (OSOs, Skylab,.etc..until SMM and HRTS) and such ground-based instruments as the VLA (and coronagraphs). More progress relies on improved instrumentation : the next step will be achieved with the ESA/NASA mission SOHO. It will include a euv multilines telescope, a set of coronagraphs which includes a Fabry-Perot working very close to the limb (1.1 R_o), normal and grazing incidence spectrometers, with high spatial and spectral resolutions from 15 to 160 nm. These two last instruments will be ideal for the study of the PCTR since the observed uv lines span the 10^4-10^6 K interval, some line ratios are sensitive to density or temperature, and H and He main lines and continua will be accessible (in conjonction with MSDP Hα observations). Further steps include the Orbiting Solar Observatory and the proposed uv interferometer SUN.

V. SUMMARY and CONCLUSIONS

We can summarize what we know about the PCTR as follows :
The pressure range is 0.01-0.2 dyn cm^{-2}, some average value being 0.1 dyn cm^{-2}.
The temperature gradient is less steep than in the CCTR (10^{-2}-10^{-3} K/cm at 10^5 K).
The PCTR is certainly highly structured (as the cool regions probably are) with a (surface) filling factor between a few 0.01 and a few 0.1.
Up and down flows less than 5 km/s have been measured (in quiescent prominences) although impulsive events exist at the edges. The turbulence is lower than in the CCTR (30 km/s) but probably results from the superimposition of different velocity fields.
As for densities, an intriguing result should be noted : the electron density around a prominence may be higher than in the prominence itself (Noens et al 1988 and this issue). How can both the gas and the

kinetic pressure be higher in the PCTR than in the prominence ? Would the magnetic field lines be concentrated in the PCTR ?

We must conclude with what we do not know : * the heating process (especially below 10^5K) * the geometry * the location of the material versus the magnetic field (direction of the temperature gradient, of the flows as compared to the field) and * the field itself in the PCTR.

A multiwavelength (multitemperature) analysis would help to fill the gap.

Acknowledgements: I would like to express my thanks to the many people who sent me their papers or their results in advance, especially Franca Chiuderi-Drago for discussion.

References :

Athay, R.G., 1986, Ap. J. **308**, 975

Athay, R.G., and Illing, R.M.E., 1986, Journal of Geophys. Res. **91**, 10961.

Ballester, J.L., and Priest, E.R., 1987, Solar Phys. **109**, 335

Ballester, J.L., and Priest, E.R., 1989, "Model for fibril structure of solar prominences", preprint

Balthasar, H., Knölker, M., Stellmacher, G. and Wiehr, E., 1986, Astron. Astrophys., **163**, 343.

Engvold, O., 1988, "The Prominence-Corona Transition Region" in Proceedings of the 9th Sacramento Peak Summer Symposium, ed. R.C. Altrock, 151

Engvold, O., 1989, "Prominence Environment" in Dynamics and Structure of Quiescent Solar Prominences, ed. E.R. Priest, Kluwer Academic Publishers, p.47

Engvold, O., Wiehr,E. and Wittmann, A., 1980, Astron. Astrophys. 85, 326

Engvold, O., Tandberg-Hanssen, E., Reichman, E., 1985, Solar Phys. 96, 35

Engvold, O., Kjeldseth-Moe, O., Bartoe, J.-D.F., and Brueckner, G.E., 1987, 21st Eslab Symposium, ESA SP 275, 21

Fontenla, J.M. and Rovira, M., 1985, Solar Phys. **96**, 53

Fontenla, J.M., and Poland, A.I., 1989, Solar Phys. **123**, 143

Fontenla, J.M., Avrett, E.H., and Loeser, R., 1990, "Hydrostatic Thermal Models with Ambipolar Diffusion", to be published in Astrophys. J.

Gaizauskas, V., 1989, in "Physics of Magnetic Flux Ropes", Proceedings of AGU Chapman Conference.

Harrison, R.A., Rompolt, B., and Garczynska, I., 1988, Solar Phys. **116**, 61

Heinzel, P., Gouttebroze, P. and Vial, J.- C. , 1988, "Non LTE modelling of prominences", Workshop on Dynamics and Structure of Solar Prominences, ed. E. Priest and J.L. Ballester, Mallorca

Hirayama, T., 1985, Solar Phys., **100**, 415

Koutchmy, S., 1981, Space Sci. Rev. **29**, 375

Kundu, M.R., 1986, "Coronal Prominence Plasma" Workshop, NASA 2442, A. Poland Editor, p.117

Kundu, M.R., Melozzi, M., and Shevgaonkar, R.K., 1986, Astron. & Astrophys. **167**, 166

Kundu, M.R., Schmahl, E.J., and Fu, 1988, Ap. J. **325** 905

Lang, K.R., and Willson, R.F., 1989, Ap. J. Letters **344**, L73

Lantos, P., and Raoult, A., 1980, Solar Phys. **66**, 275

Leroy, J.L., Ratier, G., and Bommier, V., 1977, Astron. & Astrophys. **54**, 811

Malherbe, J.M., and Priest, E.R., 1983, Astron. & Astrophys. **123**, 80

Malherbe, J.M., Schmieder, B., Mein, P., and Tandberg-Hanssen, E., 1987, Astron. Astrophys., **172**, 316.

Mouradian, Z., Martres, M.J., and Soru-Escaut, I., 1980, Proc. Japan-France Seminar on Solar Phys. (ed F. Moriyama and J-C Henoux), p. 195

Mouradian, Z., Martres, M.J., and Soru-Escaut, I., 1989,

Kjeldseth-Moe, O., Cook, J.W., and Mango, S.A., 1979, Solar Phys. **61**, 319

Noens, J.C., Schmieder, B., and Mein, P., 1988, Proceedings of the Workshop "Dynamics and structure of solar prominences", ed. Ballester and Priest, p. 177

Orrall, E.Q., and Schmahl, E.J., 1976, Solar Phys. **50**, 365

Orrall, E.Q., and Schmahl, E.J., 1979, Ap. J. **240**, 908

Schmahl, E.J. and Orrall, E.Q., 1986, "Coronal Prominence Plasma" Workshop, NASA 2442, A. Poland Editor, p.127

Simon, G., Schmieder, B., Demoulin, P. and Poland, A., 1986, Astron. and Astrophys. **166**, 319.

Orrall, E.Q., and Speer, R., 1974, in "Chromospheric Fine Structure", ed. R.G. Athay, Reidel, 193
Orrall, E.Q., and Zirker, J.B., 1961, Ap. J. **134**, 72
Poland, A.I, Tandberg-Hanssen, E., 1983, Solar Phys. **84**, 63.
Poland, A.I., and Mariska, J.T., 1986, Solar Phys. **104**, 303
Rabin, D., 1986, "Coronal Prominence Plasma" Workshop, NASA 2442, A. Poland Editor, p. 135
Saito, K., and Tandberg-Hanssen E., 1973, Solar Phys. **31**, 105
Schmahl, E.J., Foukal, P.V., Huber, M.C., Noyes, R.W., Reeves, E.M., Timothy, J.G., Vernazza, J.E., and Withbroe, G.L., 1974, Solar Phys. **39**, 337
Schmahl, E.J., and Orrall, E.Q., 1979, Ap. J. Letters **231**, L41
Schmahl, E.J. and Orrall, E.Q., 1986, "Coronal Prominence Plasma" Workshop, NASA 2442, A. Poland Editor, p.127.
Schmieder, B., Poland, A.I., Tompson, B., and Demoulin, P., 1988, Astron. Astrophys., **197**, 281.
Schmieder, B., Raadu, M.A., Demoulin, P., and Dere, K.P., 1989, to be published in Astron. Astrophys.
Steele,C.D.C., and Priest, E.R., 1989, "Thermal equilibria of coronal magnetic loops", preprint
Stellmacher, G., Koutchmy, S. and Lebecq, C., 1986, Astron. & Astrophys. **167**, 351.
Toot, G.D., and Malville, J.M., 1987, Solar Phys. **112**, 67
Tsubaki, T., 1977, Solar Phys. **51**, 121
Tsubaki, T., and Takeuchi, A., 1986, Solar Phys. **104**, 313.
Tsubaki, T., Ohnishi, Y.O., and Suematsu, Y., 1987, Publ. Astron. Soc. Japan **39**, 179.
Tsubaki, T., Oyoda, M., and Suematsu, Y., 1988, Publ. Astron. Soc. Japan **40**, 121.
Vial, J.C., 1988, "The O VI (103.2 nm) prominence profile and the prominence-corona interface", Proceedings of the Workshop "Dynamics and structure of sclar prominences", ed. Ballester and Priest, p.181.
Vial, J.-C., Lemaire, P., Artzner, G. and Gouttebroze, P., 1980, Solar Phys. **68**, 187
Vrsnak, B., 1984, Solar Phys. **94**, 289
Widing, K.G., Feldman, U., and Bhatia, A.K., 1986, Ap. J. **308**, 982.
Yang, C., Nicholls, R., and Morgan, F., 1975, Solar Phys. **45**, 351
Zirker, J.B., 1989, "Physics of prominences" in The Solar Interior and Atmosphere, eds. A. Cox, W. Livingston and M. Matthews, University of Arizona

DISCUSSION

VAN HOVEN: Would you comment on the effect on the DEM observations of a configuration in which the cool region around the prominence is much larger in volume than the dense region of the prominence. Such a configuration arises in the radiative-condensation instability in a sheared magnetic field because the heat flow is supressed over a much larger layer than that in which parallel mass flow is allowed. We have made numerical simulations of these effectes (Van Hoven et al. in these Proceedings).

VIAL: Such a model will imply a low gas pressure which would decrease the DEM even if the temperature gradient is small. It will be interesting to compute the expected DEM.

MASS AND ENERGY FLOW IN PROMINENCES

Arthur I. Poland
Laboratory for Astronomy and Solar Physics
NASA/Goddard Space Flight Center
Greenbelt, Md 20771 USA

I. Introduction

In this paper I will discuss the flow of mass and energy in quiescent prominences. It is my opinion that active region prominences have a different structure, and thus different mass and energy flow characteristics. I will first discuss the observational characteristics that provide the frame in which model calculations are accomplished. I will then discuss some non-LTE radiative transfer calculations and their significance for understanding the mass and energy flow problem. In section IV I will present a discussion of hydrodynamic model calculations. In the final section I will present the material velocities and relative energy transport efficiencies from one of the "typical" model calculations.

II. Observations

There are two particularly impressive pictures of prominences that have been made in recent years that demonstrate many of their important characteristics. These pictures are presented elsewhere in this publication so I will not duplicate them here. I will discuss some of their important aspects as they relate to this paper, and present a Ly_α picture made by UVSP on SMM (Woodgate et al 1980).

The first is a filament picture made by Dr. O. Engvold at very high resolution and showing prominence fine structure. The prominence can be seen to be made up of many very fine threads that appear to be aligned along the prominence axis (neutral line). It was noted in similar data in a paper by Simon et al (1986) that in the denser regions of the prominence, which are usually associated with the feet, the loops do not appear to be well aligned. From these data we get the impression that higher, longer threads lie along the field, while shorter, lower threads form the feet.

The second picture is of a limb prominence observed by Dr. M.J. Martres. This picture (used on the cover of the Coronal and Prominence Plasmas Proceedings, Poland, 1986) shows a large quiescent prominence with "feet" at each end. The feet seem to be made up of large loops, while the central section is made up of many small apparently twisted threads. The loops and even the small threads extend in height over many thousands of kilometers. This relatively small gradient with height presents a serious problem for material support, since the pressure scale height at even 10,000K is only ~100km.

In Figure 1 we present an image of prominence type loops made at high resolution in Ly_α by the UVSP instrument on the SMM satellite. The image area is 4'x4' and the resolution is better than 3". That the intensity remains fairly constant over many pressure scale heights indicates that we are not simply looking at the outline of magnetic loops filled with static material. It also can not be a simple case of material flowing through the loop at the Ly_α temperature of ~10,000K since the observed homogeneity would require that these velocities be supersonic. Typical observed velocities are only on the order of 1 to 10 km/sec.

We also note that time lapse photographs of prominences show that individual threads and structures within a prominence have lifetimes of only ~5 to 8 min. (Schmieder et al 1988).

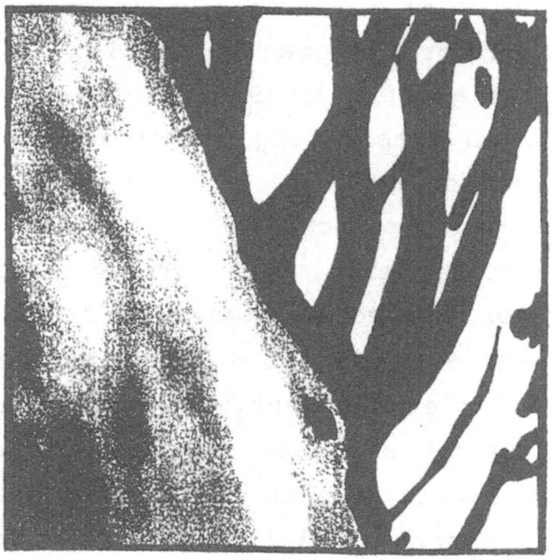

Figure 1. Ly_α image of the solar limb showing fine threads approximately 3" in diameter. The image was made by UVSP on SMM with a size of 4'x4'.

The above observations suggest that prominence support is not a simple static problem. The actual magnetic geometry is most likely quite complex. It is also evident from the observed flows and short lifetimes of individual features that the flow of mass and energy are important considerations for understanding prominence structure.

III. Radiative Transfer

One of the most important aspects of energy flow in prominences is radiative transfer. The prominence consists of cool gas suspended in a radiative "bath" of coronal, transition region, chromospheric, and photospheric light. One must consider non-LTE radiative transfer for hydrogen and to a much lesser extent helium. From a geometrical consideration one would prefer to do the calculations for many cylindrical interacting loops as shown in Figure 1, but this involves multi-dimensional radiative transfer.

The complex problem of multi-dimensional radiative transfer has been investigated by Avery and House (1969) and Jones and Skumanich (1980). These works have studied the problems of cylindrical geometry and the radiative interaction between nearby radiating structures. They discuss the important issues of length scales of interaction and, most importantly, the appropriate scaling between cylindrical geometries and plane parallel slabs. An important relation is that a plane parallel slab of thickness D is very close to a cylinder of radius R if R=D.

Energy balance work for hydrogen and helium using this result has been done by Heasley et al. (1974, 1976), Poland and Anzer (1971), and Poland et al (1988). An interesting result on energy balance, taken from Figure 3 by Poland et al 1988, is shown in Figure 2. This figure shows the non-LTE temperature achieved by a model slab irradiated by the photospheric, chromospheric, and coronal radiation fields. The curve labeled 1 is for the normal incident field, 1000 is for 1000 times the normal field, 5×10^5 is for 5×10^5 times the normal field, and 3×10^7x is for that factor enhancement, but only in the x-ray region of the spectrum. These latter examples are for active region and flare conditions. The point I wish to make with this graph is that the outer part of the slab (prominence) equilibrates to the radiative temperature shortward of the Lyman continuum (~7,000K), and the inner part equilibrates to the Balmer and Paschen continuua temperature (~4,500K). We thus see that the cool parts (T<10,000K) of prominences are probably predominantly radiatively energy balanced. If the prominence is optically thin in the Lyman continuum it will have a minimum temperature of ~7,000K, while if it is thick in the Lyman continuum it will have a minimum temperature of ~4,700K with an outer shell of ~7,000K.

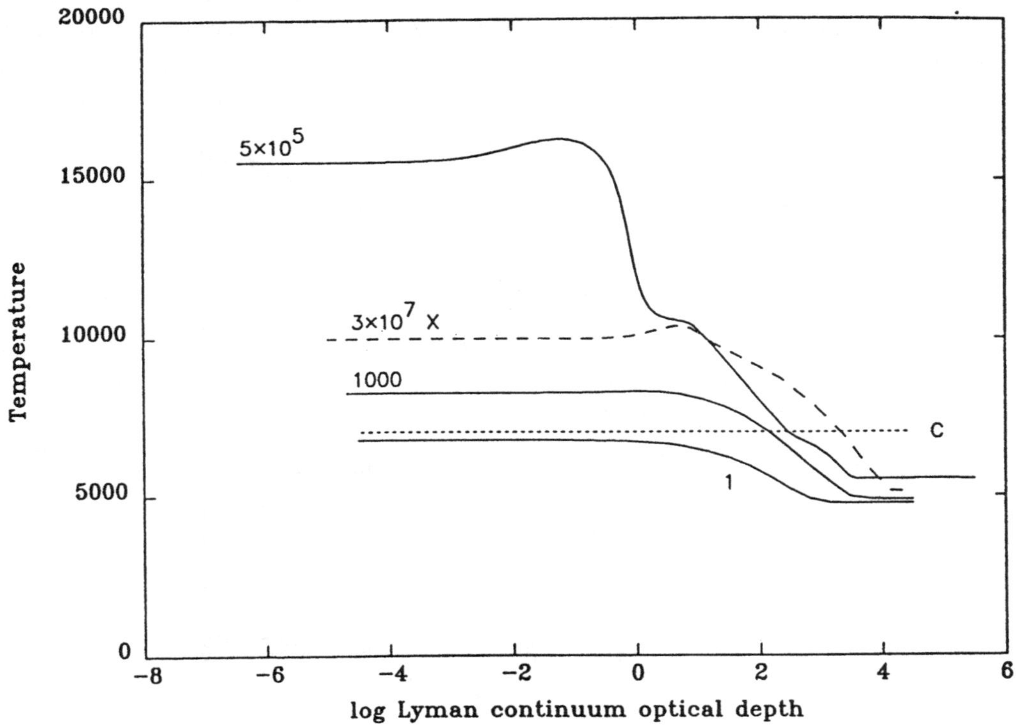

Figure 2. Electron temperature as a function of Lyman continuum optical depth in a model prominence. Each curve is for a different enhancement over normal EUV exciting radiation. The dashed line with a C shows a constant temperature of 7000°K. A normal quiescent prominence is represented by curve 1.

When one is investigating energy balance, including other sources besides radiation, the radiative losses must be calculated for a range of temperatures and pressures. An example of this is the radiative loss functions presented by Cox and Tucker (1969) who assumed optically thin radiative losses for all transitions of all species. In a recent work, Kuin and Poland (1989) calculated the losses for Hydrogen and Helium including optically thick effects and an incident radiation field of the normal photosphere, chromosphere, and corona. The results for a model slab of 1000km thickness are shown in Figure 3. Note that at sufficiently low temperatures and pressures the slab experiences a net heating from the incident radiation field. Also note that the local cooling maximum near log(T)=4.2 is reduced by including optical depth effects. The calculations by Kuin and Poland were designed to be used with hydrodynamic codes to calculate models of coronal loops and prominences.

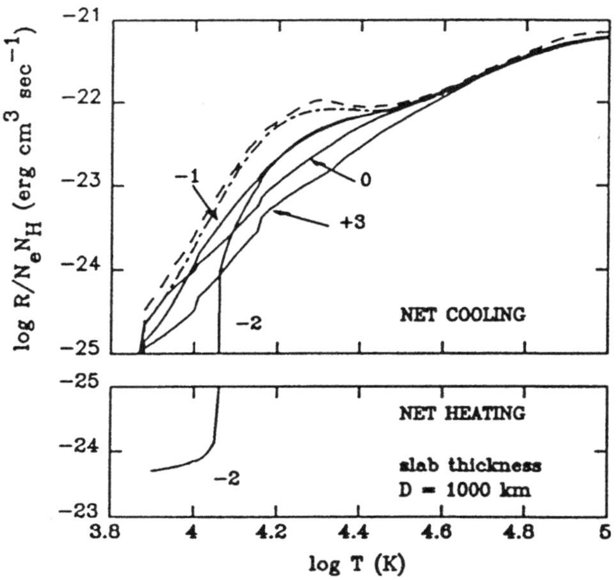

Figure 3. Radiative loss coefficient for a slab 1000 km thick (i.e. for a cell with radius of 1000 km). The full curves connect points of equal pressure. The value of log p is indicated. The dashed curve shows the results of Cox and Tucker (1969), while the dot-dashed curve shows our results for an optically thin slab with no incident radiation field.

IV. Hydrodynamic models

In order understand the energy and mass flow in prominences one must model them using hydrodynamic calculations. The first step in this process is to define a geometric model in which to confine the matter. The geometric model we use is shown in Figure 4 (see Poland and Mariska, 1987). It requires a magnetic loop with a "dip" in the center to hold the material; this is essentially the model proposed by Kippenhahn and Schluter (1957). Material will condense in the dip. We suggest that extended horizontal "threads" are regions where the field is stretched, as shown in the figure, while vertical structures are regions that are not stretched. The vertical structures are seen as a series of horizontal loops where the dips are aligned along the direction of gravity. This model is discussed further in the paper by Poland and Mariska (1987). A more physical discussion of the formation of dips by magnetic twisting is given by Priest et al 1989.

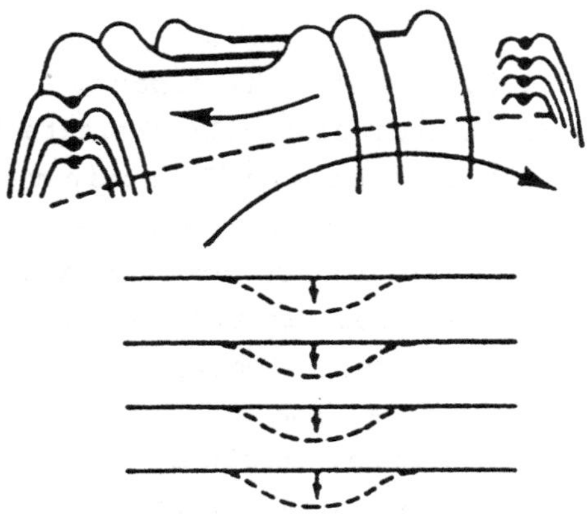

Figure 4. Sketch of magnetic configuration needed for prominence formation.

Given the magnetic structure of a dip, Poland and Mariska (1986) demonstrated with hydrodynamic calculations how a prominence condensation could form with a syphon mechanism. A major weakness of this work is that it required that the energy input to the prominence loop be almost completely shut off for a period before the prominence formed.

A significant improvement has been made to this work by Antiochos and Klimchuck (1989). They used a model and loop similar to that described by Poland and Mariska. However, they showed that a prominence could form by the following process: 1) start with a coronal loop with a dip in the center, 2) add extra heat concentrated much nearer to the chromosphere than to loop center, and 3) heating should be approximately symmetric from loop center. As in the model described by Poland and Mariska, the heat drives the material from the chromosphere into the prominence. However, this process requires less changes in the energy input and less time than the process described by Poland and Mariska. The conclusions drawn by Antiochos and Klimchuck are: 1) in order to form cool mass one needs heating concentrated away from loop center; 2) this implies that long loops are favored; and 3) they have yet to consider the effect of heating asymmetry. This work holds much promise for forming prominences in a fairly short time under reasonable conditions. It may be that the only difference between a coronal loop and a prominence is that the prominence has a dip in the center.

V. Conclusions

Once a prominence is formed it is important to understand the flow of energy into it. While radiation will dominate the energy flow in the center, convection and conduction will dominate in the outer layers. Using the results of Poland and Mariska for the computational model that represented the evolutionary process after the prominence had formed, we have plotted several important physical parameters in Figure 5. Here we see the temperature, velocity, conductive flux, and enthalpy flux plotted against distance from the highest point in the loop to the coolest part of the prominence. It can be seen from this graph that for this model the maximum velocity is only on the order of 5 km/s. However, even for this low velocity the enthalpy flux dominates the conductive flux for temperatures below approximately 200,000°K. Thus, the temperature gradient and thus thickness of the cooler part of the transition zone will not be determined by conduction alone.

From the above model calculations we see that the transition region of prominences is dominated by complex processes. For temperatures below 200,000°K we need to include the effects of mass flow, while for temperatures below 30,000°K we need to include both mass flow and optical depth effects in hydrogen. Both of these effects lead to a less steep temperature gradient through the prominence corona interface than one would get from conduction alone.

To verify these predictions one needs simultaneous observations of prominence images using spectral lines over a wide temperature range. One also needs line profiles in these lines to determine material velocities. The CDS and SUMER instruments on SOHO should provide us with such observations. Similarly, more detailed calculations should be made using various model parameters to predict the observable lines and show how the observations can be used to limit the models.

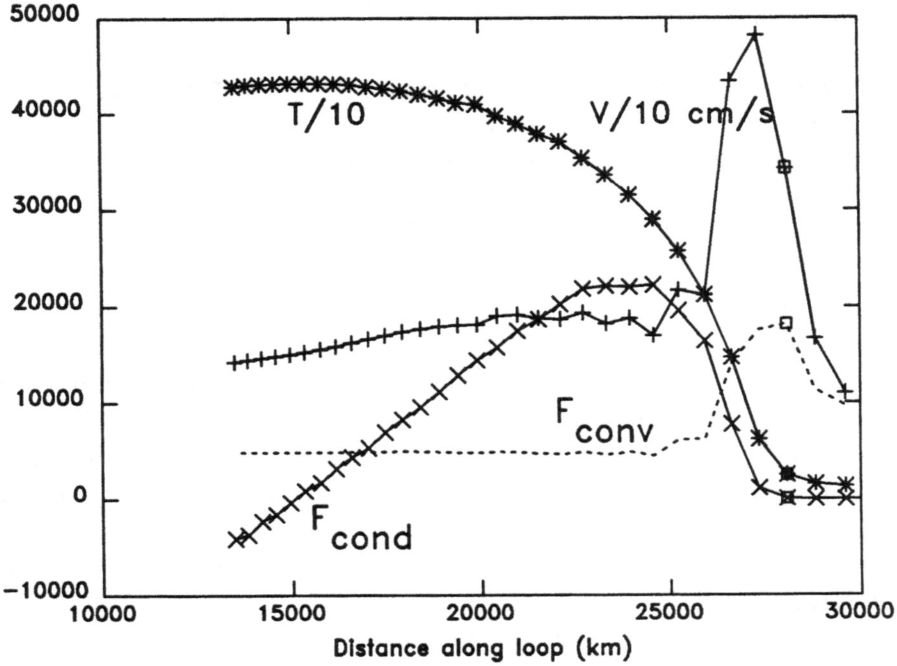

Figure 5. Physical quantities related to energy flow as a function of position along a magnetic loop having a condensation at the center. *) T/10., +) V/10., X) Conductive flux, and ---) Convective flux.

References:

Antiochos, S., and Klimchuck, J., 1989, (private communication)

Avery, L.W. and House, L.L., 1969, Solar Phys. 10, 88.

Cox, D.P., and Tucker, W.H., 1969, Ap.J. 157, 1157.

Heasley, J.N., Mihalas, D., and Poland, A.I., 1974, Ap. J. 192, 97.

Heasley, J.N., and Mihalas, D., 1976, Ap. J. 205, 181.

Jones, H.P., and Skumanich, A., 1980, Ap.J. Suppl. 42, 221.

Kippenhahn, R. and Schluter, A., 1957, Z. Astrophys. 43, 36.

Kuin, N.P.M., and Poland, A.I., 1989,(Ap.J. submitted).

Poland, A.I., and Anzer, U., 1971, Solar Phys. 19, 401.

Poland, A.I., Milkey, R.W., and Thompson, W.T., 1988, Sol. Phys. 115, 277.

Poland, A.I., and Mariska, J.T., 1987, "Dynamics and Structure of Solar Prominences" Proceedings of a Workshop held at Palma de Mallorca, November 18-20, 1987, J.L.Ballester and E.R. Priest Eds., pg 133.

Priest, E.R., Hood, A.W., and Anzer, U., 1989, Ap.J. 344, 1010.

Simon, G., Schmieder, B., Demoulin, P., and Poland, A.I., 1986, Astron. Astrophys. 166, 319.
Schmieder, B., Poland, A., Thompson, B., and Demoulin, P., 1988, Astron. Astrophys. 197, 281.
Woodgate, B.E., et al, 1980, Sol. Phys. 65, 73.

Questions:

E. Priest: Is the Lyα prominence picture of 10 March 1980 an erupting or quiescent prominence?
A. Poland: It is my understanding that this is a post flare prominence, but I do not think these fine structures are significantly different than prominence loops.
T. Hirayama: How do you compare your work with that done by An and people at Marshall?
T. Forbes: I think An's model is an injection of mass into the legs of a loop. Art's and Antiocho's model is different in that the prominence is formed by a thermal or heating process.

SUPPORT OF QUIESCENT PROMINENCES

Eberhart Jensen
Institute of Theoretical Astrophysics, University of Oslo
P.O. Box 1029, Blindern, N-0315 Oslo 3, Norway

I. INTRODUCTION

The interest for the physics of prominences has grown at an impressive rate in recent years and has resulted in a large number of publications on observations as well as on theory. A number of excellent review articles on prominences has appeared recently. In particular I will like to mention those by Hirayama (1985), Zirker (1989) and the monograph edited by Priest, "Dynamics and structure of solar prominences", Priest (1989).

The increase in resolution recently achieved by better instruments and by choosing observing sites with the very best seeing conditions, has added new aspects to our picture of quiescent prominences. As our observational data improve, confirmation or rejection of what the theories predict should be more easy to achieve as the constraints get stricter. But still the classical two-dimensional slab models of prominences are discussed in the literature and further elaborated upon. They are generalized in various ways f.inst. by adding velocity fields. Energy exchange by radiation and conduction is taken into account in more realistic ways. The mathematical descriptions have reached a high degree of sophistication and are used as a base for extensive numerical simulations.

But the lack of consent between the various models is striking. And when confronted with results of high resolution observations that reveal dynamic features like turbulent velocity fields and an intricate fine-structure composed of knots and fine threads, some of them loop-like, and changing with timeconstants down to minutes, the models fall far short of giving an adequate description.

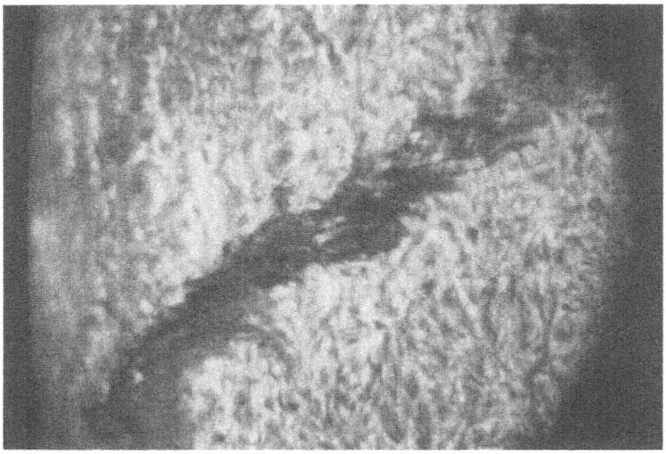

Figure 1. Filtergram of filament, Engvold and Darvann, Swedish solar Station, La Palma.

The topic we shall discuss, "Support of quiescent prominences" is in my oppinion not a problem in magnetohydrostatics. Because what we see is matter moving in all directions with velocities up to that of free fall. The supporting force must be of a stochastic nature, varying both in space and time, but capable of keeping "the ball in the air" in a statistical sense.

In the following we shall first take a critical view on some common models for quiescent prominences in confronting the observational aspects they imply with the results of high resolution observations.

II. MODELS FOR QUIESCENT PROMINENCES

In Chapter 6 of the monograph already mentioned, Anzer gives an excellent review of prominence models. Here we shall only comment shortly on a few of them, beginning with the Kippenhahn-Schlüter model.

1. The Kippenhahn-Schlüter model.

The paper describing this classic among magnetohydrostatic models was published more than 30 years ago (Kippenhahn and Schlüter 1957). It has been the subject of numerous mathematical refinements and generalizations. In the original paper the supporting magnetic field was in a direction perpendicular to the main axis of the prominence, pictured as an infinitely thin vertical sheet of cool matter.

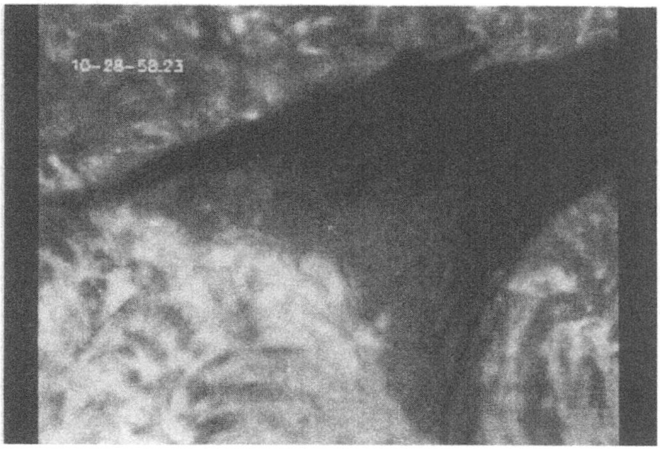

Figure 2. Filtergram of filament, Jensen and Kusoffsky, Swedish Solar Station, La Palma.

The sagging of the lines of force under the weight of the prominence situated in a central pit in the field, created an electric current which multiplied by the magnetic field gave the supporting force. Later shear of the field corresponding to the observed angle of about 15 degrees between the magnetic field and the prominence axis was introduced (Tandberg-Hanssen and Anzer 1970). In such a model the lines of force threading the prominence are connected to the photosphere. Thus there is no scarcity of matter once an injection mechanism is found. To specify an injection mechanism is clearly an important question and not only for prominences, since filling of flux tubes goes on everywhere on the sun at all times. Pikelner's "siphon model" (Pikelner 1971) operates in a symmetrical tube of force with a "pit" at the center. In focusing his attention on a single tube of force and applying the result to the two-dimensional slab model of Kippenhahn-Schlüter, Pikelner anticipated the present view that individual loops, more or less deformed, are the structure elements in quiescent prominences. By postulating an initial vertical velocity at the endpoints at the coronal base, and by choosing suitable values for the parameters in the loop, he found that a stationary solution existed that would bring matter into the central pit. More detailed calculations (Engvold and Jensen 1976, 1977) showed his solution to be highly sensitive to tube geometry. But surprisingly a change in injection velocity at the footpoints by a factor of 30, from 30 to 1 km/s altered the physical parametrs in the tube only by a few percent. These values of the injection velocity correspond more or less to respectively a ballistic and an evaporative model (Forbes 1986, Poland and Mariska 1986) in Forbes terminology. The temperature decreased monotonically with height and showed a sharp drop close to the pit. At a temperature of 20 000 degrees the optically thin approximation broke down as hydrogen started to recombine.

But the loops observed as structure elements in quiescent prominences are certainly not stationary, and the observations show no trace of pits with positive curvature on tops of loops nicely lined up in a row. The Kippenhahn-Schlüter model is thus, being two-dimensional, even with later modifications highly idealized as compared to the very complicated three-dimensional structure with its vivid dynamics that is revealed by modern observations.

2. The Kuperus-Raadu model.

The Kuperus-Raadu model was proposed in 1974 (Kuperus and Raadu 1974). Here the prominence is represented by a line-current. The supporting

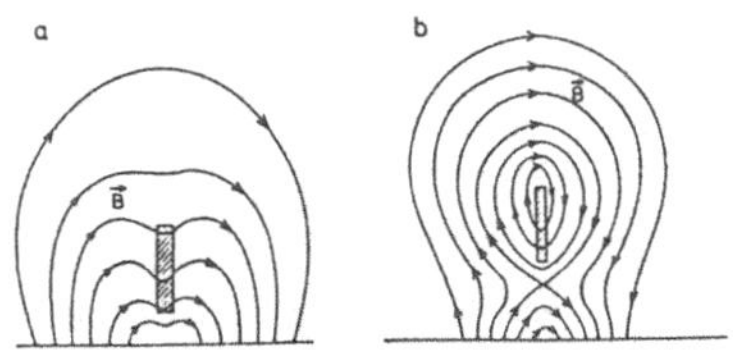

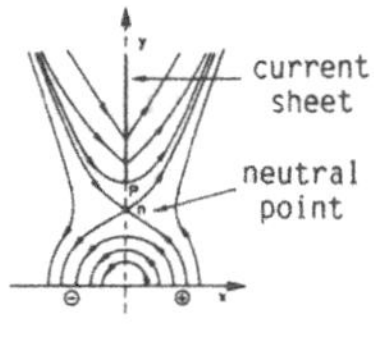

Figure 3. a. Normal, b. Inverse magnetic configuration. Below Kuperus-Raadu model. (Anzer 1989).

magnetic field has the form of a "hammoc" and the lines of force connect to the corona. The prominence is supposed to form by a continous condensation process caused by connection in the coronal plasma. Thus matter is supplied only from the corona.

The magnetic lines of force carrying the prominence have to connect to the solar wind to maintain the supporting force. It was pointed out by Anzer and Priest (1985) that in a Kuperus-Raadu configuration it is difficult to support prominences of sufficient height to match the observations. Another objection is that the current along the prominence leads to self-pinching that is not easy to reconcile with observations. Also since matter is steadily draining from the corona to the chromosphere below, a red-shift should be observed in filaments. This is not in accordance with the large body of observations that has been accumulated in particular by the Meudon group, showing that both blue and red shifts are observed, but that blue shifts tend to dominate i filaments (Maltby 1976, Schmieder et al. 1984, Malherbe et al. 1983, You and Engvold 1989).

The Kuperus-Raadu model has the property that the magnetic field threading the prominence has the opposite direction to that of a potential field connected to footpoints with similar polarity in the photosphere. This is what observations of polar crown prominences

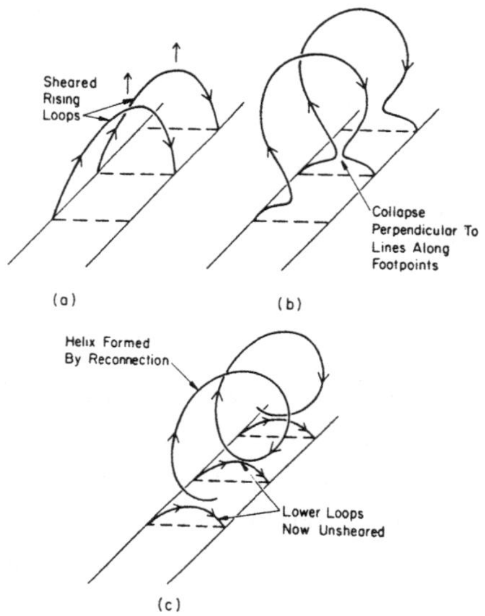

Figure 4. Pneumans mechanism of formation (Pneuman 1983).

using the Hanle effect indicate in the majority of cases, according to Athay et al. (1983) and Leroy et al. (1984). It has for this reason been suggested that these prominences may be examples of a Kuperus-Raadu configuration.

3. Hydrostatic models with helical magnetic fields.

Configurations with helical magnetic fields where matter is supported in the lower parts of a helical structure has been suggested by several authors, notably Pneuman (1983). He considered this structure to be formed from a rising collapsing loop undergoing reconnection. This model also results in an inverse magnetic configuration with downflow of matter.

Hirayama (1985) proposed this kind of models as part of a scheme describing the development in time of quiescent prominences. The starting stage is a Kippenhahn-Schlüter model.

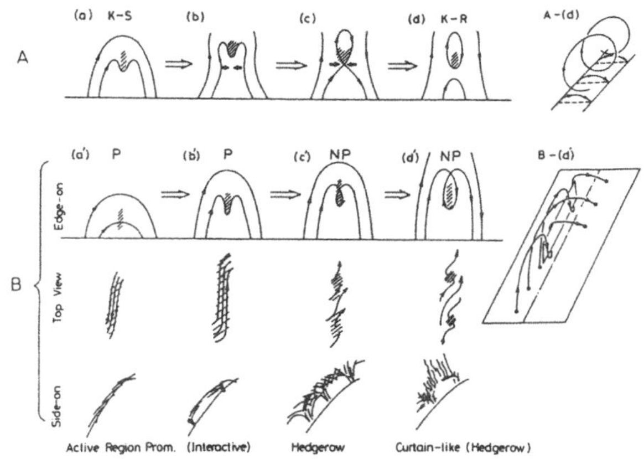

Figure 5. Hirayamas scheme of evolution (Hirayama 1985).

Clearly outlined helical structure of large scale are occasionally observed in erupting prominences. But helical structures are not easily distinguished, neither in prominences at the limb, nor in filaments. If the dynamic "dispartition brusque", where the prominence is not regenerated, is due to an electric discharge, the trigging current would mainly flow along the magnetic lines of force where the conductivity has its maximum. Then the resulting lines of force would be spirals. Thus the original configuration does not necessarily posses a helical topology.

III. COMPARISON OF PROMINENCE MODELS WITH OBSERVATIONS

We have mentioned a few out of a large number of static slab models that have been proposed for quiescent prominences. We shall now make an attempt at a more detailed comparison with observations.

The Kippenhahn-Schlüter model can account for static support of cold matter in a magnetic field and the long time stability, but it leaves most aspects of modern observations unexplained. Even one of the basic features of quiescent prominences, the existence of "feet" bridged by arches forming the lower boundary, is difficult to explain within this framework. First of all this model is two-dimensional, while high-resolution filtergrams show a complicated fine-structure. And this fine-structure is neither stationary nor static, changing with time in a matter of minutes. With the Kuperus-Raadu model we meet similar predicaments, only more so (Simon et al. 1986).

1. The magnetic field observations and the paradox of the vertical fine structure.

The magnetic field in quiescent prominences is a crucial parameter in all models for support. Both Zeeman measurements (Rust 1967, Harvey 1969, Tandberg-Hanssen 1970, 1974, Kim et al. 1982) and the Hanle data (Athay et al. 1983, Leroy et al. 1984) give comparable values for the magnetic field strength being in the range from 5 to 40 gauss. When assessing the value of the magnetic field measurements it should be borne in mind that the resolution of present magnetographs is far inferior to the resolution of filtergrams and spectra. The linear resolution of current magnetographs as applied to quiescent prominences is typically 4" to 6", while the best filtergrams go down to 0.3-0.4". Within the area corresponding to one resolution point of current magnetograms of prominences, a filtergrams thus posesses 100 information-points. This gives reason to suspect that the magnetic field strength may locally reach values considerably higher than the averages found by taking low resolution magnetic observations at their face value. And we may also infer that there is room for substantial local deviations from the directions measured. The extensive observations of prominence magnetic fields interpreted using the Hanle-effect, presents in striking contrast to the chaotic picture shown in filtergrams and spectra, a smooth configuration with predominantly horizontal magnetic fields throughout the main body of the prominence.

Also in the prominence feet the Hanle data indicate horizontal fields. As has been pointed out by several authors (Malherbe 1987) this result appears rather strange as the fine-structure in the prominence feet in high resolution filtergrams is often seen as well-defined slightly twisted strands with a direction approximately vertical. However, due to scattered photospheric light the Hanle observations are applicable only to altitudes exceeding 10 000 km, and with the limited spatial resolution contamination by scattered light from lower layer cannot be ruled out.

Also in the upper parts of quiescent prominences observed at the limb, where the fine structure in the form of threads often is mainly vertical, the direction of the magnetic field comes out to be horizontal.

Figure 6. Quiescent prominences at the limb, Dunn, Sacramento Peak Observatory.

Here structure elements are often observed to move both up and down with velocities of a few km per second (Engvold 1976). Spectroscopic observations on the disk show that these are mostly true mass motions and not excitation or ionization effects (Schmieder 1989, and reference therein). Thus horizontal magnetic fields imply motions across the magnetic field 5 orders of magnitude higher than the velocity of diffusion, if the classical value of the electrical conductivity is used (Kippenhahn and Schlüter 1957). Since observations show the prominence plasma to be turbulent, it could mean that the Böhm conductivity determines the diffusivity. If so, the diffusion velocity could be of the same order of magnitude as the observed velocity. But this implies that the approximately vertical strands of the fine-structure do not outline the direction of the magnetic field, but are perpendicular to it. In such a configuration the slightest perturbation would set up motions along the magnetic lines of force i.e. in the horizontal direction. However, high resolution observations only very rarely show the presence of features of this kind.

The lack of evidence for vertical fields in the upper parts from the Hanle observations may probably be explained by the fact that the vertical threads in the prominences are not particularly bright and have a small filling factor. This combined with the low resolution of the magnetic observations could mean that radiation from the main body still dominates in the aperture of the magnetographs.

2. Prominences with normal and inverse magnetic topology.

Another result of the Hanle observations is that the majority of prominences in the polar crown has a topology of their magnetic fields different from quiescent prominences at lower latitudes. In 75% of these prominences it was found that the magnetic field has a direction opposite to that of a potential field with similar polarity at the footpoints (Athay et al. 1983, Leroy et al. 1984). They posess "inverse" magnetic polarity. Still the magnetic field is found to be mainly horizontal. For this reason it has been suggested that polar crown prominences are built according to the Kuperus-Raadu model with the X-type singular line below the well containing prominence matter. The direction of the magnetic field in this model is opposite below and above the singular line. In contrast, quiescent prominences at low latitude are found to exhibit the same direction as a potensial field, thus having a "normal" magnetic polarity as predicted by of the Kippenhahn-Schlüter model.

But if slab models are dismissed altogether as incompatible with newer observations that resolve the prominence into a complex tangle of thin loops and other fine structure elements, how may the existence of prominences with inverse polarity be interpreted?

Could it be that the observed difference in polarity only reflects the existence of two types of neutral lines, and the preference of quiescent prominences to form over one of them? Frances Tang (1987) has compared the number of prominences that form over neutral lines separating magnetic polarities within one and the same bipolar region, to the number formed over neutral lines between different, but neighbouring bipolar regions. Her statistics covered the years 1973 and 1979. The first year was in the declining phase of cycle 20, while the second was at the maximum of cycle 21. Altogether 330 prominences were investigated. For the maximum year 66% of the prominences were formed on neutral lines between different bipolar regions, against 34% within one and the same region. In 1973 the tendency was the same, with the corresponding numbers 55 and 45% respectively. Thus quiescent prominences prefer neutral lines between different bipolar regions.

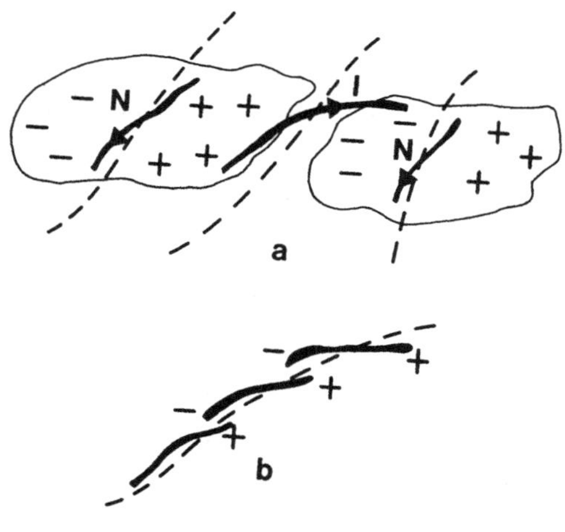

Figure 7. a. Filaments and neutral lines. b. Double magnetic polarity at footpoints.

The directions of the magnetic field in prominences is found from limb observations and it is then a question of how the magnetic polarity at the subjascent photosphere has been determined. If it has been assumed that the prominence span neutral lines inside bipolar regions, while most of the prominences appear at neutral lines between different regions, the observations confirm Tang's statistics (Fig. 7a). Then there is no room for prominences with inverse magnetic polarity involving a complicated magnetic topology, simple loops will do.

If on the other hand the magnetic polarity at the footpoints has been determined with reasonable certainty from extrapolation of magnetograms of photospheric fields, then the existence of prominences with inverse fields represents a real enigma. In that case quiescent prominences cannot be considered a collection of simple loops at the best observations show them to be.

It has been amply demonstrated by many observers that high-resolution observations both in Hα and in Lyα reveal a fine-structure whose elements have life-times of the order of minutes (Dunn 1960, Engvold 1976, Engvold 1980, Bonnet et al. 1980). Models suposed to represent the physical conditions in a prominence cannot ignore this fact and two dimensional models cannot be used to reproduce these properties. It is also apparent from high resolution observations that the morphology of the fine structure is different in different parts of a prominence. Let us first take a look at the connection to the photosphere, the lower parts of a prominence and its "feet".

3. Magnetic topology in the lower parts of prominences.

Observations support the view that the prominence feet are connected with the supergranulation network, although it is difficult to determine the position of the intersections with the photosphere with any precision. In some cases the feet do not reach down to the photosphere at all, we have the phenomenon of "suspended feet". In well-developed "hedgerows" the "feet" consist of more or less parallel often quite smooth vertical threads.

The distance between the footpoints is on the average roughly 30 000 km (Plocieniak and Rompolt 1973). The results from magnetic observations both using the Zeeman- and the Hanle-effect, show conclusively that the approximately horizontal magnetic field in the main body does not change sign along the prominence. If we accept that the "feet" are the ancherpoints for magnetic arches in the photosphere, we must therefore conclude that the footpoints have double polarity (Fig. 7b).

This was first pointed out by Kleczek (1980). Rompolt and Bogdan (1986) added the important suggestion that since in the footpoints magnetic fields of opposite polarity are in juxaposition, reconnection ought to occur. The observations that filaments are formed where photospheric magnetic fields of opposite polarities approach each other, may be taken as a confirmation of this view (Martin, 1973, 1986). If reconnection takes place the "suspended feet" phenomenon could be a temperature effect. One would then predict that the lower parts of such feet would be observable in high temperature lines. It has been demonstrated recently that the "dispartition brusque" when the whole or parts of a prominence, as observed in Hα, disappears to reappear later in a similar shape as before, is due to heating (Mouradian et al. 1986, Malherbe 1989).

The double polarity of the prominence "feet" has a further consequence. With the angle between the magnetic field in a prominence and its main axis being about 15 degrees, the minimum "width" of the feet should be of the order of 30 000 sin 15° = 8000km or roughly 10 seconds of arc. This should be compared to the observations of filament thickness, giving 6000-10 000km (L. d'Azambuja and M. d'Azambuja 1948). The basic magnetic configuration in a quiescent prominence will therefore consist of loops running from one "footpoint" to the next. This is also what is observed at the limb as lower bounderies of the arches that connect adjascent footpoints. These arches often have high intensities and apparently consist of several strands that may run parallel over large distances. Smooth flows of long duration have been observed in these regions (Engvold and Livingston 1971).

In filaments seen off the center of the disk these boundaries appear as lower bright rims to the dark filaments on filtergrams, often with conspicuous contrast (L. d'Azambuja and M. d'Azambuja 1948).
The formation time of a section of a filament from one footpoint to the next is of the order of hours (Forbes 1986). Even on this relatively large scale the structure of the prominences thus change with time scales some two orders of magnitude shorter than the life-time of

the prominence as a whole. Also from this point of view a prominence is truly dynamic.

Is there any observational evidence for matter entering or leaving the prominence through the foot-points? In filaments it is often difficult to identify the feet on spectrograms, but Kubota and Uesugi (1986) found upward directed flows in four regions that apparently coincided with footpoints. Schmieder et al. (1985) on the other hand observed one case when matter drained down at a footpoint with a velocity of 10 km/s. Malherbe et al. (1983) concluded from their analysis of the velocity-field in a filament that, "these motions correspond likely to the rise of material along magnetic loops closely related to the prominence structure". A detailed study by Kubota et al. (1988) confirms such an interpretation. The "feet" apparently play a key role for the mass balance in quiescent prominences.

4. The main body of the prominences.

In the main body of quiescent prominences we often see a completely chaotic picture both on filtergrams and on high resolution spectra. In many cases the velocities measured in the best spectra are in the supersonic range relative to the sonic velocity in the cool component i.e. at temperatures of 6-7000 K, which amount to about 8 km/s. In the prominence plasma the electrical conductivity is sufficiently high to make the magnetic field "frozen-in" at the scale of the observed fine-structure. If equipartition between the kinetic energy density and the turbulent part of the magnetic field is established, we arrive at an amplitude for the magnetic fluctuations of the order of 2-3 gauss. Local fluctuations in the magnetic field of this order is impossible to discern with present technique. Both with the fluctuations due to the presence of Alfven-waves and in the case of fully developed MHD turbulence, the kinetic and magnetic energy densities will be in equipartion. In any case a realistic model of quiescent prominences must account for the presence of this turbulent velocity field.

Until recently the problem of prominence support was to keep "a thin, vertical sheets of cool matter" from falling. As we have seen, high resolution observations have changed this picture completely. Prominences do not bear any resemblance to static models. We have to deal with the problem of supporting matter in individual tubes of force that undergo complicated changes with time-scales of minutes or less (Jensen et al. 1989). Observational evidence suggests that the supporting force at work is of a stochastic nature. Locally matter may ascend or descend for shorter or longer intervals of time. There is no observational evidence for any sagging of the lines of force under the weight of matter collecting in a centerally located pit.

A turbulent "spagetti" model is what is required to describe observations of high resolution. The tubes of force are seemingly more or less deformed loops. Some are smoth and may sustain flows that may be traced over several thousand kilometers, others look completely broken up.

The resulting morphology is different in different parts of a fully developed "hedgerow". In the uppermost parts the fine-structure may appear somewhat more ordered, often characterized by hanging "draperies" with the fine-structure consisting of knots and vertical threads. It is here observations often show structure elements to fall with velocities mostly in the range 2-5 km/s (Engvold 1976). What we observe is apparently an impeded fall, as if the support mechanism were insufficient to carry the whole load.

With the incapability of the "classical" prominence models to account for the crucial observational fact that we have to do with a turbulent plasma, a hypothesis was proposed that stresses the dynamical aspects of the problem and suggests a more flexible way of supporting matter. The keyword is Alfven-waves (Jensen 1983, 1986).

IV. WAVE SUPPORTED PROMINENCES

Taking the observed value of mass-density and magnetic field in prominences at their face values the Alfvén-velocity in prominence matter becomes 50-100 km/s. If it is postulated that Alfvén-waves may be exited anywhere on the sun where magnetic fields are present and the flux density is taken to be of a reasonable value for heating the corona and accounting for the energy in the solar wind field, an Alfvén-flux of the order of F = E5-E6 erg $cm^{-2}s^{-1}$ is required. If a wave-flux of this order is channelled into a quiescent prominence from below, it turns out that the waves become non-linear (Jensen 1983).

The damping length of Alfven-waves is reduced drastically in prominence matter as compared to the surrounding corona. In a way quiescent prominences may be said to act as a trap for Alfvén-waves. In a non-linear wave field compression effects are induced and mode conversion may result. With a stohastic generation of waves in the convection zone a turbulent velocity-field would be expected, thus accounting for the quasistatic "mess" that is observed. In the dissipation process momentum is transferred from the waves to matter. With standard values for the physical parameters in the prominence plasma, the resulting force appears to be of the right order to counteract gravity. Let us look at some equations;

The Alfvén flux-density may be written;

$$F_A = \frac{1}{2}\rho(\Delta V)^2 V_A = \frac{1}{8\pi}(\Delta B)^2 V_A \tag{1}$$

Here ΔV and ΔB denotes respectively the fluctuations in velocity and magnetic field in the wave, whose carrier field is B. V_A is the Alfvén-velocity,

$$V_A = \frac{B}{\sqrt{4\pi\rho}} \tag{2}$$

For the fluctuations in the wave we get the expressions;

$$\begin{aligned} \Delta V &= 2\pi^{1/4}\rho^{-1/4}F_A{}^{1/2}B^{-1/2} \\ \Delta B &= 4\pi^{3/4}\rho^{1/4}F_A{}^{1/2}B^{-1/2} \end{aligned} \tag{3}$$

The damping length is (Wentzel 1977);

$$L_A \sim \frac{B^4}{F_A\rho} \tag{4}$$

The order of magnitude of the supporting force becomes;

$$K_A \sim \frac{F_A}{L_A V_A} \sim \frac{F_A{}^2\rho^{3/2}}{B^5} \tag{5}$$

The dissipation also gives in additional contribution to the heating in the prominence of the order:

$$Q \sim \frac{F_A}{L_A} \tag{6}$$

The heating effect has been used by Anzer (1989) as an argument against support by waves. Let us estimate what this additional heating represents in terms of radiated energy. It may be written;

$$\Delta I \sim \frac{Qnl}{4\pi} \tag{7}$$

Here l is a characteristic dimension of a fine-structure element and n the number of elements along the line of sight, cf. Schmahl and Orrall (1986) and Hirayama (1986). With $F_A = 5x10^5$, $L_A = 2x10^7$cm, we get $\Delta I = 2x10^{-3}$ nl. Choosing $l = 2x10^7$cm we get $\Delta I = 4x10^4 n$.

This should be compared to the total radiation from the main body of the prominence. This quantity shows considerable variation from prominence to prominence, but a reasonable average seems to be in the range from $5x10^4$ to $2x10^6$erg $cm^{-2}s^{-1}str^{-1}$ (Engvold 1989). The value of n is quite uncertain, but we may conclude that the contribution of wave heating could be of importance for the energy balance in prominences, but will not dominate the energy budget. That the heating term is proporsional to the density, while the loss term goes as the density squared suppresses excessive heating and wakens Anzer's argument against support by waves.
Since we have to do with a non-stationary mechanism operating in a highly inhomogeneous medium, the heating by waves may explain some of the time changes observed in the fine structure.

The hypothesis of wave support rests on the assumption that the main characteristic of quiescent prominences is the presence of turbulence created by non-linear Alfvén-waves. If the carrier field of the waves becomes too strong the non-linear case is not reached for a given wave-flux and turbulence fails to develop. This condition defines an upper limit for the magnetic field in quiescent prominences. Similarily a lower limit follows from the condition that the magnetic field must be sufficiently strong to be able to carry the waves without excessive dissipation. This explains why quiescent prominences exist only with magnetic fields within a limited interval. The order of magnitude estimates of the limits for the magnetic field were shown to be in reasonable agreement with observations some years ago (Jensen 1983). With newer data the numbers used then have to be modified as the values both for the magnetic fields and the density must be increased.

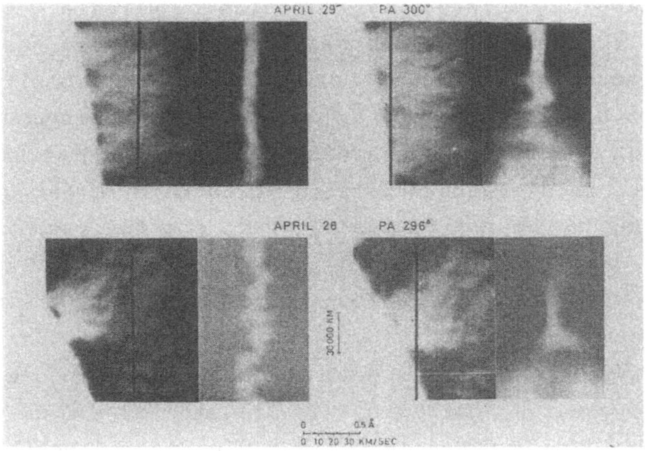

Figure 8. Edge effects, Engvold and Livingston (1971), Kitt Peak National Observatory.

Edge effects (Engvold and Livingston 1977, Engvold et al. 1978) fits in well with the effects of waves or turbulence. When either the density or the carrier field is reduced, velocity-fluctuations increase cf. equ. (3). If on the other hand the edge effects is a result of prominence reconnection and subsequent flare-like explosions, such events are in themselves wave-generators.

According to Leroy et al. (1983) the magnetic field strength for polar crown prominences varies with phase in the solar cycle. This is predicted from the hypothesis of wave support as a consequence of variations in the Alfvén-wave flux with solar activity.

V. FUTURE ASPECTS

To identify the wave-modes that may be transmitted from the convection zone, the conversion they are subject to and the interactions that create and sustain the observed plasma turbulence is of great importance for the hypothesis of wave support. For this problem a critical parameter is the magnetic field strength in the prominence feet, that have to act as a guide for waves from below. Since the damping length for Alfvén-waves is proportional to the fourth power of the field strength a field three times higher than in the prominence itself, would increase the damping length by two orders of magnitude to some ten thousand kilometers. But this also requires a vertical magnetic field in the footpoints. A horizontal field here, as some of the Hanle observations indicate, would rule out the possibility for Alfvén-waves to be of importance in quiescent prominences. To eliminate the present uncertainty regarding the magnetic parameters in prominence feet, reliable observations would be of great value. This is difficult, but maybe not impossible with present technique.

Many observers have reported periodic oscillations in the velocity-field in quiescent prominences (Schmieder 1989). Solovjev (1985) showed that Alfvén-waves in magnetic loops could reproduce the observed periods. This indication of the presence of Alfvén-waves in prominences should be further investigated using the best high resolution observations.

Further observational data which may be obtained with present instrumentation, are time-series of filtergrams with optimal spatial resolution simultaneous with spectral data in Balmer lines, D3 from He and H and K from CaII. Correlation analysis between the turbulent velocity field and intensity fluctuations should be carried out for different parts of prominences, both as functions of time and heliographic position. In an Alfvén wave-field such correlation will show up when non-linear effects become of importance, while these waves are incompressible in the linear approximation. A crucial observation for identification of Alfvén-waves in prominences would be the existence of local correlations between fluctuations in velocity and in the magnetic field. However, due to the low resolution of present magnetographs this is not yet possible. The deplorable situation for magnetic data will apparently not be substantially improved until THEMIS (Rayrole 1987) or LEST (Engvold 1989) become operative.

At that future date we could also hope to get an independent determination of mass density through the equipartition between kinetic and magnetic energy of the fluctuations.

The properties of filament channels should be further investigated. We have already seen some very interesting results at this meeting regarding chromospheric structure in filament channels (Martin 1989).

With growing evidence for sub-arc second structures the first attempts at three dimensional modelling are appearing in the literature (Poland and Tandberg-Hanssen 1983, Fontenla and Rovira 1985, Ballester and Priest 1989). This is a difficult generalization, but obviously a necessary one if we want to understand the physics operating in quiescent prominences.

I am indebted to dr. O. Engvold for numerous discussions and valuable comments.

VI. REFERENCES

Anzer, U. and Priest, E., 1985, Solar Phys. 95, 263.
Anzer, U., 1989, in "Dynamics and Structure of Quiescent Solar Prominences", ed. E.R. Priest, Kluwer Academic Publishers, 143.
Athay, R.G., Querfield, C.W., Smartt, R.N., Landi degl'Innocenti, E., Bommier, V., 1983, Solar Phys. 89, 3
Ballester, J.L. and Priest, E.R., 1989, These procedings.
Bonnet, R.M., Bruner, E.C. Jr., Acton, L.W., Brown, W.A., Decaudin, M.; 1980, Ap.J., 237, L47.
d'Azambuja, L. and d'Azambuja, M., 1948, Ann. Obs. Meudon 6
Dunn, R.; 1960, Ph.D. Thesis, Harvard Univ.
Engvold, O. and Livingston, W.; 1971, Solar Physics, 20, 375.
Engvold, O. and Jensen, E.; 1976, Institute of Theoretical Astrophysics, Report No. 47.
Engvold, O.; 1976, Solar Physics, 49, 283.
Engvold, O. and Jensen, E.; 1977, Solar Physics, 52, 37.
Engvold, O. and Malville, J.M., 1977, Solar Physics, 52, 369.
Engvold, O., Malville, J.M., Livingston, W., 1978, Solar Physics, 60, 57.
Engvold, O., 1980, Solar Physics, 67, 351.
Engvold, O., 1989, private communication.
Fontenla, J.M. and Rovira, M., 1985, Solar Physics, 96, 53.
Forbes, T.G.; 1986, NASA Conf. Pub. No. 2442, ed. A.I. Poland, p.21.
Harvey, J.W., 1969, Ph.D. thesis, University of Colorado.
Hirayama, T., 1985, Solar Physics, 100, 415.
Hirayama, T., 1986, NASA Conf. Pub. No. 2442, ed. A.I. Poland, p. 149.
Jensen, E.; 1983, Solar Physics, 89, 27.
Jensen, E.; 1986, Coronal and prominence plasmas, NASA Conf. Pub. 2442, ed. A.I. Poland, p.63.

Jensen, E., Engvold, O. and Scharmer, G.B.; 1989, in E. Leer and P. Maltby (eds) Proc. of Miniworkshop on Flux Tubes in the Solar Atmosphere, 67.
Kim, I.S., Koutchmy, S., Nikolsky, G.M., Stellmacher, G., 1982, Astron. and Astrophys. 114, 347.
Kippenhahn, R. and Schlüter, A. 1957, Z.f. Astrophysik, 43, 36.
Kleczek, J., 1980, Hvar Obs. Bull. 4, 35.
Kuperus, M. and Raadu, M., 1974, Astron. and Astrophys. 31, 185.
Kubota, J. and Uesugi, A., 1986, Publ. Astron. Soc. Japan, 38, 903.
Kubota, J. Tohmura, T., Uesugi, A., 1988, Vistas in Astronomy, 31, 39.
Leroy, J.L., Bonnier, V., Sahal-Brechot, S., 1984, Astron. and Astrophys. 131, 33.
Malherbe, J.-M., 1987, These de Doctorat d'Etat, Université de Paris.
Malherbe, J.-M., Mein, P., Schmieder, B., 1983, Advances in Space Res., 2, 57.
Malherbe, J.-M., 1989, in "Dynamics and Structure of Quiescent Solar Prominences" ed. E.R. Priest, Kluwer Academic Publishers, 115.
Maltby, P., 1976, Solar Physics, 46, 149.
Martin, S.F., 1973, Solar Physics, 31, 3.
Martin, S.F., 1986, Coronal and Prominence Plasmas, NASA Conf. Pub. No. 2442, ed. A.I. Poland, p.73.
Martin, S.F., 1989, In these proceedings.
Mouradian, Z., Martres, M.J., Soru-Escaut, I., 1986, Coronal and Prominence PLasmas, NASA Conf. Pub. No. 2442, ed. A.I. Poland, 221.
Pikelner, S.B., 1971, Solar Physics, 17, 44.
Plocieniak, S. and Rompholt, B., 1973, Solar Physics, 29, 399.
Poland, A.I. and Tandberg-Hanssen, E., 1983, Solar Physics, 84, 63.
Poland, A.I. and Mariska, J.T., 1986, Solar Physics, 104, 303.
Pneuman, G.W., 1983, Solar Physics, 88, 219.
Priest, E.R., 1989, Dynamics and Structure of Quiescent Solar Prominences, Kluwer Academic Publishers, Dordrecht, Boston, London.
Rayrole, J., 1987, The Role of Fine-Scale Magnetic Fields on the Structure of the Solar Atmosphere, ed. Schröter, Vazquez, Wyller, p. 367.
Rompolt, B. and Bogdan, T., 1986, Coronal and Prominence Plasmas, ed. A.I. Poland, NASA Conf. Rep. No. 2442, p. 81.
Rust, D., 1967, Ap.J., 150, 313.

Schmahl, E.J. and Orrall, F.Q., 1986, Coronal and Prominence Plasmas, ed. A.I. Poland, NASA Conf. Rep. No. 2442, p. 127.
Schmieder, B., Malherbe, J.M., Poland, A.I., Simon, G., 1985, Astron. and Astrophys., 153, 64.
Schmieder, B., 1989, Dynamics and Structure of Quiescent Solar Prominences, Kluwer Academic Publ. ed. E.R. Priest, Chapter 2, 15.
Simon, G., Schmieder, B., Demoulin, P., and Poland, A.I., 1986, Astron. Astrophys. 166, 319.
Solovjev, A.A., 1985, Solnecnye Dannye, No. 9, 65.
Tandberg-Hanssen, E. and Anzer, U., 1970, Solar Physics, 15, 158.
Tandberg-Hanssen, E., 1970, Solar Physics, 15, 359.
Tandberg-Hanssen, E., 1974, Solar Prominences, Dordrecht, Reidel.
Tang, F., 1987, Solar Physics, 107, 233.
Wentzel, D.G., 1977, Solar Physics, 52, 163.
You, Jianqi and Engvold, O., 1989, These proceedings.
Zirker, J.B., 1989, Solar Physics, 119, 341.

DISCUSSION

KOUTCHMI: Concerning the Alfven waves which are needed to produce the Alfven pressure in the mechanisum you described, could you make a kind of prediction which could be useful for observers, like the range of periods we should look at.

JENSEN: My guess would be 50-150 seconds.

FORBES: Comment. I Don`t see any mechanism in your model which explains why thermal condensations occur at a particular location in the solar atmosphere.

JENSEN: This hypothesis aims at describing what is going on in fully developed quiescent prominences. But since the prominences are composed of loops it is a matter of filling loops with matter. A thermal instability is not needed.

PRIEST: 1) You mentioned that there is no evidence for a dip, but quantitatively an extremely slight dip would be needed and so would not be observed.

2) Regarding Tang´s observation of many quiescent prominences forming between 2 bipolar regions, Demoulin has a good model for such prominences.

3) Perhaps there <u>are</u> strong small scale variations in magnetic field strength in prominences, but in a low-beta plasma you can produce strong small-scale variations in plasma pressure with only weak magnetic variations. Also Leroy has some arguments for feeling that the field is rather uniform at small scales.

LEROY: It is true that the interpretation of avaible magnetic field measurements gives the picture of a more or less homogeneous magnetic structure in quiescent prominences, this is certainly not the case in active prominences as illustrated well on the poster by Koutchmy and Zirker. However, I must confess that, as an observer, I am not happy with this conclusion which I feel contradictory with the very appearance of prominences. The measurement of prominences magnetic field with an improved resolution (say less than 1") is urgently needed!

MAGNETIC STRUCTURE OF PROMINENCES

(Invited Review)

E R Priest

Mathematical Sciences Dept, The University

St Andrews KY16 9SS, Scotland

1. Introduction

Prominences have been observed for many years but it was only 30 years ago that the first models for magnetic structure were proposed. In particular, the Kippenhahn-Schlüter model (1957) has the magnetic topology indicated in Figure 1. The prominence is represented here as a sheet with current I at a height h and directed out of the plane, since it produces a change in the direction of the vertical component of the magnetic field from down on the left- hand side to up on the right-hand side. If the photospheric footpoints are line-tied during the formation of the prominence, the preservation of the footpoint position can be modelled by adding an image current (-I) a distance h below the photosphere to the original arcade and the prominence sheet. Thus the prominence of mass m is supported against gravity both by the line tying (the repulsion $\mu I^2/(4\pi h)$ between I and -I) and also by the Lorentz force IB acting on I in the original background field B at height h.

An alternative magnetic topology was proposed by Kuperus and Raadu (1974) with the magnetic field passing through the prominence in the opposite direction, as shown in Figure 2, where the current I is now directed into the plane. The basic topology is shown on the left-hand side with the outwards spreading of the field lines from the footpoints providing a magnetic tension force upwards. When the lowest field lines are in the form of magnetic loops straddling directly across the polarity inversion line, as on the right-hand side, there is an X-type neutral point below the prominence sheet. Support in the Kuperus-Raadu model is only by line tying ($\mu I^2/(4\pi h)$, as before) since the Lorentz force IB now acts downwards.

During the past 5 years, there has been a renewed interest in refining these classical models and in developing new ones, and so today I want to try and summarise this activity. Because there

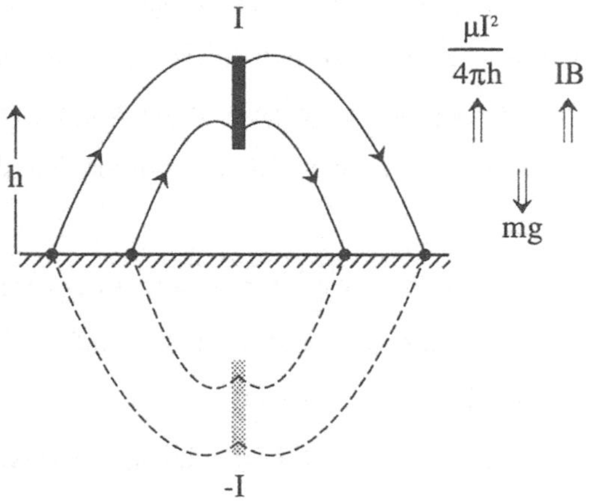

Figure 1. Magnetic topology for prominences of Normal Polarity

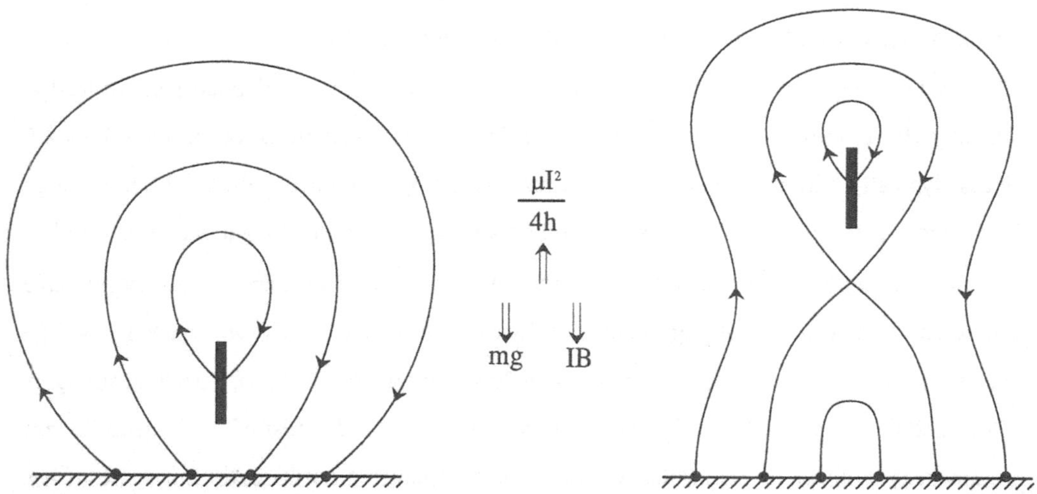

Figure 2. Magnetic topology for prominences of Inverse Polarity

are now several other models of the above types, it was suggested at the Prominence Workshop in Mallorca (Priest, 1988; Ballester and Priest, 1988) that both prominences and models in which the magnetic field goes through the prominence like a normal magnetic arcade (Figure 1) be said to possess Normal Polarity, whereas those in which it goes through in the opposite direction be said to possess Inverse Polarity.

The main observational features of quiescent prominences have been reviewed by Tandberg -Hanssen (1974) in his outstanding text-book and also in the conference proceedings by Jensen *et al* (1977) and Poland (1986). They are as follows

(1) The geometry is that of a thin vertical sheet, lying above a reversal in the line-of-sight magnetic field in the photosphere (Babcock and Babcock, 1955). The length of the sheet lies between 60 and 600 Mm (typically 200 Mm), while its height ranges between 10 and 100 Mm (typically 50 Mm) and its width is between 4 and 15 Mm (typically 6 Mm).

(2) The density is between 10^{16} and $10^{17} m^{-3}$, and the temperature is usually between 5000 K and 8000 K (Vial, 1986; Bommier, 1986; Hirayama, 1986; Engvold, 1986). They both tend to decrease with height (Bommier, 1990; Fang Cheng, 1990).

(3) The magnetic field above 10 Mm is rather homogeneous, horizontal and inclined at typically $20°$ to the prominence axis (Leroy, 1979). It has a magnitude of between 3 and 30 Gauss (typically 5G) and usually increases with height by about 50% (Rust, 1967), although Kim (1990) finds a decrease for those which lie roughly north-south along a line of longitude.

(4) In Leroy's (1985) study he found that the high-latitude quiescent prominences have heights above 30 Mm, field strengths between 5 and 10 Gauss and are all of Inverse Polarity. The low-latitude prominences near active regions are low-lying (below 30 Mm), have field strengths of about 20G and are of Normal Polarity. Kim (1990) also finds most of the large quiescent prominences to be of Inverse Polarity (except those orientated north-south) but some of her active-region prominences, which include ones of lower altitude than Leroy, are Normal and some are Inverse.

(5) Prominences are long-lived, lasting for 1-300 days, with high-latitude ones enduring typically 140 days. The polarity inversion zones tend to migrate slowly towards the poles and to become stretched in an east-west direction.

(6) The observed flows in prominences are much smaller than the free-fall speed$\sqrt{(gh)}$

of about $100\,kms^{-1}$ (Engvold, 1976; Mein, 1977; Martres *et al*, 1981), and so the plasma is essentially in equilibrium with a rough balance between the magnetic, pressure and gravitational forces as it very slowly dribbles through the magnetic field. Typically one sees a downflow of $0.5\,kms^{-1}$ when observed on the limb and an upflow of $0.5 - 3\,kms^{-1}$ when viewed on the disc. Prominences tend to form where there are converging flows (Martin, 1990) and shearing flows (Gaizauskas, 1990).

(7) Plage or active-region prominences are smaller than their quiescent cousins by a factor of 3 or 4. They are lower in height and have larger densities ($\geq 10^{17}\,m^{-3}$) and field strengths ($\sim 20 - 100\,G$) and strong horizontal flows (Schmieder, 1990).

(8) On the limb one sees that a prominence consists of vertical threads, typically 5-7 Mm long and 0.3-1 Mm wide, with a filling factor of between 0.01 and 0.1 (Engvold, 1976; Simon *et al*, 1986; Rabin, 1986).

(9) Quiescent prominences appear to reach down to the surface in a series of "feet", spaced by about 30 Mm and located at supergranule boundaries (Plocienak and Rompolt, 1973).

When confronted by the above observational facts, what is the aim of theory? It is certainly not to reproduce on the computer all the features of a beautiful $H\alpha$ picture of a prominence. Rather, the aim is to understand the basic physics and causes of the overall structure and the main properties. Why is it essentially sheet-like? How is it supported? What is the global magnetic structure? In order to make progress, one starts with a simple physical model and tries to understand it. For example, a great advance in understanding can be obtained by using a rectangular slab to explain the global properties as outlined in (1)-(7) above. Then later one can attempt to understand the feet (property (9)) and the microstructure in the form of threads (property (8)). But without such slab models one would not have obtained an overall basic understanding within which to consider such complexities.

Using the philosophy of mathematical modelling, we shall therefore see how different authors have made theoretical progress by taking different assumptions. Starting with the internal structure of the prominence sheet (Section 2), we shall move on to consider the external prominence magnetic fields of Normal Polarity (Section 3) and of Inverse Polarity (Section 4), followed by brief comments on fibril structure and feet (Section 5). A new Flux Tube Model is summarised in Section 6 and the conclusions are presented in Section 7, including some comments about the longitudinal magnetic component and the long-term evolution of prominences.

Before embarking on this programme let us remind ourselves of the magnetohydrostatic equation for equilibrium under a balance between magnetic, pressure and gravitational forces

$$\mathbf{j} \times \mathbf{B} - \nabla p + \rho \mathbf{g} = \mathbf{0} \tag{1}$$

where

$$\mathbf{j} = \nabla \times \mathbf{B}/\mu \tag{2}$$

$$\nabla \cdot \mathbf{B} = 0 \tag{3}$$

and

$$P = \frac{R}{\tilde{\mu}} \rho T \tag{4}$$

in the usual notation. Using (2) the magnetic force may be decomposed into the sum

$$\mathbf{j} \times \mathbf{B} = -\nabla(B^2/2\mu) + (\mathbf{B} \cdot \nabla)\mathbf{B}/\mu$$

of two terms, the first representing a gradient in magnetic pressure and the second a magnetic tension force.

Along a magnetic field line the magnetic force vanishes and so (1) reduces to

$$-\nabla p + \rho \mathbf{g} = 0$$

or

$$-\frac{dp}{dy} - \rho g = 0$$

if the y-axis is directed upwards in the opposite direction to gravity. Using (4), the solution to this is

$$p = p_0 \, exp - \int_0^y \frac{dy}{H(y)}$$

where p_0 is the pressure at y = 0 and

$$H(y) = \frac{RT(y)}{\tilde{\mu} g} \tag{5}$$

is the pressure scale-height. Thus for a uniform temperature we recover the familiar exponential decay

$$p = p_0 \, e^{-y/H}$$

of the pressure with height. Since H is only about 180 km for a temperature of 6000 K this explains the narrow width of a cool prominence sheet. If the prominence plasma is sitting at rest in a curved magnetic flux tube, supported against gravity by the magnetic tension (Figure 3) the width of the plasma structure depends on the inclination to the horizontal of the field lines. For a small inclination one needs a large horizontal distance (ℓ) to move a distance H vertically. Thus the half-width (ℓ) may be estimated to be

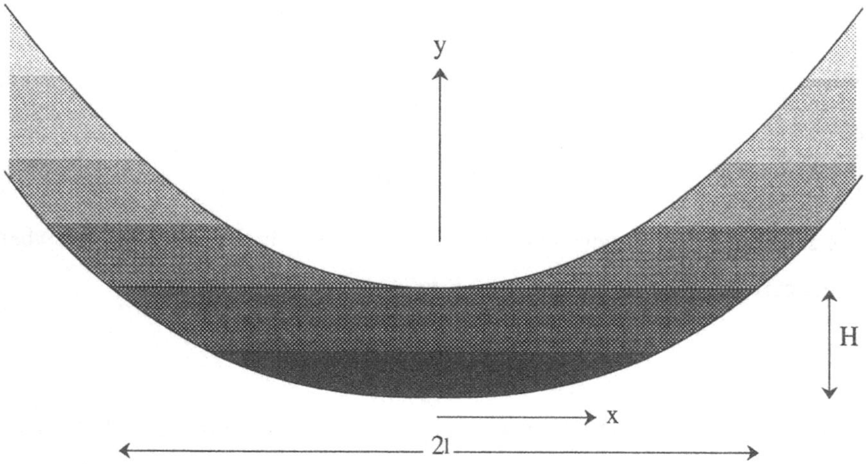

Figure 3. Support of plasma in a curved flux tube

$$\ell = 2\frac{B_x}{B_{y0}}H, \tag{6}$$

where B_{y0} is the vertical field at a distance (ℓ) from the centre, the factor 2 arising because the field lines are curved rather than straight. This relation may be used to estimate the inclination of the field. With H = 180 km and ℓ = 3 Mm, say, one finds a value

$$\frac{B_x}{B_{y0}} \approx 8,$$

corresponding to an inclination to the horizontal at the edge of the sheet of only a few degrees. In other words the required dip in the magnetic field is extremely small.

Outside the prominence the pressure and gravitational forces are dominated by the magnetic field and so (1) reduces to

$$\mathbf{j} \times \mathbf{B} = \mathbf{0}. \tag{7}$$

The magnetic field is force-free with the electric current given by (2) being parallel to the magnetic field, so that

$$\nabla \times \mathbf{B} = \alpha \mathbf{B} \tag{8}$$

where α is a scalar function of position. The divergence of (8) implies that

$$0 = \mathbf{B} \cdot \nabla \alpha$$

so that α is constant along a magnetic field line. When α takes the same value on every field line the curl of (8) yields

$$(\nabla^2 + \alpha^2)\mathbf{B} = \mathbf{0}, \tag{9}$$

the basic equation for a constant-α or linear force-free field. In the particular case when $\alpha = 0$, the electric current vanishes and this reduces to

$$\nabla^2 \mathbf{B} = \mathbf{0} \tag{10}$$

for a potential field. In the models that follow sometimes the external field is assumed to be potential and sometimes force-free.

2. Internal Structure of the Prominence Sheet

Whether the prominence is of Normal or Inverse polarity it may be modelled as a vertical sheet. Kippenhahn and Schluter (1957) set up a simple model for the magnetohydrostatic support of such a sheet by assuming that the temperature (T) and horizontal field components (B_x, B_z) are constant while the vertical field (B_y), pressure (p) and density (ρ) depend on x alone. The horizontal and vertical components of force balance (1) then reduce to

$$p + \frac{B_y^2}{2\mu} = \frac{B_{y0}^2}{2\mu} \tag{11}$$

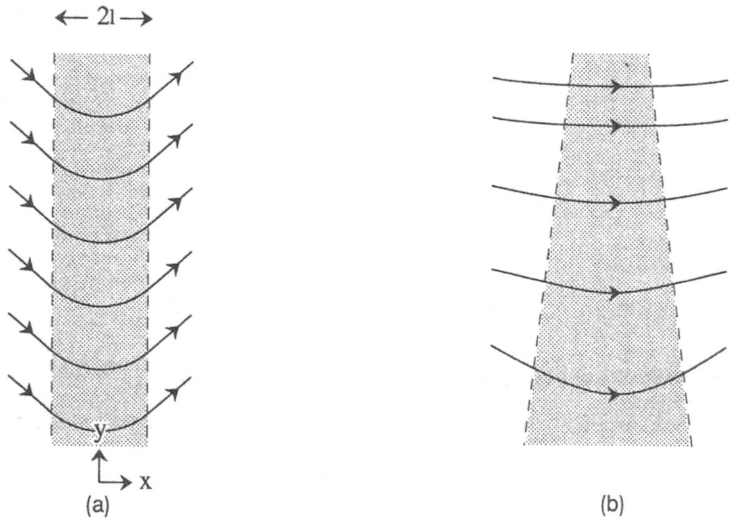

Figure 4. A prominence sheet that is (a) uniform or (b) narrows with height

and

$$\frac{dB_y}{dx}\frac{B_x}{\mu} - \rho g = 0 \qquad (12)$$

where B_y approaches $\pm B_{y0}$ as x tends to $\pm\infty$ and p approaches zero. The magnetic field therefore plays two roles. According to (11) it compresses the plasma sheet laterally and increases the plasma pressure in the sheet by a small amount equal to the external magnetic pressure $(B_{y0}^2/2\mu)$ associated with the vertical field. In addition (12) expresses the support of the plasma by the magnetic tension against gravity.

The solution of (11) and (12) is

$$B_y = B_{y0}\tanh(x/\ell), \quad p = (B_{y0}^2/2\mu)\,\mathrm{sech}^2(x/\ell), \qquad (13)$$

where the prominence half-width (ℓ) is given by (6) as expected.

Several generalisations of this solution have been constructed. Poland and Anzer (1971) allowed for imposed spatial variations T(x) of the temperature so that

$$B_y = B_{y0}\tanh\int_0^x dx/\ell(x).$$

Milne *et al* (1979) coupled the magnetohydrostatics to a very simple energy balance equation

$$\frac{d}{dx}(\kappa \frac{dT}{dx}\frac{B_x^2}{B^2}) = \rho^2 Q(T) - h\rho$$

where

$$p = p_e \qquad \text{and} \qquad T = T_e \quad \text{at} \quad x = \pm a$$

and

$$B_y = dT/dx = 0 \qquad \text{at} \qquad x = 0\,.$$

The resulting solutions depend on the plasma beta ($\beta = 2\mu p_e/B_x^2$) and the shear angle $\Phi = tan^{-1}(B_z/B_x)$. Prominence-like solutions are found when β is smaller than a critical value, but it will be important in future to try and include radiative transfer effects in a better manner.

Another modification due to Ballester and Priest (1987) is to allow slow variations with height by writing the magnetic field as

$$\mathbf{B} = \mathbf{B}_0(x) + \epsilon \mathbf{B}_1(x, y)$$

where $\mathbf{B}_0(x)$ is the Kippenhahn-Schlüter solution (13). The result is that the width decreases slowly with height, while the field lines become less curved and the field strength increases, in agreement with observations (Figure 4b).

3. External Fields of Normal Polarity

Menzel (1951) proposed a model, which has been rather overlooked in favour of the Kippenhahn -Schlüter model, but which as we shall see later has been generalised by Hood and Anzer (1990) in a most interesting manner. The Menzel model assumes that the temperature and, therefore the scale-height $H = RT/\tilde{\mu}g$, is constant and it considers a force-balance in which the pressure (p) and two field components (B_x, B_y) are separable functions of x and y in the forms

$$p = P(x)e^{-y/H}, \quad B_x = X(x)e^{-y/(2H)}, \quad B_y = Y(x)e^{-y/(2H)} \qquad (14)$$

so that the pressure and magnetic field decay exponentially with height. Solutions were found which are periodic in x (Figure 5a), a feature which, together with the lack of a strong component along the prominence axis, is a disadvantage of the model.

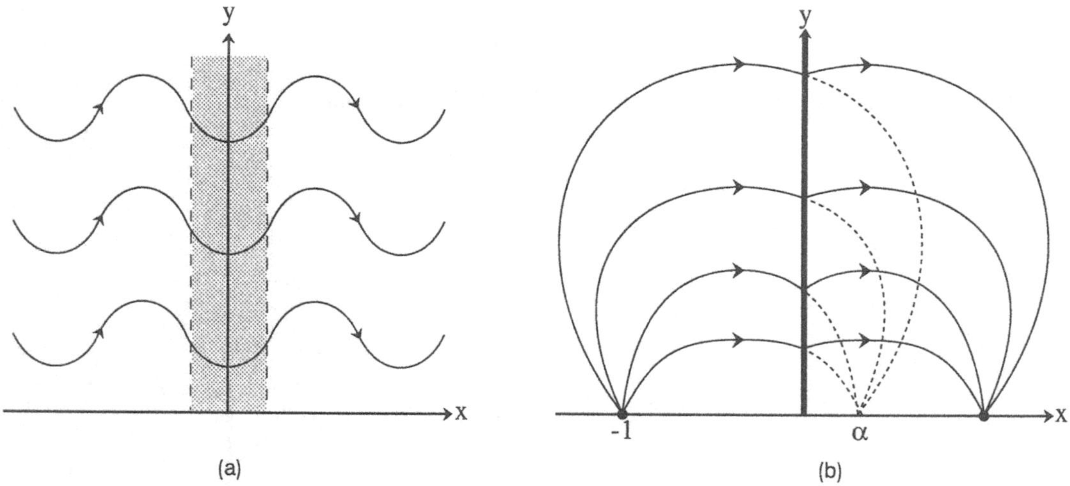

Figure 5. The magnetic fields for the models of (a) Menzel and (b) Kippenhahn and Schlüter

Kippenhahn and Schlüter (1957) modelled the external field as a potential field outside an infinite vertical current sheet (Figure 5b). The field is symmetric in x, so that for example the field in the left-hand plane is due to a positive line dipole at x = -1 on the x-axis together with an equal and opposite line dipole at x = α. Thus with

$$\mathbf{B} = \nabla\phi$$

the potential in the left-hand plane ($x < 0$) is

$$\phi = log\ r_1 - log\ r_\alpha$$

where

$$r_1^2 = (x + 1)^2 + y^2$$

and

$$r_\alpha^2 = (x - \alpha)^2 + y^2 .$$

The variation of B_x and B_y along the y-axis can be calculated and used to deduce the variation with height of the mass (m) that can be supported in the prominence sheet by magnetic tension from

$$mg = [B_y]\frac{B_x}{\mu}$$

where $[B_y]$ is the jump in the vertical field B_y in crossing the sheet.

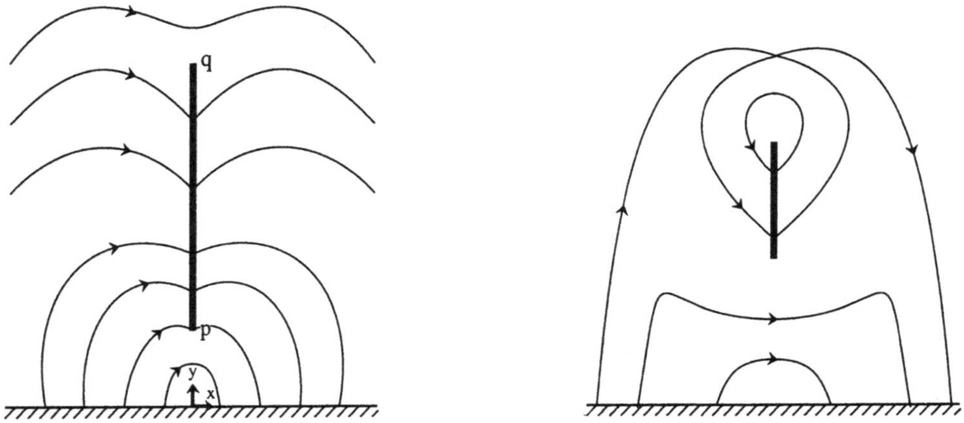

Figure 6. Malherbe-Priest models of Normal Polarity

More recently, Malherbe and Priest (1983) used complex variable theory to model the prominence as a current sheet of finite height (Figure 6). They treated the sheet as a cut in the complex plane from $z = ip$ to $z = iq$, where $z = x + iy$. Then, if the combination $(B_y + iB_x)$ is written as an analytical function of z outside the prominence, the magnetic field components are potential, satisfying Laplace's equation. For example, the forms

$$B_y + iB_x = -\frac{B_0[(p^2 + z^2)(q^2 + z^2)]^{1/2}}{z(z + ih)^2} - \frac{B_1}{z}$$

and

$$B_y + iB_x = -\frac{B_0[(p^2 + z^2)(q^2 + z^2)]^{1/2}}{z(z + ih)^2} + \frac{B_1(z + iH)}{z(z - iq)}$$

give field configurations of Normal Polarity that are sketched on the left and right of Figure 6. As well as the prominence cut, they also both have a dipole source at $z = -ih$. The first is similar to the usual Kippenhahn-Schlüter type of field, whereas the second has an X-point above the prominence.

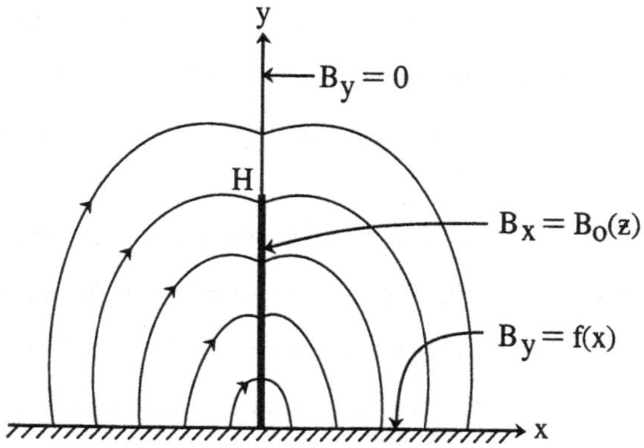

Figure 7. Anzer's model

In a more general, but not analytical, analysis Anzer (1972) had solved

$$\nabla^2 \mathbf{B} = 0$$

numerically for the (symmetric) potential field around a prominence (Figure 7). He imposed the normal component $B_y = f(x)$ at the photosphere ($y = 0$), the normal component $B_x = g(y)$ in the prominence ($x = 0, 0 \leq y \leq H$) and the condition $B_y = 0$ on $x = 0$ above the prominence ($y \geq H$). In principle, the functional forms $f(x)$ and $g(y)$ could be taken from observations. For the forms adopted, he found that the curvature at the prominence sheet was upwards for $0.4H \leq y \leq H$ and so the prominence could be supported there, but below $0.4H$ the model gave a downwards curvature and so failed. Recently, Demoulin, Malherbe and Priest (1989) have extended the model by allowing for magnetic flux to exist below the prominence where Anzer's boundary condition is replaced there by $B_y = 0$.

Anzer (1969) also considered the MHD stability of a prominence sheet by applying Bernstein's principle. He found stability if and only if

$$J\frac{dg}{dy} \geq 0$$

and

$$g\frac{dJ}{dy} \leq 0 \qquad or \qquad \frac{dJ^2}{dy} \leq 0$$

where $J = 2B_y(0+, y)/\mu$ is the current in the prominence sheet. Although the first condition is satisfied by observations which show the horizontal field strength increasing with height in a prominence, the observations are not precise enough to test the second condition.

Aly and Amari (1988) have considered a current sheet of general shape in a two- dimensional potential field $\mathbf{B}(x, y)$. They derive general relations for the current, mass and stored magnetic energy in terms of the values of the normal field $B_y(x, 0)$ at the photosphere ($y = 0$). The analysis has been extended to the case of an external force-free field by Amari and Aly (1988).

Recently, Priest (1988) and Priest *et al* (1989) have suggested that one needs to create a dip in the magnetic field before a prominence can form by thermal condensation or chromospheric injection. The reason is that usually the freefall time is much smaller than the cooling time and so in a low-beta plasma the plasma will tend to drain down before it collects, as shown in the numerical experiments of An *et al* (1988). Thus the question is: how can one create a magnetic dip in a coronal arcade?

Demoulin and Priest (1989) pointed out that a quadrupolar field naturally tends to possess a dip (Figure 8). This may be formed between two active regions, a place where prominences are observed to be present, or due to the presence of parasite polarity in the photosphere. A linear force-free model for such a configuration may be set up by superposing a harmonic on the fundamental arcade field

$$B_x = -\frac{\ell}{k}\cos kx \cdot e^{-\ell y},$$

$$B_y = \sin kx \cdot e^{-\ell y},$$

$$B_z = -\frac{\alpha}{k}\cos kx \cdot e^{-\ell y},$$

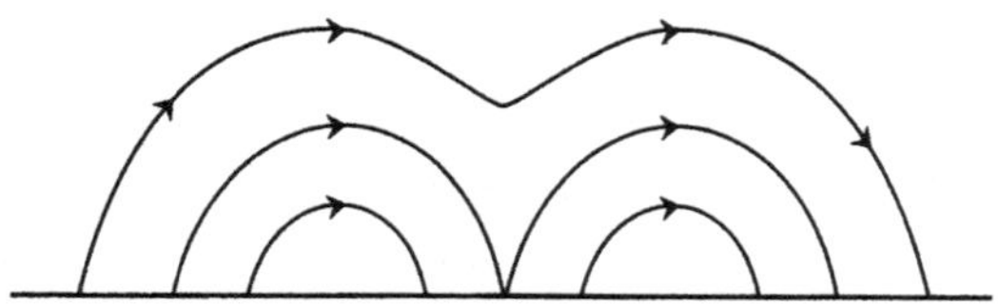
(a)

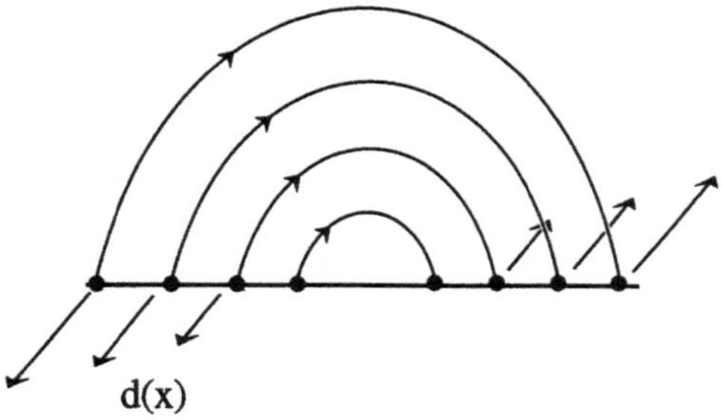

(b)

Figure 8. (a) A dip in a quadrupolar field. (b) Shearing of a 2D arcade.

where $\ell^2 = k^2 - \alpha^2$, so that for example the vertical field component becomes

$$B_y = \sin kx \cdot e^{-\ell y} + B_n \sin nkx \cdot e^{-\ell_n y},$$

where $\ell_n^2 = n^2 k^2 - \alpha^2$. For such a model, parasite polarity at the base ($y = 0$) is necessary in order to create a dip in the overlying field. More generally, Demoulin, Amari, Browning and Priest (1989) proved a theorem which states that, for a two-dimensional force-free arcade without parasitic flux, there is no shear profile $d(x)\hat{\mathbf{z}}$ of the photospheric footpoints which can create a dip.

Recently, Hood and Anzer (1990) have produced a generalisation of Menzel's (1951) largely forgotten model which combines the results of this and the previous section by including both the internal and external structure of a prominence. They assume that the field components have the form

$$(B_x, B_y, B_z) = (X(x), Y(x), Z(x))\, exp(-y/(2H_c))$$

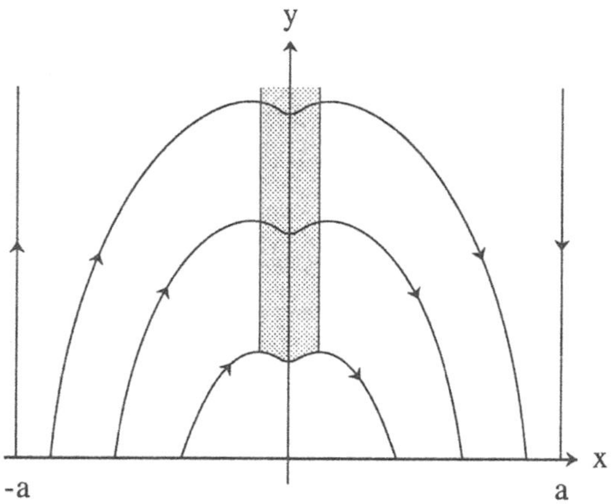

Figure 9. Hood and Anzer's model.

while the pressure and scale height are

$$p = \frac{P(x)}{2\mu} exp(-y/H_c), \quad H(x) = \frac{RT(x)}{\tilde{\mu} g},$$

where

$$Z(x) = \alpha X(x), \quad Y(x) = 2\frac{dX}{dx}H_c.$$

The configuration is in the form of an arcade of width 2a within which there is a prominence of width 2a. For simplicity, the temperature is assumed to have a uniform value of T_p in the prominence ($|x| \leq \ell$) and T_c in the corona ($\ell \leq x \leq a$), which gives a corresponding step-function for the scale-height H(x) (Figure 10a). The horizontal and vertical components of the force balance give the following two equations for P(x) and X(x):

$$P + X^2 + Y^2 + Z^2 = P_T,$$

$$2H_c XY' = [(1 - H_c/H)Y^2 + (H_c/H)(P_T - X^2 - Y^2) - P_T],$$

where P_T is constant and $Y = 2X'H_c$. In the corona P(x) is roughly uniform and the horizontal

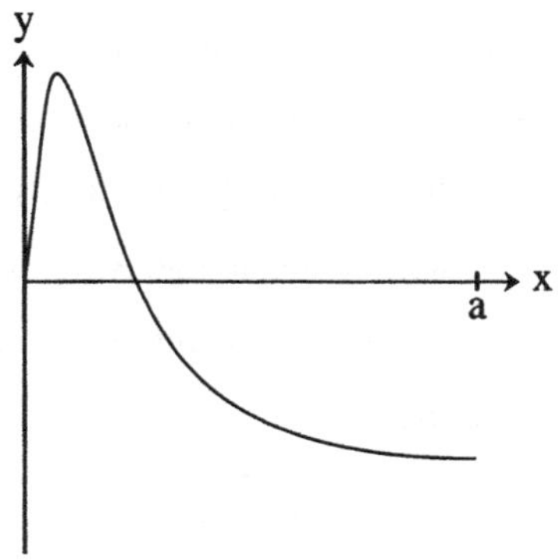

Figure 10. The variation of the vertical field with distance from the prominence axis in the Hood-Anzer model.

variation of the vertical field is given approximately by

$$Y = cos[(1 + \alpha^2)^{1/2}(x - a)]$$

for $\ell \leq x \leq a$. The full numerical solution is sketched in Figure 10b. Hood and Anzer were able to deduce reasonable values for the corona for given prominence conditions. They also extended the model to allow a potential field below the prominence base.

4. External Fields of Inverse Polarity

Kuperus and Tandberg-Hanssen (1967) suggested that a prominence may form in an open field - i.e. a current sheet - created when a closed arcade erupts and blows open as a result of a flare or a prominence eruption. Kuperus and Raadu (1974) suggested that the current of such a sheet would coalesce to form the prominence, which they modelled as a line current of strength I at height h and containing a mass m (Figure 11). Such a magnetic configuration can be regarded as the sum of an open field with straight field lines plus a set of closed field lines which do not intersect the photospheric boundary. The latter may in turn be considered as the field due to the sum of a line current (I) at height h and an image current -(I) at a depth h below the photosphere. Thus the support from magnetic tension due to line tying, if the footpoints do not move during the formation, is the repulsion between I and the image current -I. If this balances the downwards force of gravity we have for equilibrium

$$\frac{\mu I^2}{4\pi h} = mg \tag{4.1}$$

where $I = 2B_\phi \pi/\mu$ and $m = \pi R^2 \rho$ in terms of the field B_ϕ at the radius R of the filament. If the prominence density is $\rho = 10^{-10}\, kgm^{-3}$ and the prominence height is h = 10 Mm, this implies a reasonable field strength of $B_\phi = 6G$.

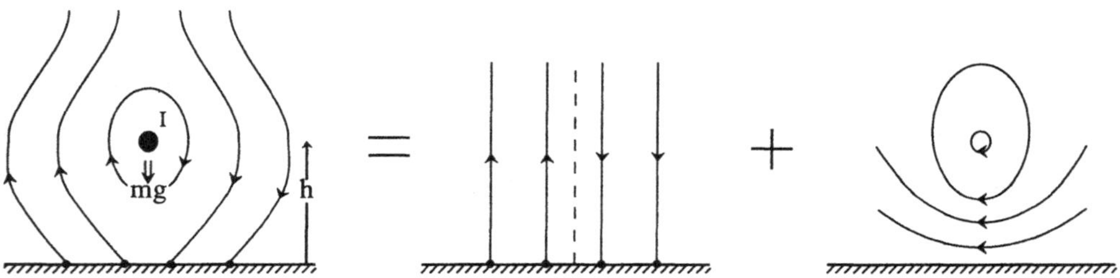

Figure 11. The Kuperus-Raadu configuration

Later Van Tend and Kuperus (1978), Kuperus and Van Tend (1981) and Kaastra (1985) extended this model by adding a background field B(h) which modifies (4.1) to

$$\frac{\mu I^2}{4\pi h} = mg + IB. \tag{4.2}$$

The last term represents the Lorentz force between the prominence and the background field. For a given function B(h) this equation can be solved for I, and it is found that for some such functions when I is too large there is no equilibrium and the force imbalance is such as to make the prominence erupt.

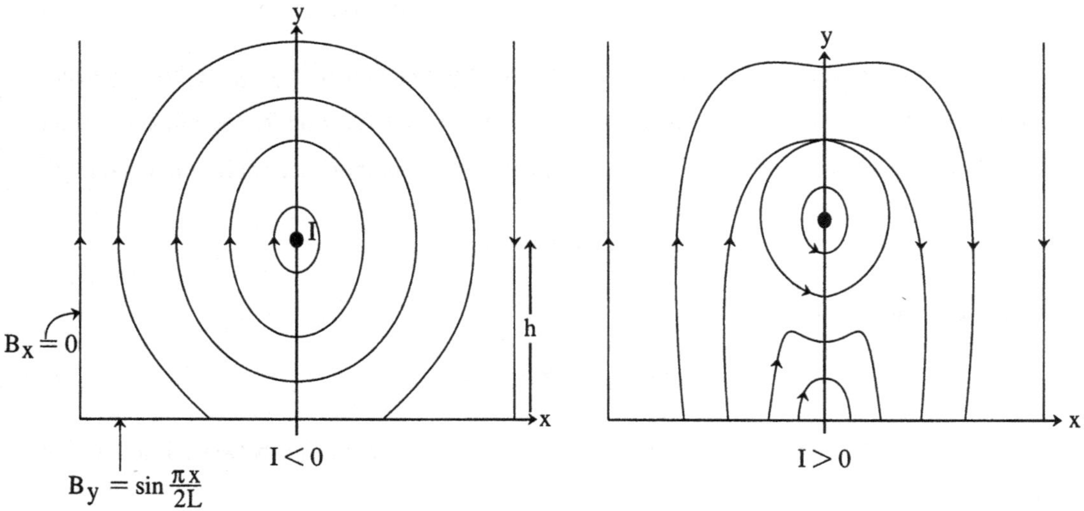

Figure 12. The configuration of Amari and Aly for (a) Inverse and (b) Normal Polarity.

However, the weak feature of the above order-of-magnitude model is that B(h) is imposed in an adhoc manner, and so Amari and Aly gave a more detailed analysis including the ambient field in a self-consistent manner. They modelled a prominence as a line of mass and current supported in a two-dimensional, linear force-free field (Figure 12) of the form

$$B_x = \frac{\partial A}{\partial y}, \quad B_y = -\frac{\partial A}{\partial x}, \quad B_z = B_z(A) \tag{4.3}$$

where the flux function A(x,y) satisfies

$$\nabla^2 A + \alpha^2 A = \delta(x)\delta(y-h). \tag{4.4}$$

The delta-functions on the right-hand side represent the line current at x=0, y=h. When the current I is negative the field is of Inverse Polarity and when I is positive it is of Normal Polarity.

The boundary conditions were to set

$$B_x = 0 \quad \text{on} \quad x = \pm L \quad \text{(the side boundaries)}$$

$$B \to 0 \quad \text{as} \quad y \to +\infty$$

$$B_y = \sin\frac{\pi x}{2L} \quad \text{on} \quad y = 0 \quad \text{(the base).}$$

The solution is written as the sum of a complementary function satisfying the homogeneous version of (4.4) and a particular integral which is the Green's function for the problem. After calculating the field components from (4.3) the equation for prominence equilibrium, namely,

$$IB_x(h) = mg$$

may then be written in the form

$$I^2 R(h) - IB(h) = mg. \tag{4.5}$$

This is remarkably similar to (4.2) except that now R is the repulsion between I and all the images of the force-free field. It is given by

$$R(h) = \frac{\mu}{L}\sum_0^\infty exp(-2h\gamma_{2n+1})$$

$$\text{where} \quad \gamma_{2n+1}^2 = (2n+1)^2\pi^2/L^2 - \alpha^2.$$

Also B(h) is now deduced from the force-free field for the imposed boundary conditions, and is

$$B(h) = \frac{\mu}{L}\sum_0^\infty b_{2n+1} L\gamma_{2n+1}\, exp(-h\gamma_{2n+1}).$$

Amari and Aly (1989) found from (4.5) that I increases monotonically with h and so the prominence is always in equilibrium, unlike the result of Van Tend and Kuperus! However, Demoulin and Priest (1988) generalised their analysis to allow for extra harmonics on the base of the region and found that the prominence could erupt because of a magnetic nonequilibrium when the current and shear parameter (α) are too large since the functional form of I(h) changes from monotonic to one with a maximum and minimum (Figure 13).

Lerche and Low (1980) and Low (1981) have modelled a prominence as a cylinder of finite radius surrounded by a potential field.

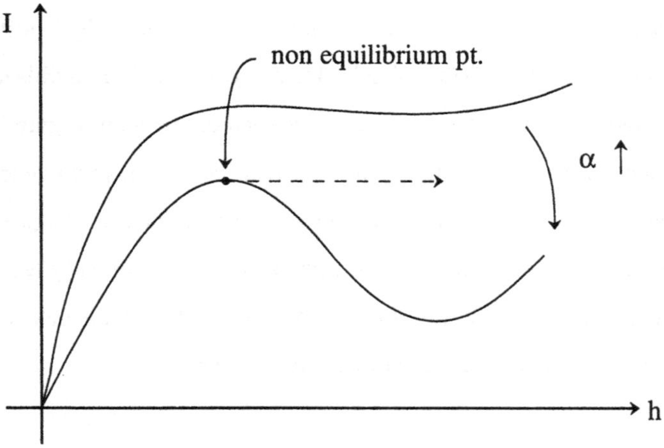

Figure 13. The current (I) as a function of prominence height (h) for different values of α (After Demoulin and Priest, 1988b).

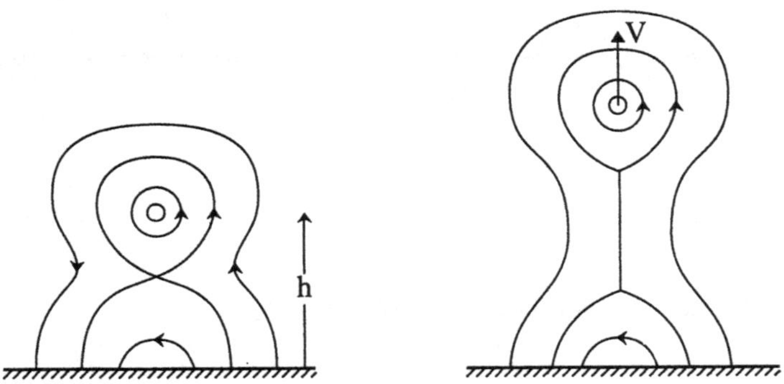

Figure 14. The magnetic field (a) before and (b) during eruption (After Priest and Forbes, 1990).

The process of eruption has been considered in more detail by Priest and Forbes (1990) and Isenberg and Forbes (1990) (see also Molodensky, 1990). The equilibrium is modelled as a line current supported against gravity in a background dipole field (Figure 14). When the twist is too great a state of magnetic nonequilibrium is reached and the prominence erupts. As it does so, it drives the formation of a current sheet below the prominence. The prominence would still erupt if there were no reconnection in the current sheet, but normally reconnection would be expected to be driven by the eruption and so would allow the prominence to erupt more rapidly as the anchoring of field lines to the photosphere is broken.

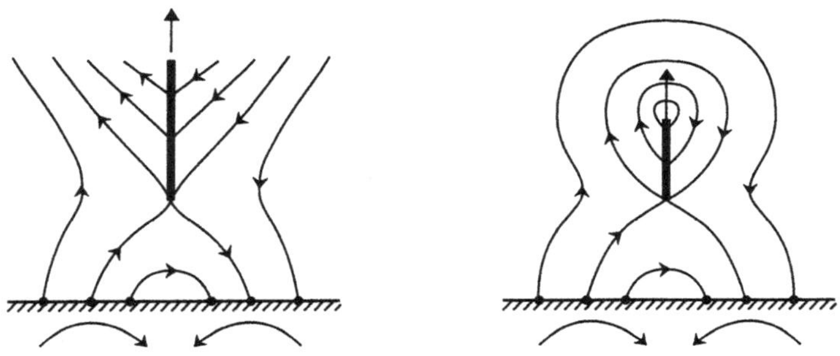

Figure 15. Malherbe-Priest models of Inverse Polarity.

Malherbe and Priest (1983) were able to construct complex variable models of Inverse Polarity prominences with finite current sheets stretching from $z = ip$ to $z = iq$. For example, the forms

$$B_y + iB_x = \frac{B_0}{z}[(p^2 + z^2)(q^2 + z^2)]^{1/2} + B_1(z - ip)$$

and

$$B_y + iB_x = -\frac{B_0[(p^2 + z^2)(q^2 + z^2)]^{1/2}}{z(z + ih)^2} + \frac{B_1(z - ip)}{z(z - iq)}$$

give the types of configuration shown in Figure 15. They suggested that the slow upflow of $0.5 - 3\,kms^{-1}$ seen in filaments viewed from above on the disc could be a response to slow converging footpoint motions if the prominences lie at the boundaries of giant cells on giant fault lines in the solar surface.

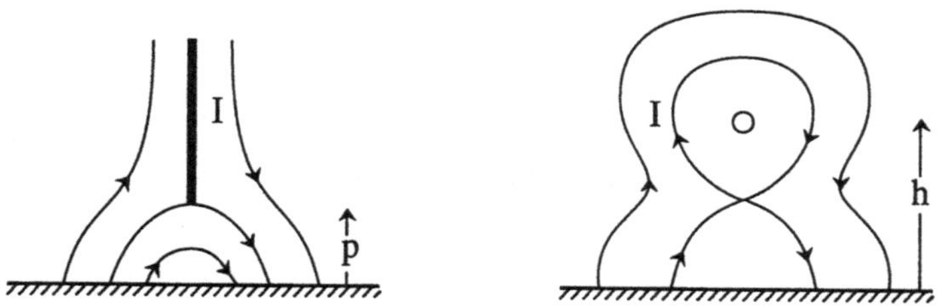

Figure 16. (a) A field with an open current sheet (b) The field that results after current coalescence.

Inverse-polarity models possess several difficulties. Anzer (1984, 1988) pointed out that a current sheet tends to possess a self-pinching force which is upwards in the lower part of the sheet but downwards in the upper part. This implies that the upper part cannot be in equilibrium since for Inverse-Polarity models the external IB force is downward too. Also Anzer and Priest (1985) found that it is hard to form an Inverse-Polarity prominence from a stretched-out sheet by a process that conserves the current as the current coalesces (Figure 16). Their calculation showed that if the base of the initial sheet is a height p = 30 Mm, say, above the photosphere then, after the coalescence, the prominence height (h) is of order 4 Mm, much lower than observed.

Amari and Aly (1990) considered a sheet prominence (Γ) in a linear force-free field (Figure 17), so that the field components are of the form given in Equation (4.3), where the flux

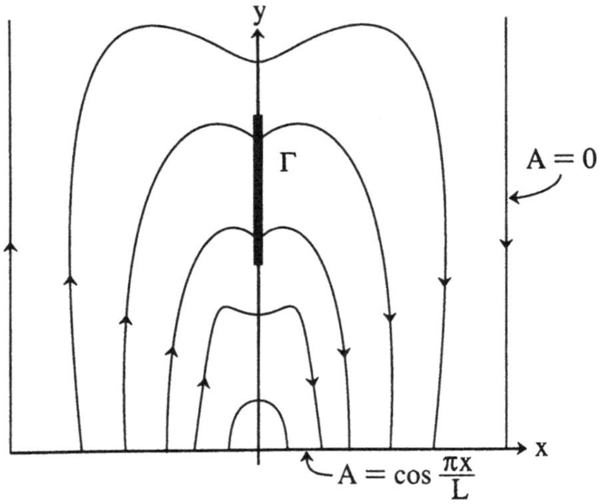

Figure 17. The model of Amari and Aly (1990).

function (A) satisfies

$$\nabla^2 A + \frac{d}{dA}(\frac{1}{2}B_z^2) = j\delta(\Gamma)$$

and the boundary conditions are that $A = cos\pi x/(2L)$ on the base (y=0) while A = 0 on the sides ($x = \pm L$). They suppose the current in the sheet behaves like

$$j = \frac{I}{h^3}(y-p)(q-y)$$

and write the solution as the sum of a complementary function and particular integral where the latter is

$$A_{PI} = \int Gj\delta(\Gamma)dx'dy'$$

in terms of the Green's function. They find that, whereas a current filament can always obtain an equilibrium, a current sheet of Inverse Polarity is never in equilibrium and one of Normal Polarity possesses an equilibrium if the current (I) and prominence mass (m) are less than critical values.

5. Prominence Structure

5.1 Fibril Structure

Having modelled the global sheet structure we need to turn to a consideration of the fibril or thread-like microstructure of a prominence. The cause for the thread-like nature of a prominence has not yet been isolated but it may well be due to the effects of the magnetic field, magnetic diffusivity and gravity on the thermal instability. The nature of thermal instability in a magnetic field has been studied in detail by Steinolfson and Van Hoven (1984), Van Hoven and Mok (1984), Sparks and Van Hoven (1985,1987), Van Hoven *et al* (1986).

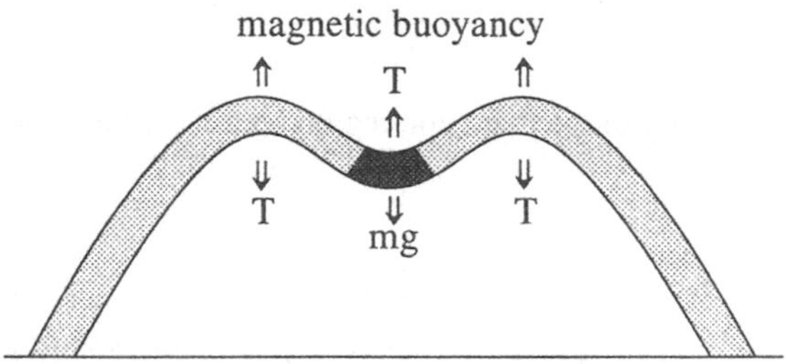

Figure 18. A slender loop with a cool condensation at its summit.

Ballester and Priest (1989) have modelled fibrils as slender loops in a coronal arcade in equilibrium under a balance between magnetic tension, magnetic buoyancy and gravity (Figure 18). They determine the tube shape from such a force balance and at the moment the effect of energy balance is being incorporated (Degenhart, 1990).

Poland and Mariska (1988) suggested that a local condensation sags down and tends to create dips on neighbouring field lines above and below. Also Steele and Priest (1989) have modelled different types of cool and hot-cool loops.

Furthermore, the radiative transfer of prominence threads has been studied in detail by Fontenla and Rovira (1985), Heinzel *et al* (1988) and Vial *et al* (1989).

5.2 Prominence Feet

One of the most puzzling features of prominences is the way they reach down towards the solar surface in a series of feet-like or tree-trunk-like structures. The cause is not well- established and they have not yet been well-modelled. Nakagawa and Malville (1969) considered the Rayleigh-Taylor instability of cool plasma of density ρ_p supported against gravity by a magnetic field. The growth-rate is given by

$$\omega^2 = -gk\left(\frac{\rho_p - \rho_c}{\rho_p + \rho_c}\right) + \frac{2B_0^2 k_x}{\mu(\rho_p + \rho_c)}$$

where ρ_c is the coronal density in the magnetic region below the prominence. The fastest mode has a wavenumber

$$k = \frac{\rho_p - \rho_c}{2B_0^2} g\mu$$

and so, for a prominence density $\rho_p = 2.10^{11}\, kgm^{-3} >> \rho_c$ and a field strength of $B_0 = 10G$, they find a wavelength of 30 Mm, the same as the observed footpoint separation.

Milne, Priest and Roberts (1979) found that when the plasma beta is greater than about unity there is no prominence equilibrium. They suggested that the prominence magnetic field is then too weak to support the plasma which pushes the magnetic field down towards the photosphere to form the feet. Martin (1986) found that the feet occur at the junctions of supergranular cells where the flow converges and the magnetic field is concentrating and cancelling.

Demoulin, Priest and Anzer (1989) set up a three-dimensional force-free model for the field around a prominence and tried to incorporate the foot phenomena. For a linear force- free field the curl of the equation

$$\nabla \times \mathbf{B} = \alpha \mathbf{B} \tag{4.6}$$

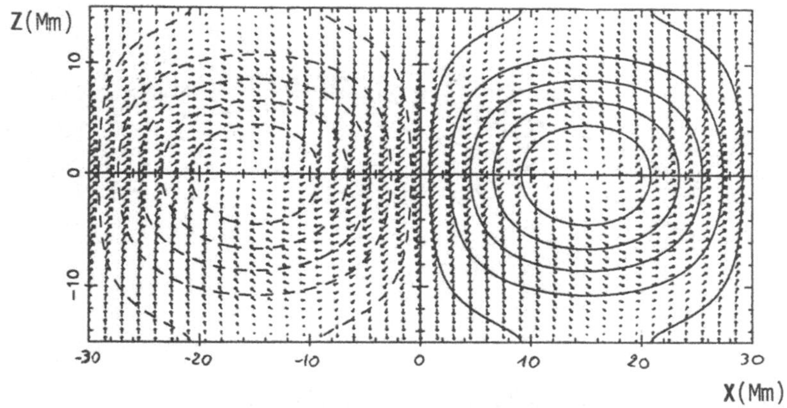

Figure 19. The magnetic configuration of Demoulin *et al* (1988) in the photosphere. The z-axis is along the prominence.

gives

$$(\nabla^2 + \alpha^2)\mathbf{B} = 0.$$

Thus the usual arcade solution may be generalised to allow periodic variations along the prominence axis (the z-direction) by choosing

$$B_y = \cos k_z z \;\; \sin k_x x \cdot e^{-\ell y}$$

where $\ell^2 = k^2 - \alpha^2$ and (4.6) implies

$$B_x = k^{-2}(\alpha k_z \sin k_x x \cdot \sin k_z z - \ell k_x \cos k_x x \cdot \cos k_z z)e^{-\ell y}$$

$$B_z = k^{-2}(\ell k_z \sin k_x x \cdot \sin k_z z + \alpha k_x \cos k_x x \cdot \cos k_z z)e^{-\ell y}$$

Higher harmonics are added to give a concentration of flux at the photosphere (y=0) and also a line of current and mass is added to represent the prominence itself. The resulting configuration in the plane y=0 is sketched in Figure 19, with the arrows indicating the direction of the horizontal field and the dashed and solid contours giving the contours of positive and negative B_y. The result is that a prominence of Normal Polarity has its feet at supergranule centres while one of Inverse Polarity (the usual case) has its feet at supergranule boundaries (in agreement with

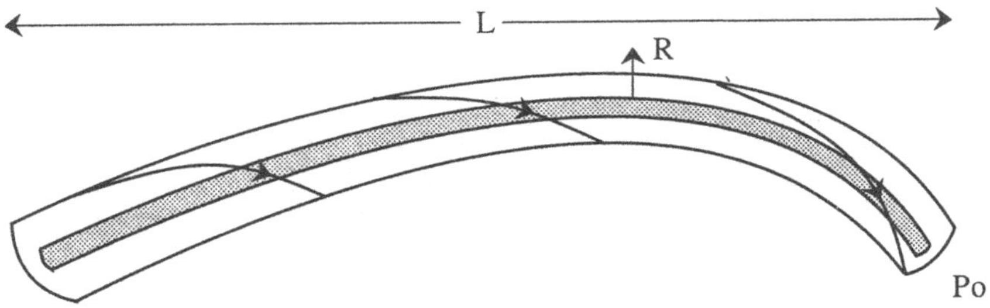

Figure 20. An active-region flux tube with a cool core.

observations).

6. A Flux Tube Model

Because of the fact that many prominences are of Inverse Polarity and all the previous inverse- polarity models have severe difficulties, Hood, Anzer and I sat down at the Mallorca Prominence Workshop to try and propose a new flux tube model which could have Inverse Polarity and which could also agree better with many other observational features. Previously, Hood and Priest (1979) had suggested that a *plage filament* may be a low-lying twisted flux tube (Figure 20), since one can often have motions along such an active-region prominence and it can sometimes end in a sunspot, both of which features contradict the Kippenhahn-Schlüter model. They modelled the flux tube as a cylinder and solved the force balance

$$\mathbf{j} \times \mathbf{B} = \nabla p$$

in the radial direction. Along each field line the pressure is uniform and they solved an energy balance equation and found a hot coronal equilibrium when the tube is untwisted. However, if the loop length, pressure or twist is increased, eventually a cool filament forms in the core of the flux tube.

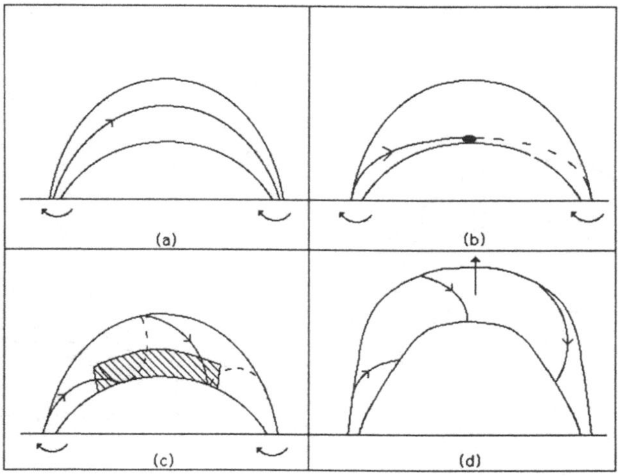

Figure 21. The Twisted Flux Tube model scenario.

Priest, Hood and Anzer (1989) suggested that for a quiescent or active-region prominence the basic geometry is a large-scale curved flux tube (Figure 21). Twist of the tube may be created in several ways - either by general evolutionary footpoint motions (which could produce inverse or normal polarity) or by Coriolis forces (which produce inverse polarity) or by flux cancellation (which produces inverse polarity). Evidence for such twist has been found by many authors (e.g. Schmieder *et al*, 1985, Mein and Schmieder, 1988) and braids seen in plage filaments are one such evidence (Gaizauskas, 1985). Coriolis forces would relentlessly tend to twist up a tube and produce one complete twist in about 35 days. As the twist increases, eventually a dip with upwards curvature is created at the summit (Figure 21b) and at this point the prominence can begin to form either by condensation (especially for quiescents) or by chromospheric injection (only likely for active-region prominences). The suggestion is therefore that the magnetic mould with the right environment for a prominence to be born (in particular a state of upwards curvature) needs to be created. As the twisting or flux cancellation continues, the prominence grows in length (Figure 21c) and eventually, when the twist or prominence length is too large (Hood and Priest, 1980; Einaudi and Van Hoven, 1981), the prominence erupts. It undergoes a metamorphis, like a beautiful giant butterfly, and reveals its true form as a flux tube for the first time in its life!

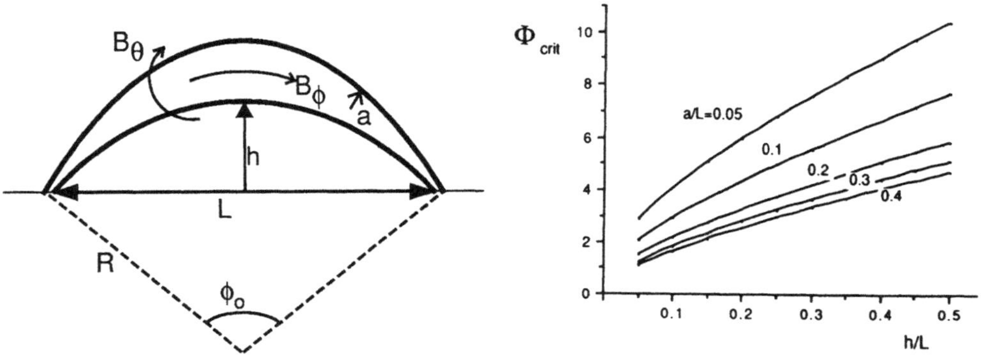

Figure 22. (a) The notation for a flux tube. (b) The critical twist for prominence formation as a function of flux tube height (h) and radius (a).

The critical twist for support for a large aspect-ratio tube is

$$\Phi_{crit} = \phi_0 \left(\frac{R}{a}\right)^{1/2}$$

in terms of the major (R) and minor (a) radii of curvature (Figure 22a). The way in which this increases with the summit height (h) and flux tube radius (a) is shown in Figure 22b in terms of the footpoint separation L. The way the prominence length increases with twist has also been estimated.

The force-free structure may be modelled by neglecting the large-scale curvature and writing in cylindrical geometry

$$B_r = \frac{1}{r}\frac{\partial A}{\partial \theta}, \quad B_\theta = -\frac{\partial A}{\partial r}, \quad B_z = B_z(A)$$

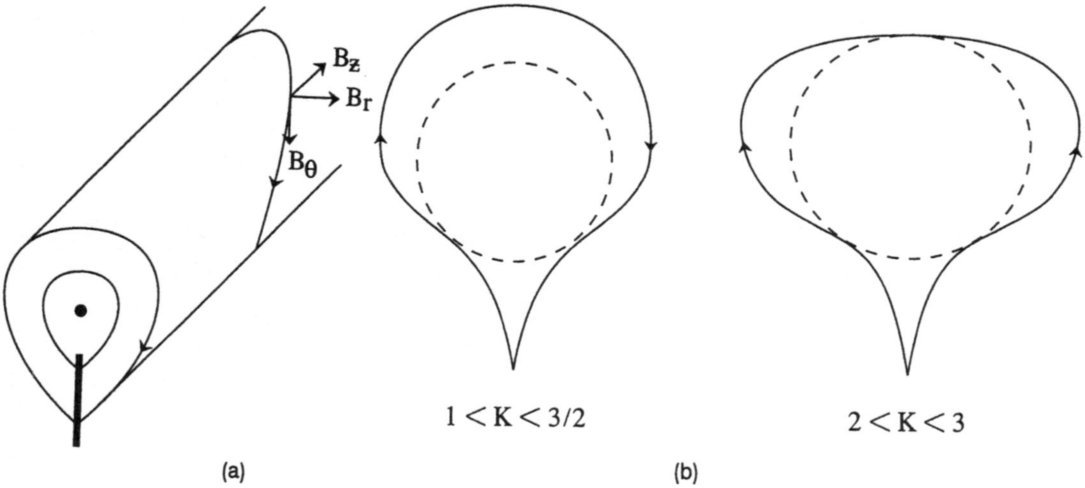

Figure 23. (a) The notation for a flux tube supporting a prominence sheet. (b) Field line shapes.

where the z-axis is directed along the flux tube (Figure 23a) and the flux function satisfies

$$\nabla^2 A + \frac{d}{dA}\left(\frac{1}{2}B_z^2\right) = 0.$$

Before the prominence forms one may consider a tube with

$$B_{0\theta} = \frac{B_0 r}{\sqrt{2}\ell}, \quad B_{0z} = B_0\left(1 - \frac{r^2}{\ell^2}\right), \quad 0 \leq r \leq a,$$

where B_0, ℓ and a may be found in terms of the flux (F), twist (Φ) and external pressure (P_e). After the prominence forms it may be treated as a sheet at $\theta = \pi$ and the field may be written as $\mathbf{B}_0(r) + \mathbf{B}_1(r,\theta)$, where

$$B_{1r} = -Kr^{K-1} sinK\theta, \qquad B_{1\theta} = -Kr^{K-1} cosK\theta.$$

The shapes of the field lines when $1 < K < 1.5$ and $2 < K < 3$ are shown in Figure 23b on the left and right, respectively, and the variation of the prominence mass with K has been calculated. Unlike the previous inverse-polarity solutions, there are no difficulties at the flux tubes axis where there is an O-type neutral point.

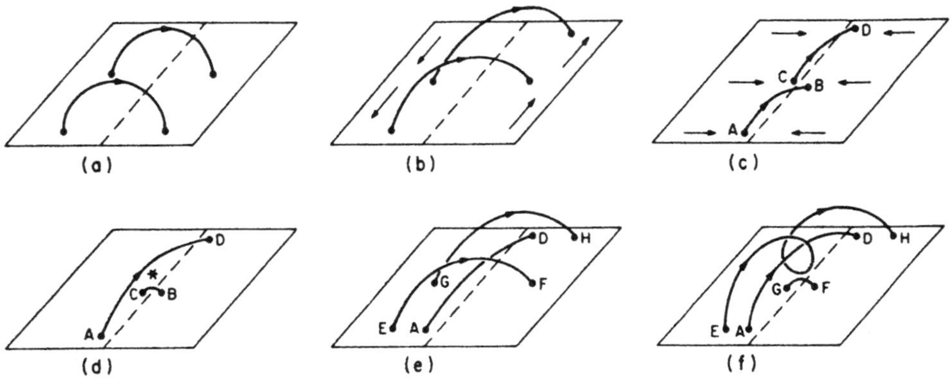

Figure 24. The formation of a twisted flux tube by flux cancellation in the photosphere (Van Ballegooijen and Martens, 1989).

The suggestion by Pneuman (1983), Rompolt (1986) and, most recently, van Ballegooijen and Martens (1989) that a helical flux tube may be created by flux cancellation at the polarity inversion line (Figure 24) is particularly appealing since it agrees with the observations of cancelling magnetic features in videomagnetographs by Martin (1986, 1990). As more flux cancels, so the helix tends to rise.

Hyder (1965) and Rust (1967) pointed out that the longitudinal component of polar crown prominences is opposite in direction from what one would expect from differential rotation acting on an inverse-polarity arcade. Martens and van Ballegooijen (1989) suggest that differential rotation plus cancellation acting on an arcade with its polarity inversion line originally orientated in a north-south direction can indeed produce a flux tube with the correct longitudinal field (Figure 25). However, if the inversion line is originally inclined substantially to the north-south direction, the mechanism fails and it is not clear what is happening in the polarity inversion zone to produce the correct field against the action of differential rotation. Clearly, more observations are needed.

7. Conclusion

We have seen the effects of several themes running through the attempts to model prominences. One is the use of potential (or, better, force-free) magnetic fields as external fields to create the magnetic environment of a prominence. Another is regarding the prominence as a line current, or, better, as a sheet current. A third is assuming the temperature is uniform, or, better, a function of position.

As the symphony has developed, we have observed how, through ingenuity and mathematical cunning, it has gradually become more realistic, although the earlier simple forms have laid the necessary foundation for later developments. But now we are at a stage where new themes are beginning to be heard, involving fibrils, feet and flux tubes, and I expect them to play a much more prominent part in future.

The flux tube model is particularly promising since it contains five new features that account for previously puzzling observations, namely the formation of a magnetic dip in the corona, the flux cancellation in the photosphere, the generation of the correct longitudinal component, the growth of the prominence as its twist increases and consistency with the struture of a prominence during its eruption.

Acknowledgement

I am most grateful to friends and colleagues for helpful suggestions and to the UK SERC for financial support.

References

Aly, J J and Amari, T (1988) *Astron. Astrophys.* **207**, 154.
Amari, T and Aly, J J (1989) *Astron. Astrophys.* **208**, 261.
Amari, T and Aly, J J (1988) *Astron. Astrophys.* **193**, 291.
Amari, T and Aly, J J (1990) *Astron. Astrophys.* , in press.

An, C, Wu, S and Bao, J (1988) in Ballester and Priest (1988) p.89.
Anzer, U (1969) *Solar Phys.* **8**, 37.
Anzer, U (1972) *Solar Phys.* **24**, 324.
Anzer, U (1978) *Solar Phys.* **57**, 111.
Anzer, U (1984) *Measurements of Solar Vector Magnetic Fields* (ed M Hagyard) NASA Conf. Pub. 2374, p. 101.
Anzer, U (1988) in Ballester and Priest (1988), p.99.
Anzer, U and Priest E R (1985) *Solar Phys.* **95**, 263-268.
Babcock, H and Babcock, H (1955) *Astrophys. J.* **121**, 349.
Ballester, J L and Priest, E R (1987) *Solar Phys.* **109**, 335-350.
Ballester, J L and Priest, E R (1988) *Dynamics and Structure of Solar Prominences*, Univ. les Illes Balears Press.
Ballester, J L and Priest, E R (1989) *Astrophys. J.*, in press.
Bommier, V (1986) *Coronal and Prominence Plasmas* (ed A Poland) NASA Conf. Pub.
Bommier, V (1990) this book.
Degenhart, U (1990) in preparation.
Demoulin, P, Malherbe, J M and Priest, E R (1989) *Astron. Astrophys.* **211**, 428-440.
Demoulin, P and Priest, E R (1988) *Astron. Astrophys.* **206**, 336-347.
Demoulin, P and Priest, E R (1989) *Astron. Astrophys.* **214**, 360-368.
Demoulin, P, Amari, T, Browning, P and Priest, E R (1989) *Astron. Astrophys.*, in press.
Demoulin, P, Priest, E R and Anzer, U (1989) *Astron. Astrophys.* **221**, 326-337.
Einaudi, G and Van Hoven, G (1981) *Phys. Fluids* **24**, 1092.
Engvold, O (1976) *Solar Phys.* **49**, , 283.
Engvold, O (1986) *Coronal and Prominence Plasmas* (ed A Poland) NASA Conf. Pub. 2442, p.97.
Fang Cheng (1990) this book.
Fontenla, J and Rovira, M (1985) *Solar Phys.* **96**, 53.
Gaizauskas, V (1985) Proc. Workshop on Solar Phys. and Interplan. Trav. Phen. (ed C de Jager and Chen Biao), p.710.
Gaizauskas, V (1990) this book.
Heinzel, P, Gouttebroze, P and Vial, J C (1988) *Dynamics and Structure of Solar Prominences* (ed J Ballester and E Priest), p.71.
Hirayama, T (1985) *Solar Phys.* **100**, 415.
Hirayama, T (1986) *Coronal and Prominence Plasmas* (ed A Poland) NASA Conf. Pub. 2442, p.149.
Hood, A W and Anzer, U (1990) *Solar Phys.*, in press.
Hood, A W and Priest, E R (1979) *Astron. Astrophys.* **77**, 233.
Hood, A W and Priest, E R (1980) *Solar Phys.* **66**, 113-134.
Hyder, C (1965) *Astrophys. J.* **141**, 1374.
Isenberg, P and Forbes, T G (1990) submitted.
Jensen, E, Maltby, P and Orrall, F (1979) *Phys. of Solar Prominences*, IAU Colloq No.44.
Jockers, K and Engvold, O (1975)
Kaastra, J (1985) Ph.D. thesis, Utrecht.
Kim, I (1990) this book.
Kippenhahn, R and Schlüter, A (1957) *Zs. Ap.* **43**, 36.

Kuperus, M and Raadu, M A (1974) *Astron. Astrophys.* **31**, 189.
Kuperus, M and Tandberg-Hanssen, E (1967) *Solar Phys.* **2**, 39.
Kuperus, M and Van Tend, W (1981) *Solar Phys.* **71**, 125.
Lerche, I and Low, B C (1977) *Solar Phys.* **53**, 385.
Leroy, J L (1979) in Jensen *et al* (1979), p.56.
Leroy, J L (1985) NASA Conf. Pub. 2374 (ed M Hagyard) p.121.
Low, B C (1981) *Astrophys. J.* **246**, 538.
Malherbe, J M and Priest, E R (1983) *Astron. Astrophys.* **123**, 80-88.
Martens, P and Van Vallegooijen, A (1989) Preprint.
Martin, S (1986) *Coronal and Prominence Plasmas* (ed A Poland) NASA Conf. Pub. 2442, p.73.
Martin, S (1990) this book.
Martres, M, Mein, P, Schmieder, B and Soru-Escaut, I (1981) *Solar Phys.* **69**, 301.
Mein, P (1977) *Solar Phys.* **54**, 45.
Mein, P and Schmieder, B (1988) *Dynamics and Structure of Solar Prominences* (ed J Ballester and E Priest) p.17.
Menzel, D (1951) Proc. Conf. on Dynamics of Ionized Media, London.
Milne, A, Priest, E R and Roberts, B (1979) *Astrophys. J.* **232**, 304.
Molodensky, M (1990) this book.
Nakagawa, Y and Malville, J (1969) *Solar Phys.* **9**, 102.
Plocienak and Rampolt, B (1973) *Solar Phys.* **29**, , 399.
Pneuman, G (1983) *Solar Phys.* **88**, 219.
Poland, A (1986) *Coronal and Prominence Plasmas*, NASA Conf. Pub. 2442.
Poland, A and Anzer, U (1971) *Solar Phys.* **19**, 401.
Poland, A and Mariska, J (1986) *Solar Phys.* **104**, , 303.
Priest, E R (1989(*Dynamics and Structure of Quiescent Solar Prominences*, Kluwer Academic.
Priest, E R and Forbes, T G (1990) *Solar Phys.* , in press.
Priest, E R, Hood, A W and Anzer, U (1989) *Astrophys. J.* **344**, 1010-1025.
Rabin, D (1986) in Poland (1986) p.135.
Rampolt, B (1986) in Poland (1986) p.81.
Rust, D (1967) *Astrophys. J.* **150**, 313.
Schmieder, B (1990) in this book.
Schmieder, B, Malherbe, J, Poland, A and Simon, G (1985) *Astron. Astrophys.* **153**, 64.
Simon, G, Gesztelyi, Schmieder, B and Mein, N (1986) in Poland (1986) p.229.
Sparks, L and Van Hoven, G (1985) *Solar Phys.* **97**, 283.
Sparks, L and Van Hoven, G (1987) *Phys. Fluids* **26**, 2590.
Steele, C and Priest, E R (1989) *Solar Phys.* **119**, 157-196.
Steinolfson, R and Van Hoven, G (1984) *Astrophys. J.* **276**, 391.
Tandberg-Hanssen, E (1974) *Solar Prominences*, D Reidel.
Van Ballegooijen, A and Martens, P (1989) Preprint.
Van Hoven, G, Sparks, L and Tachi, T (1986) *Astrophys. J.* **300**, 249.
Van Hoven, G and Mok, Y (1984) *Astrophys. J.* **282**, 267.
Van Tend, W and Kuperus, M (1978) *Solar Phys.* **59**, 115.
Vial, J (1986) in Poland (1986) p.89.
Vial, J (1990) in this book.

DISCUSSION

MARTIN: The great disparity between observations and the theories of prominence formation continues partly because we observers fail to adequately communicate the few facts that we know about prominence formation and partly because theorists do not recognize the significance of published observations (or the real circumstances are too difficult to model). For example, many high quality observations of prominences at the limb and filaments on the disk, taken over the last two decades, show no evidence of dips in either prominence structure, in related coronal and chromospheric structure or by inference from combinations of structures seen at different wavelengths. Nevertheless, you have just said that dips in magnetic field lines are a necessary condition for prominence formation. Why do you insist on having dips in field lines when the observations show that the primary requirement is for a continuous mass flow into and out of prominences with no requirement for retaining stationary mass at any location within a prominence?

PRIEST: I agree strongly that we need better communication between theorists and observers, so that theorists can attempt relevant models also that observers can attempt relevant observations, useful to advance our physical understanding.

Regarding dips, the theory considers dips with angles of only 5 degrees or so to the horizontal and so this is not inconsistent with observations. Furthermore the general theoretical assumption is that the observed vertical threads are <u>not</u> indicating the field line directions but represent the sum of a series of beads supported one above the other in a series of field lines.

The reason for needing support against gravity in large quiescent prominences is that plasma is observed to be sitting up above the surface, essentially in equilibrium, i.e. to lowest order in the Alfven Mach number (the ratio of plasma to Alfven speed). Since the observed vertical speeds in such prominences are only a few kilometers per second they are very much less than the Alfven speed and so the inertial or acceleration term in the equation of motion is much less than the Lorentz force-i.e. to a high degree of approximation the plasma is in equilibrium! Thus the observed flows in quiescent prominences represent a slow emptying through small holes, say, of a bucket of water and at the same time its refilling - i.e. it is likely that plasma is slowly dribbling out of an essentially equilibrium prominence and is at the same time being slowly replenished. I believe a study of such an inflow and outflow - i.e. the mass balance - is very important to pursue, and it is likely also to be crucial for the energy balance.

Active-region prominences may well be different. I do not know of any systematic study of their flows - do they always have flows? What is their magnitude? Are such flows essential for their existence? If such flows along active-region prominences - i.e. along the flux tube of which I suspect them to exist - are much less than the sound speed then they are essentially in hydrostatic equilibrium. If the flows are sonic or greater then a genuinely dynamic model will be necessary.

FORBES: In response to S. Martins comment I would like to say, that flow into and out of a prominence can easily be incorporated into the static-support models by adding the thermodynamical processes. That is, material could continually be condensing into and evaporating out of the prominence, and this would give the flows you are referring to.

MARTIN: Another disparity between observations and theory concerns the existence macroscopic arcades of field lines and structures in the prominence environment. The majority of observations show a clear absence of either coronal structures or arcades of fibrils that join opposite polarities or opposing sides of a prominence except high above prominences. Yet many models assume the existence of magnetic arcades that would lie under or pass through prominences and ignore the real overlying arcade.

PRIEST: The flux tube model does not necessarily include a coronal arcade of the classical Kippenhahn-Schlüter type, which should please you! However, the lack of observation of an arcade in H_α does not imply lack of existence, since if the arcade is filled with low density coronal plasma, you would not see it in H_α and it would be hard to see in soft X-ray pictures. However, the classical pictures of coronal cavities and helmet streamers seem to imply an arcade topology overlying a prominence. For consistency and continuity therefore it is not unreasonable to assume an arcade structure in the cavity and threading the prominence.

MARTIN: Comment: I would like to emphasise the importance of modelling the pre-prominence magnetic field configuration. There is a need to show how and why the transverse (horizontal) structures of the chromosphere lie at a large angle with regard to the overlying coronal arcade.

PRIEST: I agree. Such observed large angles together with the observed magnetic field inclination angle are the main reasons why most recent theories include a large magnetic shear as an important ingredient. However, the earlier Normal and Transverse polarity models tended to add the field component along the prominence axis almost as an afterthought, whereas in the Flux Tube Model that component is central.

ENGVOLD: You say that it will be possible in "static" models to have matter "dribbling" through the prominence. How large flows can be accepted perpendicular to the lines of force of these models? Vertical mass flows are the rule for quiescent prominences. Numerous observations show mass motions, both upwards and downwards, typically $\pm$ 5 kms^{-1}.

PRIEST: The observed velocities are very much smaller than the Alfven speed of a few hundred kilometers per second and so are a small perturbation to the magnetohydrostatic models. In other words, the plasma is very slowly dribbling through a basic magnetic mould.

ZIRKER: Isn´t the twisted tube model of prominence formation in conflict with observations? We see the feet appear first, whereas the model predicts we should see the center first.

PRIEST: The model shows how a vertical sheet of cool plasma can be formed, but it does not yet include feet and so is not in coflict with the observations. I agree, however, that a better observational and theoretical understanding of feet in future is a key problem.

PHYSICAL CONDITIONS IN PROMINENCES

T. Hirayama
National Astronomical Observatory, Mitaka, Tokyo 181, JAPAN

§1. Introduction

Since many recent review papers are available (Hirayama, 1985; Vial, 1986; Zirker, 1989; Schmieder, 1989), I will concentrate here on a few selected topics and try to make critical assessment, particularly on the electron density and hence pressure determinations, and radiative equilibrium modeling. As a result I became very critical upon more important and elaborate works, hoping to learn more from further developments in the immediate future.

Fine structures, non-random velocity fields, theoretical aspect of Lyman transfers, and UV and radio observations are not included (see other reviews), and mostly quiescent prominences are treated. At the end of each section a short summary is given, and a résumé table of various physical quantities derived there is given in the last section, including earlier results reviewed in Hirayama (1978).

§2. Temperature and Non-Thermal Velocity

Temperature can best be determined by comparing the widths of the optically thin hydrogen and metallic lines. In this case one should carefully estimate self-absorption if one uses Hα and/or Ca K lines because their optical depths in the central part of quiescents amount to several, or even more than 100 (Kubota, 1980).

Two dimentional distributions of the kinetic temperature (T_k) and non-thermal velocity (v_{nt}) for a large arch-shaped quiecent prominence were obtained by Zhang et al. (1987). They obtained on the average T_k=7500K, and v_{nt}=6km s^{-1} from two line profiles of Hα and Ca K, and deduced total hydrogen column density of 2.2×10^{18}cm^{-2} from the optical depth of Ca K. They claim that the kinetic temperature is higher and non-thermal velocity is lower in the edges and the opposite in the center part, and also that the physical parameter does not change much along the vertical directions. Although the quality of spectra seems very good and the method of analyzing spectrum is rather sophisticated, it is hoped to use at least another two lines such as Hβ and Ca H lines so as to make the derivation of the line-center optical depths, and hence T_k and v_{nt} (about 3 for Hα and Ca K) more reliable. In another study, Zhang and Fang (1987) observed H and K and Hα, Hβ lines, and fitted line profiles almost perfectly with the observation (too good!), using extensive calculations of non-LTE transfers for H and Ca with a depth dependent temperature distribution. From observers stand point, it is desirable to see what is inferred directly from the observation.

The intensity ratio can in principle be used for the temperature determination. One of the temperature sensitive ratios so far claimed is Hβ/Ca K (Heasely and Milkey, 1978), and this ratio has been calculated only for the very low pressure case, but it may also be dependent, though perhaps weakly, upon the incident ionizing radiation from the environment which does change very much from place to place. In a study of the prominence oscillations Suematsu <u>et al.</u> (1990) report that in fact this intensity ratio increased when the kinetic temperature as determined from the widths was increasing with time at certain points of a prominence.

Regarding the temperature determination using the Lyman continuum observation it has been believed that the relation $I_\nu = \int_0^\infty S_\nu e^{-\tau} d\tau = S(\tau_\nu = 1)$ gives B_ν and hence temperature, and b_1 if one uses the intensity observation at least at two wavelengths. This is because the Lyman continuum source function is given by $S_\nu = B_\nu(T_e)/b_1$, where b_1 is the departure coefficient from LTE. However as Vernazza, Avrett and Loeser (1981, VALIII) have shown that though in the chromosphere this process will give T_e=9000-9500K and b_1=814-2050, the optical depth of these temperature regions is less than 0.1. In fact the temperature is 7600K at τ_0(Lyc)=1. They concluded that "meaningful values of T_e and b_1 cannot be determined from Lyman continuum observation". I would rather state that Lyc will tell an upper limit of the temperature at τ_0(Lyc)=1, but of course if there is no temperature stratification we will obtain the correct values, and this may be the case for active region filaments lying parallel to the magnetic field without prominence-corona interface.

It is without saying that the true nature of non-thermal

motions can best be studied by high spatial observations either with tuned filtergrams, or spectrograms. One of the latter study by Engvold, Wiehr, and Wittman (1980) at 2" resolution showed that knots having larger line shifts (about 20km s^{-1}) tend to have rather smaller FWHM of 0.2A or so (10km s^{-1}) in terms of optically thin Doppler width, while knots having smaller line shifts tend to have larger FWHM. (See also Hellwig, Stellmacher, and Wiehr, 1984.)

Engvold and Brynildsen (1986) reported hydrogen Paschen line measurements of three prominences using the Fourier Transfer Spectrometer at Kitt Peak. For example, the kinetic temperature at one point of a prominence was found to be T_k=6200 $\pm$ 900K from the width of Paschen, HeI, OI, and CaII lines, while the excitation temperature among Paschen lines gives T_{ex}=8000 $\pm$ 700K. Calculation by Heasley and Milkey (1978) shows that $T_k > T_{ex}$ for a gas pressure of P_g=0.15dyn cm^{-2}, and $T_k < T_{ex}$ for P_g=0.01. So their observations would indicate that the gas pressure or the electron density in this example may have been very small.

A new method of determining temperature was suggested by Brickhouse and Landman (1987), where OI 7774/Na D_2 ratio is used. Because the 2nd ionization potential of sodium is quite large (47.3 eV), NaD_2 intensity will give the total number density of, in this case, oxygen. And they find that if T_e>9000K, the ratio will give a measure of the kinetic temperature, where the charge exchange process of $O+H^+ \leftrightarrow O^+ +H$ was claimed to be important. The observed intensity ratio of about 6 leads to just $T_e \approx$9000K. In the lower temperature range of T_e<8000K, the ratio becomes sensitive to pressure, however.

Although the kinetic temperature are generally in the range of 5500-9000K, we are sure that there are those showing the kinetic temperature as low as 4300K (Hirayama, Nakagomi, and Okamoto, 1978). Here the widths of HeI lines are found to be consistent with this low kinetic temperature and non-thermal velocity of 3.4km s^{-1} derived from hydrogen Balmer series lines and metallic lines (see Figure 1). The electron density in this case (average of three prominences, altogether 12 positions) is $n_e \approx 10^{11.4} cm^{-3}$ as inferred from Stark effect (upper left of the Figure) and the emission measure of $n_e^2 L = 9.5 \times 10^{27}$, where L(cm) is the effective length in the line of sight.

There has been a controversy whether the temperature and non-thermal velocity increase towards edges of quiecents. When the

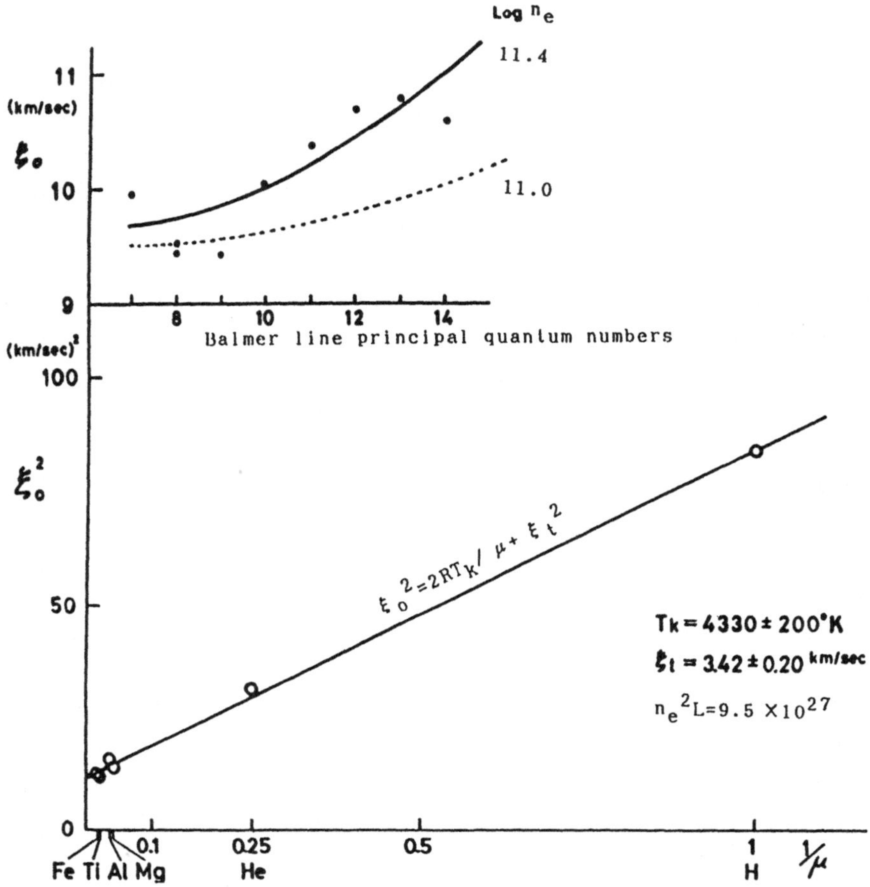

Fig. 1. Three equatorial hedgerow quiescent prominences showing very low kinetic temperature and high electron density observed with Norikura 25cm coronagraph-7m Littrow spectrograph (1.7mm A^{-1}, 80mm solar image, and altogether 280 lines to get the figure). The increase of observed widths $\xi_0=C\Delta\lambda_{1/e}/\lambda$ towards higher Balmers are real, because metallic line widths show no defocusing effect along λ and small overall rms deviations (±0.20km s^{-1}). ξ_0 (without Stark effect)=9.15km s^{-1}. The apparent thickness of L=1.5km is converted to a thread diameter of less than 60km! See section 6 (n (in measured slit length of 2.4"))>1 was assumed here).

"increase" was not found, often either usual telescopes were used and/or only CaHK and H α (or H β) lines were measured. In an earlier paper (Hirayama, 1971, Figs. 4, 5 and 6) it was shown that the fainter part of quiescents which is as faint as coronal emissions has larger T_k and v_{nt}. However even there, a prominence (Pr. C) showed that the kinetic temperature at lower heights did not rise towards the edges, so both cases exist. The increase of non-thermal velocity at the edges are more common than the temperature increase, and my interpretation is that because one finds more horizontal fine structures at the edges and at the top, horizontal motions are easily developed, and hence larger v_{nt}.

Finally it may be interesting to remark here that the height distribution of the kinetic temperature of chromospheric spicules has been determined to give that the temperature is $T_k \approx 8000$K at the height of 2200km, it decreases towards larger height to T_k=5500K at 3200km, and then it increases again to 8200K at 6000km (Matsuno and Hirayama, 1988). This has been derived from a large number of emission profiles from eclipse flash spectra. In particular, the decreasing temperature from 2200km to 3200km is imperative, because the optically thin hydrogen lines show a <u>constant</u> Doppler widths of about 16km s^{-1}, while the optically thin metallic lines show an <u>increasing</u> widths there.

§3. Electron Density

Bommier, Leroy, and Sahal-Bréchot (1986a and 1986b) in the very detailed studies have, for the first time, obtained the electron density from the depolarization of H β lines. Linear polarization of the resonance scattering of H β line is expected to be p=3.5% at about 60" above the limb. But if the magnetic field is present (Hanle effect), or the proton collisions are not too small (Stark effect), the polarization degree decreases: in fact they measured the polarization of 14 prominences and found p=0.5 ±0.1% (the change of polarization vector with respect to the Sun horizon is 14° ±10°), which leads to an average electron density of $n_e = 1 \times 10^{10} cm^{-3}$, ranging from 3×10^9 to $4 \times 10^{10} cm^{-3}$ (No. 12 was excluded), from the nearly simulataneous observations of H β and D_3.

However there is a problem in the total effective thickness. Namely since the average intensity of H β is 1.9×10^4 erg $cm^{-2}s^-$

$^{1}sr^{-1}$, corresponding to an equivalent width of 6.2mA (Leroy and Bommier, priv. comm.), the effective length becomes $L=3.6\times10^{4}$km for $n_e=10^{10}$ (for $\bar{n}_e=10^{9.7}$, it is 7.5×10^{4}km!), apparently too large a value. For this we have used the observed excitation temperature of 8000K among principlal quantum numbers 5-23 extrapolated to continuum (Engvold and Brynildsen, 1986), so that the intensity of the optically thin Hβ becomes $I(H\beta)=5.31\times10^{-26}n_e^{2}L(T_e/7000)^{-3/2}$cgs. If T_{ion}=7200K is used from Landman, Illing and Mongillo (1978), the numerical factor becomes 6.09×10^{-26}, while $I(H_9)=3.66\times10^{-27}n_e^{2}L(T_e/7000)^{-3/2}$. Moreover since the 6.2mA equivalent width is the lower limit because of possible smearing due to seeing, the true intensity would become at least a factor of two to three larger, which implies that $L=7\text{-}10\times10^{4}$km for their (logarithmic) average value of $n_e=1.0\times10^{10}$. And I think that it is too large when compared with Hα pictures in Bommier et al. (1986a), which show thread diameters of less than 10"-30". If the average value were increased, by unknown reasoning to $n_e=10^{10.5}$, the effective length in the line of sight L becomes $5\text{-}7\times10^{3}$km, which seems reasonable in view of the comparison with the electron density from Stark effect.

Altogether I am compelled to doubt the values below $n_e=5\times10^{9}$ derived from the Hanle effect, so that the decreasing electron density with height (Bommier et al., 1986a, Fig.3) may be looked at with caution. This, however, does not mean that $n_e<5\times10^{9}$ or $dn_e/dh<0$ is not seen in quiescent prominences, but simply that I cannot be confident at the present time.

Contrary to the Hanle effect which can deduce the electron density when it is less than about 6×10^{10}, the Stark effect is applicable when $n_e>2\times10^{10}$ from the width of whole series members of higher Balmer lines (H_7 to H_{32}). Therefore even with the coronagraph, in order to obtain down to $n_e=2\times10^{10}$ it is neccessary that either the prominence is very bright ($I(H_9)\approx10^{3}$cgs) or the terrestrial sky is very dark ($I(H_9)\approx10$cgs). The average electron density for six quiescent promincences, to which Stark effect is reliably applicable, was found to be $10^{10.8}$, the maximum being $10^{11.4}$ (logarithmic average, Hirayama, 1985 and 1986), while for many other prominences in various positions it is surmised that n_e is less than 10^{10}. Thus Stark effect shows that there are such low density portions, though positive determination is impossible. Higher values reaching to $10^{12.8}$ were obtained for the post flare loops (see Hirayama, 1978). New determinations for loops are reported by Hanaoka, Kurokawa, and Saito (1986), Foukal, Hoygt, and Gilliam (1986), and Heinzel and Karlický (1987).

Landeman (1986) has revised his earlier electron density determination by increasing recombination coefficient from many upper levels of Na-atom: the average electron density for his (rather bright) objects now becomes $n_e=9\times10^{10}$ instead of $n_e=2\times10^{11}$, which is in agreement with the above, and this leads to a much smaller gass pressure (see section 6). Kubota (1981) has derived the electron density of $n_e=10^{11.0}$ on the average for brighter prominences from CaHK and IR line observations with the help of the extensive non-LTE calculations.

The first systematic determination of the electron density of erupting prominences at the height of a few solar radii was reported by Athay, Low, and Rompolt (1987) using the coronagraph aboard SMM. Here the intensity of H α, roughly proportional to n_e^2L, is compared with the electron scattering continuum intensity, which is proportional to n_eL. After finding that the temperature of 2×10^4K is within the acceptable range from level calculations, they obtained on the average $n_e=10^8(r/4)^{-6}$ cm^{-3} and $L=10^9(r/4)^3$cm, where r the height in unit of solar radius. These famulae will give $n_e=4\times10^{11}$ and L=160km when extrapolated to r=1, although such extrapolation is a bit dangerous. Foukal, Little, and Gilliam (1987) determined the electron densities of two erupting prominences from Stark effect of Paschen and Balmer lines: $n_e=10^{12.0}$ (Dopper width was 60km s^{-1}), and $n_e=10^{11.3}$ (5km s^{-1}).

Koutchmy, Lebecq, and Stellmacher (1983) obtained the electron density of 3×10^9 cm^{-3} for a prominence observed at the total eclipse of July 31, 1981 using the intensity ratio of H β and the red continuum images with an assumed temperature of 10^4K (if 8000K, $n_e=10^9$, and if 1.3×10^4K, $n_e=10^{10}$). From the same eclipse and with the use of wide band filters at 5500A and 6500A, they found $n_e=2\times10^9$cm^{-3} for an erupting prominence at 10^5km above the limb, while the density of the corona neaby was 6×10^8. In Hirayama (1971, Table IV) electron densities were derived from continuum and Stark effect from the same spectrograms and using the observed T_k, and they are coincident whithin a factor two: $n_e=10^{10.2}-10^{10.6}$.

The conclusion on the electron density is that for quiescent prominences n_e ranges from a little less than 1×10^{10} cm^{-3} to 2.5×10^{11}, while in erupting prominences a maximum value of $n_e=10^{12}$

at zero height was found in one case, and the decreasing density were found with height (or with time) reaching a value of 10^8 at 4 solar radii. Problem to be pursued further may be the Hanle effect in $n_e<5\times10^9$ range, and SrII level calculations (see section 6).

§4. Radiative Transfer of Lyman Lines — Comparison with Observations

Since the hydrogen Lα and Lβ transfer is treated extensively elsewhere in this proceeding by Heinzel, Gouttebroze, Vial, and Zharkova, I will briefly report the relation with observations (an excellent short review in Heinzel, Vial, and Goutebroze, 1989; see also Heinzel, Vial, and Goutebroze, 1987). First, the observation from OSO-8 shows that the intensity ratio of Lα to Lβ is slightly less than the average normal Sun value (Vial, 1982), though I would say almost the same. Lβ observation is reported only from OSO-8. The absolute intensity of Lα so far reported is similar to the very quiet Sun value. The latest observations from SMM-UVSP show various interesting examples (Fontenla, Reichman, and Tandberg-Hanssen, 1989). However it is apparent that the faintest part of any prominences will be expected to be by orders of magnitude smaller than the average quiet value. The emission line profiles of Lα and Lβ show central depressions whose depth and peak-to-peak wavelength distance are similar to the quiet Sun.

Now the theoretical problems include 1) whether the coherent scattering is properly treated, 2) how (many) slab models are included with or without mutual interaction of radiation among them, 3) the temperature structure either of isothermal model or a model with hot transition region, and 4) whether the model is iso-baric or non-isobaric. Except for the 4)-point, the inclusion of which seems still exploratory, now it may be possible to reproduce Lα profiles and the absolute intensities with varying assumptions and parameters. However almost always the same model and the same computation predict very low Lβ line wing intensities as compared with the observation. Even the latest computation for the normal chromosphere using VALIII model and what is considered the most complete treatment of the partial redistribution process shows a factor of ten smaller intensity than the observed one at 0.3A from line center (Cooper, Ballagh, and Hubeny, 1989). It might be that the diffusion of neutral hydrogen to higher temperature regions as advocated by Fontenla, Avrett, and Loeser (1989) helps resolve the discrepancy. In case of prominences we would like to see computations with more atomic levels (hopefully n=7 or so), much higher gas pressure than 0.02 dyn cm^{-2}, and lower central temperatures. And to those of us who are familiar with the best H α pictures, the number of slabs or threads of some 40-100 being used by theoreticians seems too large.

§5. Radiative Equilibrium Model with Heat Conduction

Fontenla and Rovira (1985) have made extensive calculations of the radiative equilibrium models of quiescent prominences, which included also heat conduction. (See earlier study in Heasley and Mihalas (1976), and a critical review by the author (1978).) They treated an iso-baric slab atmosphere with pure hydrogen of 4 levels plus continuum. The full non-LTE radative transfer equations were solved using integral forms (Λ-operator) for the mean intensity. The radiation loss due to L α, L β, Lc and H α were included together with the Cox-Tucker radiation loss excluding hydrogen contributions. With a given constant gas pressure (P_g = 0.02-2 dyn cm^{-2}), and a prescribed temperature at the slab center (T_c=6500-10^4K) for each model, the temperature and all the other parameters as a function of geometrical depths were determined by iterative procedures. The result shows that L α (and L β and Lc) is radiated mainly in a steep temperature gradient region at 16000K. H α is in radiative balance, hence dark if seen on the disk (T_{ex}=3500K). Most striking fact is that the geometrical width of the slab at $<1\times10^4$K is only 2-20km, inspite of the fact that they explored cases with non-classical conductivity, or extra heating terms.

In order to be compatible with the observation, they assumed that the emergent intensities of lines and continua are emitted passing through a large number of threads (slabs) of N<50, where $I=I(\Delta\tau)\sum_{n=0}^{N} e^{-n\Delta\tau}$. Here $I(\Delta\tau)$ is the emergent intensity of each slab of the equal optical thickness of $\Delta\tau$ calculated in the above, assuming no interaction among slabs. Total optical depths are τ_{Lc}=0.16-0.36, $\tau_{L\alpha}$=1.4-3.3$\times10^4$, and $\tau_{H\alpha}$=0.05-(5). And the intensities of L α and L β are compatible with observations (Vial, 1982), while the H α intensity is incompatible with the average observed intensity unless one uses, say, ten times of the classical conductivity, or extra heating source (the case of $\tau_{H\alpha}=5$). Here we note that the ionization ($n_{HII}/n_{HI}=0.3-0.4$) seems to be determined from the incident Lyc. And the total geometrical thickness becomes in the acceptable range, though numbers of the slab seem too large.

My interpretation on the origin of very small thickness of a single slab is as follows: From their tables of radiation loss (in the standard cases of $T_c=6500K$, and $P_g=0.02-0.2 dyn\ cm^{-2}$), the energy balance equation can be written in a local form as $d(\kappa T^{5/2}dT/dh)/dh=R_0 P_g{}^2 T^{5/2}$ where $R_0=5\times10^{-11}$ (c.g.s. unit) for $T<10^4K$. (Incidentally the radiation loss term between $1-2\times10^4K$ is about 1/3 of Rosner, Tucker, and Vaiana (1978).) The temperature scale height $H_T=dh/d\ln T$ is then found to be $3\times(\kappa T_c/R_0)^{1/2}/P_g$, which becomes 18km for $P_g=0.02$ and 1.8km for $P_g=0.2$ with the Spitzer conductivity of $\kappa=1\times10^{-6}$. Evidently if one wants to get a wider slab, R_0 (or more generally the net radiation loss term) should be reduced by incorporating other loss terms than Lα, β, c and Hα, or by putting more heating sources than they tried. Here the (constant) pressure is already small so that the change of P_g would not help. It is desirable to calculate models for $T_c<6500K$, whick makes $dT/dh\approx0$, and hence thicker slab, and to extend to 5-levels atoms of H and Ca^+ (the most important in the chromosphere). Also a careful discussion of incident radiation is certainly needed (e.g. distance from active regions, see Kim (1987)).

If one combines the Fontenla-Rovira approach (including Ca, and hopefully Mg) with the earlier Heasely and Mihalas (1976) treatment of non-constant pressure and with the possible inclusion of ambipolar diffusion, there does not seem to be any fundamental difficulty in obtaining wide variety of models compatible with observations. I would think that to fit the Hα optical depth at least in the right order of magitude to the observation is more urgent than to fit the Lβ line precisely. As seen in the next section, the total optical depth of HI Lyc may be taken less than ten if seen sideways, so that the illumination due to the surrounding atmosphere will easily make the prominence temperature 6000-8000K, which I advocated in 1964.

§6. Gas Pressure, Hydrogen Ionization, and Filamentary Structure

Since Heasley and Milkey's work (1978) is still the most extensive to date, a brief discussion is given here again, though I dit it earlier in 1978. Very roughly speaking, they determined the gas pressure as 0.01 dyn cm^{-2} from the observed intensity ratio of CaII 8542 to Hβ (=0.38) assisted by the temperature of 7500-9500K which is inferred from the observed intensity ratio of CaIIK to Hβ. The Hβ intensity used is 500-1500 erg $cm^{-2}s^{-1}sr^{-1}$ (cgs), while the whole range so far reported covers from less than 100 to 3×10^5cgs (Kim, 1987). However, they obtained a large geometrical depth of L=11,000km, and if the filamentary nature is taken into account, the apparent thickness would become 50,000 km (one arc min!) or so in this very faint part of the prominence. It might be that something is wrong with the CaII photoionization process either of atomic or of the incident radiation, which changes markedly even within the quiet region (VALIII, Fig. 16)

On the other hand Hellwig, Stellmacher, and Wiehr (1984, and earlier studies therein) and Bendlin, Stellmacher, and Wiehr(1988) showed that more or less brighter part of prominences (I(Hβ)=1-5×10^4cgs) shows the intensity ratio of CaII 8542 to Hβ to be 0.1, which results in a gas pressure of 0.3-0.4 dyn cm^{-2} or more from the calculation of Heasley and Milkey. A similar study by Novocký and Heinzel (1990) shows P_g=0.12 dyn cm^{-2}. This tendensy of the decreasing ratio with increasing gas pressure can be understood if n(CaIII)>n(CaII) and n(HI)>n(HII) were assumed and if the Saha-Boltzamann equation and radiative ionization are used. For a higher electron density and hence higher gas pressure, where n(CaIII)<n(CaII), the ratio becomes insensitive to gas pressure as predicted by Heasley and Milkey.

An interesting fact is that this ratio together with some other intensity ratio is often seen constant within a prominence, but systematically varies from one prominence to another, while one would expect that the gas pressure (and the temperature) is not constant within a prominence. This might suggest that the incident UV radiation, which may well differ among prominences, is playing a crucial role on this (and other) ratio.

Now I turn to Landman's series of papers. As mentioned in section 3, Landman (1986) now revised electron density to 9×10^{10} by decreasing a factor of two, and hence gas pressure to 1.4 dyn cm^{-2} by decreasing a factor 4 for his rather intense objects of I(Hβ)=5-

15×10^4cgs with temperatures determined from optically thin line width ($\approx$6500K). Although $n_e=9 \times 10^{10}$ happens to be the same as that from Stark effect, there remains a serious problem with his treatment as I have shown earlier (1986, Fig. 2): the intensity ratio of R=I(MgI 3838)/I(SrII 4077) is expected to be proportional to n_e from his theory, while our observations show that it is almost constant from $n_e<10^{10}$ to $n_e=10^{11.4}$ which were determined from Stark effect. At $n_e=10^{11}$ his prediction coincide with our observation. I think that this is the reason why he obtained the electron density of about 10^{11} for all his objects, though in fact it might really have been so because the objects were brighter ones. (Mg and Na should be similar in this respect.)

I would think that discrepancy can only be resolved if Sr is mostly in SrIII instead of SrII. Then the ratio becomes independent of n_e as observed. This might well be the case, because photoionization cross-section seems much different among the literature, and if it is peaked at shorter wavelengths as he had presented in his earlier paper, and if this had a large absolute value, the ionization of SrII to SrIII will be much advanced. It is to be noted that if $(n_e n_{III}/n_{II})_{sr}$ is assumed to be dominated by the HI LYc radiation corresponding to T_r=6700K or so, near constancy of I(Mg3838)/I(SrII4077) as well as I(NaD_2)/I(SrII4077) can be predicted and their absolute values come close to the observed ratio.

Kim (1987) has treated NaD and Hβ lines in his non-LTE calculations, with <u>uniform</u> n_H and with multiple non-interacting slab models of one to 32 for a fixed total geometrical thickness, and successfully reproduced the observed ratio of $I(D_1)/I(D_2)=0.62$. In order to be consistent with the absolute values of NaD and Hβ lines, his results imply a gass pressure of less than 2 dyn cm^{-2} and $n_H<10^{12}$ cm^{-3} for his objects of very intense (or very thick) prominences of I(Hβ)=5-30 $\times 10^4$cgs. (His tables many be useful for interested researchers)

Finally our own determination of pressure is shown from an earlier paper (Hirayama, 1986) with a slight revision, but without criticism! Advantage in our case is that for each positions of each prominence we have already derived n_e from Stark effect, and the kinetic temperature from line widths, using whole series of Balmer lines and many (observationally confirmed) optically thin metallic lines for one and the same spectrum, which also contains MgI 3838 and 3832 lines. By combining T_e, $I(H_9) \sim n_e^2 T_e^{-3/2} L$, and the logarithmic average value of $n_e=10^{10.9}$ (the range of $n_e=10^{10.2}$-$10^{11.4}$, $n_e<10^{10}$ objects are excluded), we obtain the average total effective (line of sight) length to be L=80km, which ranges from 2km to 5000km. The intensity ratio of $I(H_9)$ to I(Mg), where $I(Mg) \sim n_e n_H T_e^{0.72} L$ is taken (see below), will give straight $n_{HII}/n_H=0.4$. Now all the other parameters are derived accordingly (corresponding $I(H\beta)$ ranges 1.6×10^2-1.6×10^4cgs). $n_H=2.1\times10^{11}$ (range: 2.5×10^{10}-6.3×10^{11}), $n_{HI}=1.3\times10^{11}$, $n_e n_{HII}/n_{HI}=5.6\times10^{10}$ corresponding to a radiative ionization temperature of 6400K, $\tau_0(Lyc)=6.6$, and gas pressure of 0.3dyn cm^{-2} (0.04-0.9), density of $\rho=4.9\times10^{-13}$g cm^{-3} with 10% helium, and average column mass of $\rho L=3.9\times10^{-6}$g cm^{-2}.

The following discussion will give the expression for MgI used above. Here Landman's (1984, Table 2) non-LTE calculation for 6-levels MgI is used, leading to $T_e^{0.72}$ dependency, but the absolute value of recombination is increased by 2.0 following the treatment for NaI and SrII (Landman, 1987). Since the upper levels are more or less hydrogenic in NaI and MgI, the use of the same factor may well be allowed. Besides, the VALIII model which used an 8-level atom for MgI predicts <u>similar</u>, but a factor of two larger value of n_{HII}/n_H than Landman's (1984) calculation in similar temperature ranges. So in fact I took the average of the two, and multiplied by 2.0. Because MgII is found to be the predominant ion, the factor n_H, total hydrogen number density, appears in the expression for I(Mg). (If $n_e<10^{10}$, MgIII cannot be neglected.) And in the case of such simple atoms as MgI, there does not seem to be problems in the photoionization cross-sections. One might argue if these non-LTE calculations of VAL or Landman are basically correct. For this purpose, I have used a simple Saha-Boltzmann equation and adopted the radiation temperature of 5800K from VALII for 7300A which is the wavelength of the ionizing radiation from the upper level of MgI 3838 and the dilution factor of 1/2. Then I obtained almost the same level population as in VALIII (Table 21) for the relevant temperatures. Altogether the uncertainty in our treatment is in the adoption of a factor 2.0, though I believe this is within an error of 50%.

We discuss in the following filamentary structure, or volume filling factor (Hirayama, 1986). Since the total intensity of an optically thin Balmer lines, whose upper level is near continuum such as H_9 (χ_{ion}=0.176eV), is very well expressed as $I(H_9)=a\times\int_{-\infty}^{\infty}n_e^2T_e^{-3/2}ds$ (a=constant;ds=geometrical distance), the effective length in the line of sight is defined and derived as $L=I(H_9)/(a<n_e>^2<T_e>^{-3/2})$, where $<n_e>$ is obtained from Stark effect, and $<T_e>$ from line width. Since we derived n_e and T_e from the segment of spectrum over 10" along the limb (raster step was also 10"), the number of threads in the line of sight should better be counted with this distance. If we assume that prominences consist of vertically suspending threads of a diameter ϕ of 300km (Dunn, 1960), the number of threads in the line of sight and within a horizontal distance of 10" becomes $n=L\times10''/(\pi\phi^2/4)$. Since the average L was 80km, n=8 is obtained, and the area filling factor (=volume filling factor in the present assumption) is $n\phi/10''=1/3$. For the cases we obtained L>200km, threads are seen overlapped ($n\phi>10''$) in the line of sight. When L<10km as was derived for some cases, the above formula gives erroneously n<1, because the fact that we could derive n_e means n>1. So we must expect the diameter was less than 300km. In case of the extreme value of L=2.4km which was obtained from 4 positions with $n_e=10^{11.4}$ for low altitude young quiescents, the diameter must have been less than 150km!

The result of this section is summarized as follows: Heasley and Milkey calculation (1978) may be by a factor 2 or more different from actuality, but general behavior seems to be in accord with what was observed, though the effect of varying incident radiation should be further examined. Landman's n_e and P_g are now in agreement with those derived from Stark effect. This fact, however, be better considered as a chance coincidence, because his method in the present form is clearly against observations. Resolution can be expected if the photionization process of SrII becomes well understood. Our derivation of n_H, and hence gas pressure P_g are based upon fewer assumptions because n_e and T_e have been derived directly, and presently no difficulty is found. In order to improve accuracy, the Mg ionization including many upper levels should be carried out.

As the acceptable range of parameters for the quiescent prominences which I take rather conservatively, namely by avoiding the extreme values, even for some values like $n_e=10^{11.4}$ in which I firmly believe their reality, the following values are obtained: $n_e=1\times10^{10}-2\times10^{11}cm^{-3}$, $n_H=1\times10^{10}-8\times10^{11}cm^{-3}$, $n_{HII}/n_{HI}=0.2-10$, $P_g=0.02-1.0$ dyn cm^{-2} (or may be $P_g>0.04$). As for the filling

Table 1

RÉSUMÉ OF THE SPECTROSCOPIC STUDY

	Quiescents Central Part	Quiescents Edges or Faint Part	Quiescents (average)	Loops Surges Eruptives
T_e (K)	4300 - 8500	7000 - 12,000	6500	7000 - 20,000<
v_{nt} (km s^{-1})	3 - 8	10 - 20	6	20
n_e (cm^{-3})	$10^{10.0}$ - $10^{11.3}$	$10^{9.7}$	$10^{10.8}$	$10^{11.0}$ - $10^{12.8}$ (PFL) $\gtrsim 10^{8}$ (eruptive)
P_g (dyn cm^{-2})	0.1 - 1	0.02	0.3	1 - 10 (loops)
n_{HII}/n_{HI}	0.2 - 10	~10?	0.7	various values
total effective thickness (km)	2 - 5000	?	100	2 - 30 (loops)
volume filling factor	0.01 - 0.5	?	0.3	?
thread diameter (km)	200 - 400	<1500 (direct photos)	200 - 400	<1000
thread length (km)	5000	?	5000	
co-spatial emission? (H, M, HeI, (HeII))	yes	(yes)	yes	{no {(yes) late stage in surges and loops

factor, if a cylindrical form of the threads standing nearly vertically to the solar surface is assumed, the volume filling factor, which is the same as the area filling factor here, is found to be 1/3 on the average, where the average total effective thickness was 80km, and where we assumed a diameter of thread to be 300km. In cases where the effective length was found to be 2.4km or so, a thread diameter of less than 150km is expected. Direct determinations of threads diameter are awaited.

§7. Summary

As a summary of this paper I am presenting a table which is in a very similar format as I have presented in the former IAU colloquium (Hirayama, 1978), and according change was made. Though the present study did not cover much about post flare loops, and erupting prominences (and nothing was said on surges and active region filaments), they are also included for comparison.

References

Athay, R. G., Low, B.C., and Rompolt, B.: 1987, Solar Phys. **110**, 359.

Bendlin, C., Stellmacher, G., and Wiehr, E.: 1988, Astron. Astrophys. **197**, 274.

Bommier, V., Leroy, J. L., and Sahal-Bréchot, S.: 1986a, Astron. Astrophys. **156**, 79.

Bommier, V., Leroy, J. L., and Sahal-Bréchot, S.:1986b, Astron. Astrophys. **156**, 90.

Brickhouse, N. S. and Landman, D. A.:1987, Astrophys. J. **313**, 463.

Cooper, J., Ballagh, R. J., and Hubeny, I.: 1989, in "Spectral Lines Shapes", Vol. **5**, Proc. 9th International Conf. Torun, Poland, in press.

Dunn, R. B.:1960, Ph. D. Thesis, Harvard University.

Engvold, O., Wiehr, E., and Wittman, A.: 1980, Astron. Astrophys. **85**, 326.

Engvold, O. and Brynildsen, N.: 1986, in <u>Prominence and Coronal Plasmas</u>, NASA CP2442 (ed. A. Poland), p.97.

Fontenla, J. M. and Rovira, M.:1985, Solar Phys. **96**, 53.

Fontenla, J.M., Avrett, E. H., and Loeser, R.: 1990, Astrophys. J. June 1, in press.

Fontenla, J.M., Reichman, E. J., and Tandberg-Hanssen, E.:1989, Astrophys. J. **329**, 464.

Foukal, P., Little, R., and Gilliam, L.:1987, Solar Phys. **114**, 65.

Foukal, P., Hoygt, C., and Gilliam, L.:1986, Astrophys. J. **303**, 86.

Hanaoka, Y., Kurokawa, H., and S. Saito: 1986, Solar Phys. **105**, 133.

Heasley, J. N. and Mihalas, D.:1976, Astropys. J. **205**, 273.

Heasley, J. N. and Milkey, R. W.: 1978, Astrophys. J. **221**, 677.

Heinzel, P., Gouttebroze, P. and Vial, J. -C.: 1987, Astron. Astrophys. **183**, 351.
Heinzel, P., Vial, I.-C., and Gouttebroze.: 1989, Radiative Transfer Workshop of the GdR Magnetism in the Solar Type Stars (eds. H. Frisch and N. Mein), Paris.
Heinzel, P. and Karlický, M.:1987, Solar Phys. **110**, 343.
Hellwig, J., Stellmacher, G., and Wiehr, E.:1984, Astron. Astrophys. **140**, 449.
Hirayama, T.: 1971, Solar Phys. **17**, 50.
Hirayama, T.: 1978, in Physics of Solar Prominences, IAU Collq. **44**, (eds. E. Jensen, P. Maltby, F. Q. Orrall), p.4.
Hirayama, T.:1985, Solar Phys. **100**, 415.
Hirayama, T.:1986, in Coronal and Prominence Plasmas, NASA CP2442 (ed. A. Poland), p.149.
Hirayama, T., Nakagomi, Y., and Okamoto, T.:1978, in Physics of Solar Prominences, IAU Colloq. **44** (eds. E. Jensen, P. Maltby, F. Q. Orrall), p.48.
Kim, K. S.: 1987, Solar Phys. **114**, 47.
Koutchmy, S., Lebecq, C., and Stellmacher, G.:1983, Astron. Astrophys. **119**, 261.
Kubota, J.: 1980, Publ. Astron. Soc. Japan, **32**, 359.
Landman, D. A.: 1984, Astrophys. J. **279**, 438.
Landman, D. A.:1986, Astrophys. J. **305**, 546.
Landman, D. A., Illing, R. M E., and Mongillo, M.: 1978, Astrophys. J. **220**, 666.
Matsuno, K. and Hirayama, T.: 1988, Solar Phys. **117**, 21.
Novocký, D. and Heinzel, P.:1989, preprint.
Rosner, R., Tucker, W. H., and Vaiana, G. S.: 1978, Astrophys. J. **220**, 643.
Schmieder, B.: 1988, in Dynamics and Structure of Quiescent Solar Prominences (ed. E. R. Priest), Kluwer Academic Publ., Dordrecht/Boston/London, p.15.
Stellmacher, G., Koutchmy, S., and Lebecq, C.:1986, Astron. Astrophys. **162**, 307.
Suematsu, Y., Yoshinaga, R., Terao, N., and Tsubaki, T.:1990, Publ. Astron. Soc. Japan, in press.
Vernazza, J. E., Avrett, E. H., and Loeser, R.: 1981, Astrophys. J. Suppl. **45**, 635, (VALIII).
Vial, J.-C.: 1982, Astron. Astrophys. **253**, 330.
Vial, J.-C.: 1986, in Prominence and Coronal Plasmas, NASA, CP2442 (ed. A. Poland), p.89.
Zhang, Q. Z. and Fang, C.:1987, Astron. Astrophys. **175**, 277.
Zhang, Q. Z., Livingston, W. C., Hu, J., and Fang, C.: 1987, Solar Phys. **114**, 245.
Zirker, J. B.: 1989, Solar Phys. **119**, 341.

Discussion

E. Priest: Your comment that we can take whatever density we like (within reason) makes me feel very free as a theorist! (1) Do your densities refer to densities in threads? If so then with a filling factor of 0.01, say, it would imply that the total mass of the prominence is any 1% of what you would estimate without threads. (2)By how much does the density decrease (presumably) with height in a quiescent prominence?

Hirayama: (1) When I speak of the density, either of electron or total hydrogen, it is the determination in the threads, and not the smoothed ones. Yes, you are right (see the text for total mass in the line of sight in unit area, and my contributed paper for the mass integrated also over height in unit length). (2)Bommier et al. (1986a) reported their finding, but if I exclude the values of $n_e=1\times10^9$ (see text), only one case remains: the electron density of $10^{10}cm^{-3}$ at the height of 3×10^4km decreases to 5×10^9 at 7×10^4km. See also my contributed paper.

J-C. Vial: Could the very low temperatures you quote be connected with Landman's very high electron densities and very low ionization degrees?
Hirayama: T_k=4300K and $n_e=2.5\times10^{11}$ (Stark effect) are found for the same prominences. Since $I(H_{12})/I(T_iII3759)(\sim n_e n_{HII}/n_{HI})$ shows ordinary values, $n_{HII}/n_{HI}=0.2$ is expected, the change being only from n_e-change. This is still high for T_k=4300K, perhaps due to the penetration of Lyc. As I noted in my talk that even with the revisions of Landman's n_e, his method will all lead to the value of $10^{11}cm^{-3}$ for any objects because of trouble in his calculations.

B. Rompolt(comment): One can see the supergranulation network in the vicinity of a filament in CaII K or H lines.

A. Poland: For lowest temperature ($T\sim$4300K) prominenses, what is approximate optical depth in H α.
Hirayama: It is probably a little less than unity because the emission measure is not large:$n_e^2L=9.5\times10^{27}cm^{-5}$ on the average for three prominences.

E. Wiehr: Several parameters may vary within the prominence as seems the case for our gas-pressure results you mentioned. How realistic is the temperature increase at the prominence edge if you consider error bars for those fainter emissions?
Hirayama: Including unpublished works, I believe they are realistic, but the uncertainty is rather large; 12000$\pm$2000K at the edges, while 7000K$\pm$800K or less in the center (see 1971 paper).

V. Bommier (comment): The impact approximation is valid at typical prominence densities (10^{10}-$10^{11}cm^{-3}$) for describing collisions of hydrogen atom with electrons and protons. This has been established in a recent work (Stehle, C., Mazure, A., Nollez, G., Feautrier, N., 1983, Astron. Astrophys. **127**, 263) by comparing results of impact theory with results of theories accounting for multi-perturbers effects (Microfield Model Method). These results (impact and MMM) are found in full agreement at these typical densities, for the first lines of the Balmer series.

Hirayama: We plan to observe Hanle effect with the new, fully automated 10cm-coronagraph, using a CCD camera, and hope to learn from your experience.

Fang Cheng: Do you think there would be differences of physical parameters determination if one takes into account the fine structures in prominences? How much the differences would be and could you give some estimations?

Hirayama: Since I have been using optically thin lines, fine structures are averaged evenly along the line of sight, and only lateral differences are found (see Hirayama, 1978).

O. Engovold: In a set of observations of two quiescent prominences in the near IR, using the FTS of the McMath, we have determined T_{kin} and ξ_t from a number of optically thin lines. A substantial variation in line opacities were noticed with position in our prominences using line ratios such as from the HeI λ10830A triplet. Some faint emission structure closer to prominence edges also showed large HeI 10830A opacity. When such cases were removed, we find only a marginal increase of T_{kin} and ξ_t towards the prominence edges. Therefore, can we safely exclude that the well-known increase of T and ξ_t in your papers also can be "contaminated" by line saturation effects?

Hirayama: No, where the increase was found, those are places of real edges whose faintness is only observable with a coronagraph or at the eclipse, and self-absorption effect has been carefully examined (see text). But I know even in my examples there are cases of no increase in T_{kin} towards edges.

DIAGNOSTIC OF PROMINENCE MAGNETIC FIELDS

Egidio Landi Degl'Innocenti
Dipartimento di Astronomia e Scienza dello Spazio
Università di Firenze, Largo E. Fermi 5, 50125 Firenze, Italia

and

Instituto de Astrofisica de Canarias
38200 La Laguna, Tenerife, España

1. INTRODUCTION

The role of magnetic fields in the complex physical phenomena that are usually described under the name of "prominences" can hardly be underestimated. Solid arguments of magneto-hydrodynamics lead indeed to the widespread belief that the structure of prominences, as well as their dynamical behavior, their evolution, and their existence itself, are intimately related to the geometrical and topological structure of the associated magnetic field vector. This fact justifies the conspicuous amount of work, both from the observational and theoretical points of view, that has been dedicated in recent years to the measurement of prominence magnetic fields.

As these measurements can be achieved only through the polarimetric signal observed in suitable spectral lines, the theoretical work on this subject has been mainly devoted to establish a sound physical basis for describing the polarization phenomena observed in prominence spectral lines and to clarify the diagnostic content of such polarimetric observations.

The aim of the present paper is the one of giving an updated review of the theoretical results that have been achieved in this field in the last decade. In the main body of the paper (Sections 2 to 6), the basic concepts underlying the physics of polarimetric observations in prominences are reviewed; the relevant formalism is summarized and applied to schematical situations in order to clarify the interplay of the different parameters in determining the signature of the observed polarization, and to point out the basic difficulties underlying the diagnostic of magnetic field vectors from observations.

In Section 7, an historical review of the theoretical work on the subject is presented and some remarks are made on the controversies raised by some of the results on prominence magnetic fields and densities that have been obtained through such theories and that seem to be in contradiction with the results obtained through different methods.

Finally, in Section 8 the main conclusions are drawn and recommendations are presented for future observations.

2. BASIC PHYSICAL CONCEPTS

The physical conditions that are typically met in prominences are extremely different from those of the underlying photosphere, the main difference being a reduction of the order of 10^7 in the density of the plasma; the low density of the prominence plasma results in the fact that the various energy levels of any given atomic species are populated and depopulated only by means of radiative transitions, collisional transitions being practically inefficient. The consequence of this fact is that prominence emission lines are linearly polarized, the linear polarization being directed (in absence of magnetic fields) along the tangent to the solar disk. This is the well-known phenomenon of resonance polarization which produces in an emission line a typical polarimetric signal which depends on many factors, like the height of the observed point in the prominence, the center-to-limb variation of the exciting photospheric radiation field and the various atomic parameters characterizing the spectral line.

The phenomenon of resonance polarization can be easily understood in terms of the concept of atomic level polarization. An atom is said to be polarized, or to show atomic polarization, when its Zeeman sublevels are unevenly populated and/or when well defined phase relationships (or coherences) exist between the same sublevels. The anisotropy of the exciting photospheric radiation (due to a combination of geometrical effects and limb darkening) is capable of introducing atomic polarization in the upper level of any given atomic transition; the linear polarization observed in emission lines from prominences simply reflects the physical fact that the emitting atom is polarized.

In the presence of a magnetic field the polarization observed in a spectral line is modified as a consequence of two different mechanisms : the Hanle effect and the Zeeman effect. The Hanle effect is due to a relaxation of coherences between different sublevels, while the Zeeman effect is due to a splitting of the sublevels themselves; the result of the Hanle effect is (in general) a depolarization of the linearly-polarized resonance radiation and a rotation of the direction of maximum polarization from the tangent to the solar limb, while the result of the Zeeman effect is the generation of a typical antisymmetrical circular polarization profile. The magnetic field can then be measured at least in principle, by two different methods : either by observing the linear polarization in suitable spectral lines or, alternatively by observing the circular polarization signal. The typical parameters that control the two different mechanisms are, for the case of the Hanle effect, the ratio ν_L/A (where ν_L is the Larmor frequency and A the Einstein coefficient for spontaneous de-excitation), and, for the Zeeman effect, the ratio $\nu_L/\Delta\nu_D$ (when $\Delta\nu_D$ is the Doppler broadening of the line). For the physical conditions typical of prominences, assuming a magnetic field of the order of 10 Gauss, we have : $\nu_L/A \cong 1$, $\nu_L/\Delta\nu_D \cong 10^{-3}$, so that both mechanisms can be used for measuring the magnetic field.

The theoretical analysis of the polarization phenomena that are observed in prominences, and the relative diagnostic of the magnetic field vector need a full description of the physical situation of the atoms that are present in the prominence plasma. This description can be achieved only through the introduction of the formalism of the density matrix and through the solution of the relevant equations that couple the atomic density matrix to the radiation field. In some cases, the role of depolarizing collisions (due to the long-range interaction of the atoms with neutral or ionized perturbers) cannot be neglected, so that their effect has also to be accounted for. The resulting problem is a typical non-equilibrium problem which can be compared with the standard non-LTE theory that has been developed in the past for stellar atmospheres. The differences lay into the fact that the atom has now to be described in terms of sublevel populations and of coherences between sublevels (instead of just in terms of level populations) and that the radiation field has to be described in terms of its Stokes parameters (instead of just in terms of its intensity). This typical non-equilibrium problem, for which the name of "Non-LTE problem of the 2nd kind" has been proposed, is schematized in Figure 1.

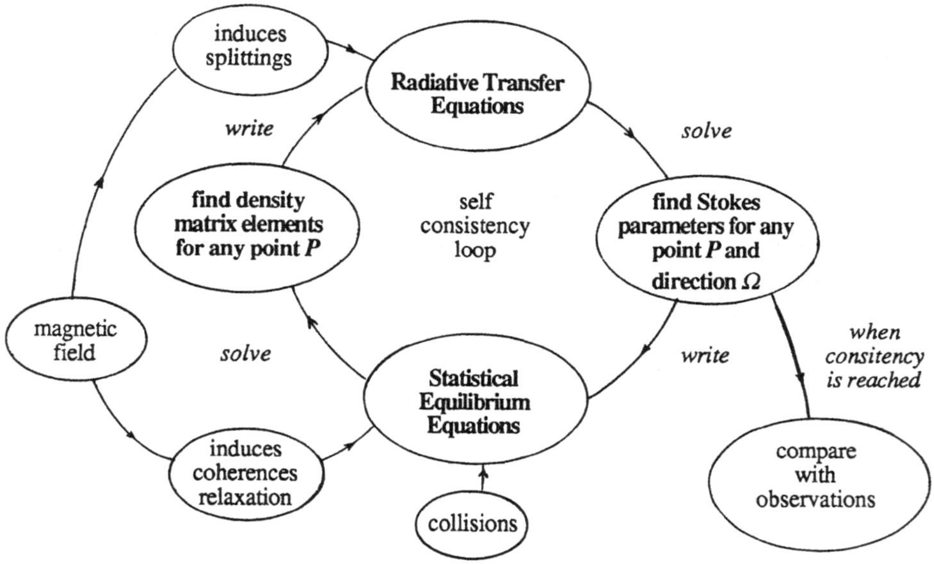

Figure 1

The non-LTE problem of the second kind is schematized by this self-consistency loop.

3. DIAGNOSTIC FROM OPTICALLY THIN LINES

Fortunately, for analyzing the polarization observed in optically thin lines from prominences, the self-consistency loop of Figure 1 considerably simplifies, as the radiation field that is entering the statistical equilibrium equations for the density matrix of the prominence atoms is nothing but the photospheric radiation field, which is well-known from observations. The problem of finding the polarization in a spectral line is then reduced to the following steps :

a) the atom responsible for the given spectral line is schematized through a suitable model consisting of several levels. Any of such levels is described through a set of parameters (the density matrix elements) that have to be considered the unknowns of the problem; such unknowns will be here denoted by the symbol $\rho^{\alpha}_{nm} = \langle \alpha n | \rho | \rho m \rangle$, where ρ is the atomic density matrix, α is a symbol that characterizes a given atomic level, and the indices m and n denote two arbitrary sublevels of the level α.

b) The statistical equilibrium equations are written for the density-matrix elements ρ^{α}_{nm} ; these equations, in stationary situations, are of the form :

$$0 = \frac{\partial}{\partial t}\,\rho^{\alpha}_{nm} = -\,2\pi i\nu_{nm}\,\rho^{\alpha}_{nm} + \sum_{\beta rs} R(\beta rs \rightarrow \alpha nm)\,\rho^{\beta}_{rs}$$

$$-\frac{1}{2}\left\{ \sum_{\beta rn'} R(\alpha n'm \rightarrow \beta rr)\,\rho^{\alpha}_{n'm} + \sum_{\beta rm'} R(\alpha nm' \rightarrow \beta rr)\,\rho^{\alpha}_{nm'} \right\}$$

where the first term in the r.h.s. is the term responsible for the Hanle effect (ν_{nm} being the energy difference, in frequency units, of the sublevels n and m), while the other terms represent the coherence-transfer rates due to radiative transitions (and collisional transitions, if important) among the various sublevels. In the radiative rates, the photospheric radiation field, and its relative center-to-limb variation, enter as known parameters.

c) The previous equations, implemented with the trace equation ($\sum_{\alpha n} \rho^{\alpha}_{nn} = 1$), are then solved to give the values of the unknowns ρ^{α}_{nm}. This is done by solving a linear system of N equations in N unknowns, where N is , generally, a rather large number counting the unknowns ρ^{α}_{nm} necessary to describe the atomic polarization of the model-atom.

d) From the values of the unknowns so obtained, one finally gets the emission coefficient in the four Stokes parameters of the prominence plasma.

The final values thus obtained for the Stokes parameters depend, in general on several parameters, that are : the height h of the observed point in the prominence; the intensity B and direction of the magnetic field (specified through the angles θ_B and χ_B defined in Figure 2), the angles θ_V and χ_V specifying the direction of the line of sight with respect to the solar radius (again defined in Figure 2); and, finally, when collisions are important, the density of the prominence plasma.Further parameters, like those describing the photospheric radiation field, or the atomic parameters (Einstein coefficients, energy shifts between fine or hyperfine structure levels, cross sections, etc ...) are, generally, rather well-known so that they can be fixed once and for all in the calculations.

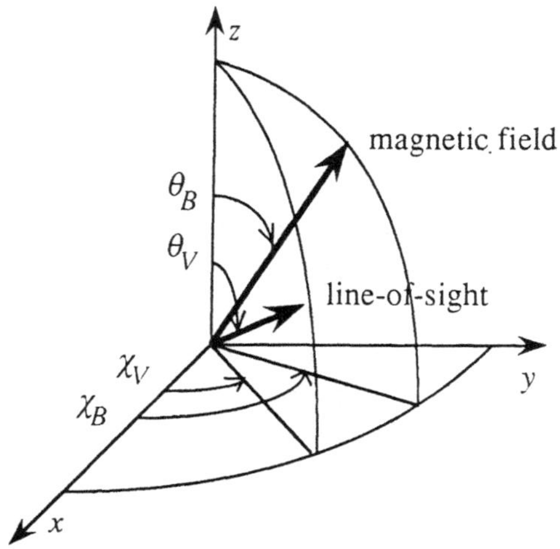

Figure 2

The scattering point inside the prominence is at the origin of the coordinate system. The meridian passing through the tip of the vector specifying the line of sight defines the positive Q-direction.

As a final result of these calculations, one gets the relative values for the Stokes parameters (Q/I, U/I, V/I) as complicated functions of the form : f (h, θ_V, B, θ_B, χ_B-χ_V); the values of h and θ_V can be derived, even with some uncertainties, from synoptic observations of the prominence crossing the solar disk as a filament; the "measurement" of the magnetic field can then be obtained by means of suitable algorithms looking for the values of B, θ_B and χ_B that best reproduce the observations.

4. SYMMETRIES AND OTHER LIMITATIONS

Unfortunately, there is an inherent symmetry in the Hanle effect (and in the Zeeman effect) which avoids the unambigous determination of the magnetic field vector from measurements of polarization in optically thin lines. Indeed, there are always two different determinations of $\vec{B}$ that give rise to the same polarization (both linear and circular); in the simplest case of a prominence in the plane of the sky, this symmetry holds between two magnetic field vectors that are symmetrical with respect to the line of sight. Another limitation typical of the Hanle effect, is that three quantities are necessary to give a full determination of the magnetic field vector. If only two quantities are measured (like for instance the linear polarization Stokes parameters in a single line), one of the three parameters (B, θ_B, χ_B) has to be independently known; only in this case the remaining two can be determined. Obviously, a full determination of the magnetic field vector can be obtained by observing the three Stokes parameters in a single line (although the circular polarization is often an order of magnitude smaller than the linear one) or by observing the linear polarization in two or more lines simultaneously.

5. SIMPLE ANALYTICAL RESULTS

The procedure that has been outlined in Section 2 requires involved calculations which imply the solution of systems of linear equations in a large number of unknowns. However, the basic characteristics of the Hanle effect, and the influence on the resulting polarization of the various parameters, can be better understood through simple analytical results that we are going to present in this section. Although the more detailed calculations previously outlined become ultimately necessary when interpreting observations, we consider important to give an intuitive grasp of the interplay of the various parameters in determining the signature of the polarization originating in a prominence spectral line.

5.a) Two-level atom with unpolarized ground level

We start from a simple two-level atom with the lower level having angular momentum J_ℓ and the upper level angular momentum J_u. If we neglect stimulated emission and we suppose that the ground level is unpolarized, the Stokes parameters of the radiation emited by our two-level atom in the direction $\vec{\Omega}_0$ are given by the expression (see Landi Degl'Innocenti (1984, 1985) for the introduction of the formalism of the irreducible spherical tensors) :

$$\varepsilon_i(\nu,\Omega_0) = \frac{h\nu_0}{4\pi}\,\phi(\nu_0-\nu)\,\mathcal{B}\,\mathcal{N}\sum_{kq} W(k,q)\,(-1)^q\,\bar{J}^k_{-q}\,\mathcal{T}^k_q(i,\Omega_0) \tag{1}$$

where :

ν_0 is the frequency of the transition; $\phi(\nu_0-\nu)$ is a profile (generally a gaussian profile in the case of prominences) centered at frequency ν_0; $\mathcal{B}$ is the Einstein coefficient for absorption; $\mathcal{N}$ is the overall population of the lower level; $W(k,q)$ is a depolarization factor which accounts for the effect of the magnetic field (Hanle effect) and is given by :

$$W(k,q) = 3(2J_u+1)\begin{Bmatrix} 1 & 1 & k \\ J_u & J_u & J_l \end{Bmatrix}^2 \left[1+2\pi i q\nu_L g_u / A\right]^{-1} \tag{2}$$

g_u being the Landé factor of the upper level, ν_L the Larmor frequency, and A the Einstein coefficient of the transition (note that $W(0,0) = 1$); $\bar{J}^k_q$ is the polarization tensor of the photospheric radiation field evaluated in the reference system of the magnetic field, and, finally, $\mathcal{T}^k_q(i,\vec{\Omega}_0)$ is the polarization tensor of the unit vectors defining the Stokes parameters of the emitted radiation. The tensor $\bar{J}^k_q$ can be first evaluated in the reference system having its z-axis directed along the solar radius. In this system, $\bar{J}^k_q$ has only two components different from zero ($\bar{J}^0_0$ and $\bar{J}^2_0$) that can be expressed, respectively, in the form $d_0 I_{ph}$ and $d_2 I_{ph}$ where I_{ph} is the photospheric radiation field at Sun's center and where d_0 and d_2 the so-called dilution factors, are given by :

$$d_0 = \frac{1}{2}\left\{(1-\cos\gamma)(1-u) + \frac{1}{2}u\,(1 - l\cos^2\gamma / \sin\gamma)\right\}$$

$$d_2 = \frac{1}{4\sqrt{2}}\left\{\cos\gamma\sin^2\gamma\,(1-u) + \frac{1}{8}u\,[3\sin^2\gamma - 1 + l\cos^2\gamma(1+3\sin^2\gamma) / \sin\gamma\,]\right\}$$

where u is the coefficient describing the limb darkening of the photospheric radiation field, γ is the aperture angle of the cone substending the Sun from the point in the prominence at height h : $\gamma = \sin^{-1}((1+h/R_\odot)^{-1})$, and finally $\ell = \ell n((1+\sin\gamma) / \cos\gamma)$(*). In the magnetic field reference system we then have :

$$\bar{J}^0_0 = d_0\, I_{ph}$$

$$\bar{J}^2_q = d_2\, I_{ph}\, \mathcal{D}^2_{0q}(\chi_B, \theta_B, 0)$$

Finally, the expression of the tensor $\mathcal{T}^k_q(i, \vec{\Omega}_0)$ depends on the convention that is used for the measurement of the Stokes parameters. If the positive Q-direction is chosen as in Figure 2 (which means that for a prominence in the plane of the sky the positive Q-direction is perpendicular to the solar limb), we have, for the only non-vanishing components of the tensor $\mathcal{T}^k_q$:

$k = 0$ $\quad \mathcal{T}^0_0(0,\Omega_0) = 1$

$k = 1$ $\quad \mathcal{T}^1_q(3,\Omega_0) = \sqrt{\frac{3}{2}}\,\mathcal{D}^1_{0q}(R)$

$k = 2$
$$\begin{cases} \mathcal{T}^2_q(0,\Omega_0) = \frac{1}{\sqrt{2}}\,\mathcal{D}^2_{0q}(R) \\ \mathcal{T}^2_q(1,\Omega_0) = -\frac{1}{2}\sqrt{3}\,[\mathcal{D}^2_{2q}(R) + \mathcal{D}^2_{-2q}(R)] \\ \mathcal{T}^2_q(2,\Omega_0) = -\frac{i}{2}\sqrt{3}\,[\mathcal{D}^2_{2q}(R) - \mathcal{D}^2_{-2q}(R)] \end{cases}$$

where $R \equiv (0, -\theta_V, -\chi_V) \times (\chi_B, \theta_B, 0)$.

(*) Note : the expressions given here are valid only in the case that the photospheric spectrum is flat around the frequency of the scattering line. Otherwise the expressions have to be generalized by considering integrals over the line profile of the form : $\int d\nu\, \phi(\nu_0-\nu)\, I_{ph}(\nu)\, d_i(\nu)$ $(i = 0,2)$.

In deriving Equation (1) we have implicitly assumed that the splitting of the sublevels of the upper level is negligible. As a consequence, Equation (1) neglects the contribution due to the Zeeman effect. This contribution can be easily evaluated if we suppose that the Zeeman splitting is very small as compared to the line broadening (an approximation that is always verified for prominences where $\nu_L/\Delta\nu_D \cong 10^{-3}$) and we further suppose that the atomic polarization induced by the anisotropy of the radiation field is weak ($\rho^k_q \ll \rho^0_0$ for k, q ≠ 0).

In this approximation, there is a further contribution that has to be added to Equation (1), namely

$$\varepsilon_3^{(1)}(\nu,\Omega_0) = \frac{h\nu_0}{4\pi} \mathcal{B} \mathcal{N} \bar{g} \cos\gamma \, \nu_L \frac{\partial}{\partial\nu} \varphi(\nu_0-\nu) \qquad (3)$$

where $\bar{g}$ is the effective Landé factor for the line considered

$$\bar{g} = \frac{1}{2}(g_u+g_l) + \frac{1}{4}(g_u-g_l)[J_u(J_u+1) - J_l(J_l+1)]$$

and where $\cos\gamma$ gives the projection of the magnetic field along the line of sight :

$$\cos\gamma = \cos\theta_B \cos\theta_V + \sin\theta_B \sin\theta_V \cos(\chi_B-\chi_V) \quad .$$

It is important to notice that, for the model atom that we have considered here, the only contribution to the emitted circular polarization comes from the Zeeman effect (as the terms $\bar{J}^1_q$ of the photospheric radiation field are zero); this results in a pure antisymmetrical profile for the circular polarization. As we will see in the following, this is no longer true for lines showing fine or hyperfine structure.

Returning now to Equation (1), we can explicitly see where the various parameters that influence the signature of the emitted polarization are contained : the intensity of the magnetic field is contained only in the depolarizing factor W, while its direction is contained in both tensors $\bar{J}^k_q$ and $\mathcal{J}^k_q$. The height of the observed point in the prominence affects only $\bar{J}^k_q$ (through the dilution factors d_0 and d_2) whereas the line-of -sight direction is contained in $\mathcal{J}^k_q$.

The behavior of the linear polarization with respect to the various parameters can be conveniently illustrated by means of suitable diagrams (called Hanle-diagrams). In Figure 3 some diagrams are presented for different values of the inclination angle, Θ_B, of the magnetic field. These diagrams, as the ones that will be presented in the following, are obtained for a height h = 70 arcsec and assuming the value u = 0.56 for the limb darkening coefficient; the transition considered is ($J_\ell = 1$, $J_u = 2$) with an Einstein coefficient $A = 7.06 \times 10^7 \, s^{-1}$ and with $g_u = 1$, that mimics the D_3 line of Helium. In each diagram, the full lines represent

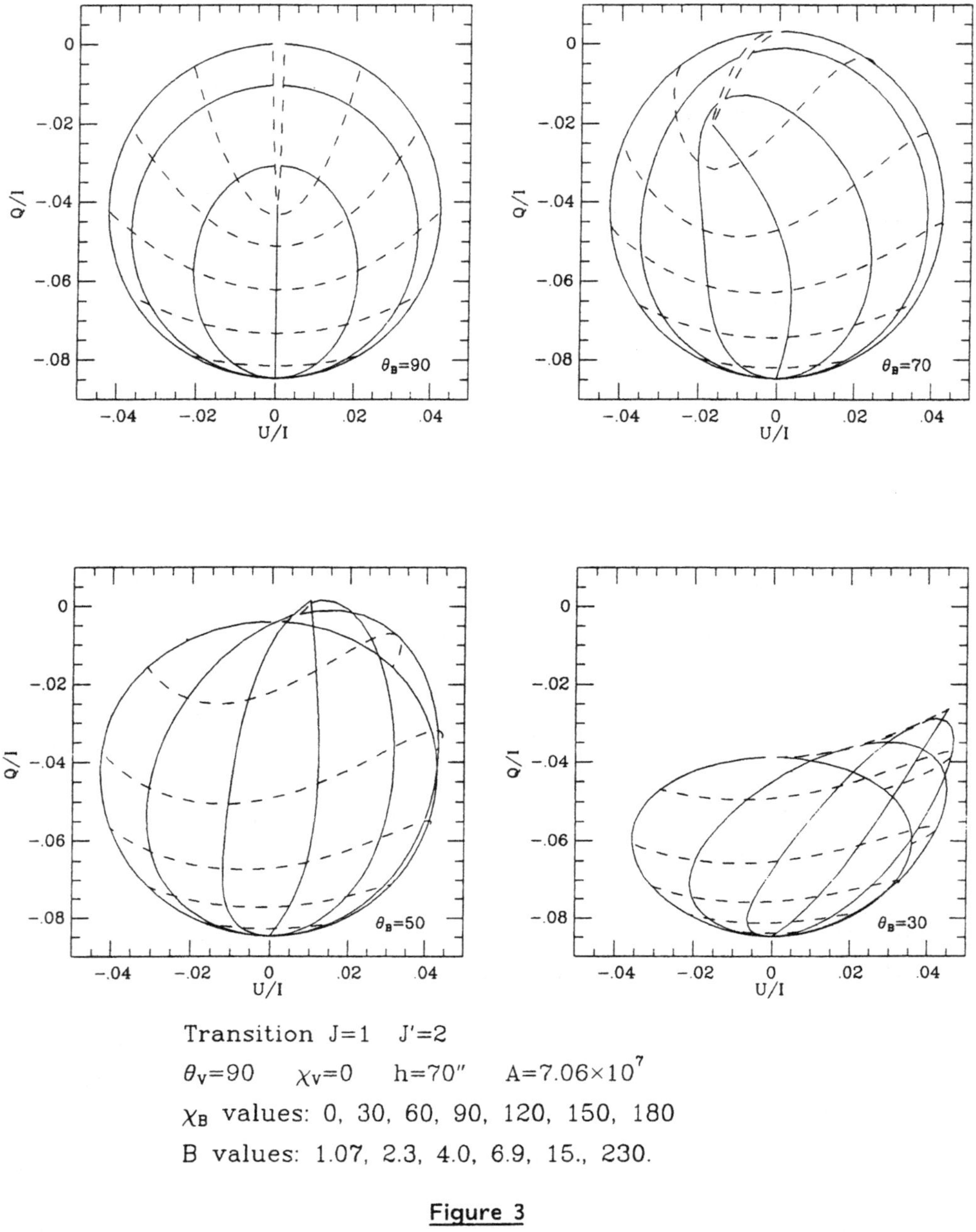

Transition J=1 J'=2

θ_V=90 χ_V=0 h=70" A=7.06×10^7

χ_B values: 0, 30, 60, 90, 120, 150, 180

B values: 1.07, 2.3, 4.0, 6.9, 15., 230.

Figure 3

Polarization diagrams for a two-level atom with unpolarized ground level. The diagrams refer to a prominence seen in the plane of the sky for different values of θ_B. The full lines represent curves at χ_B = const., with χ_B ranging from 0° (at the left end of each panel) to 180° (at the right end). The dashed lines are lines at B = const. with B increasing from the bottom to the top.

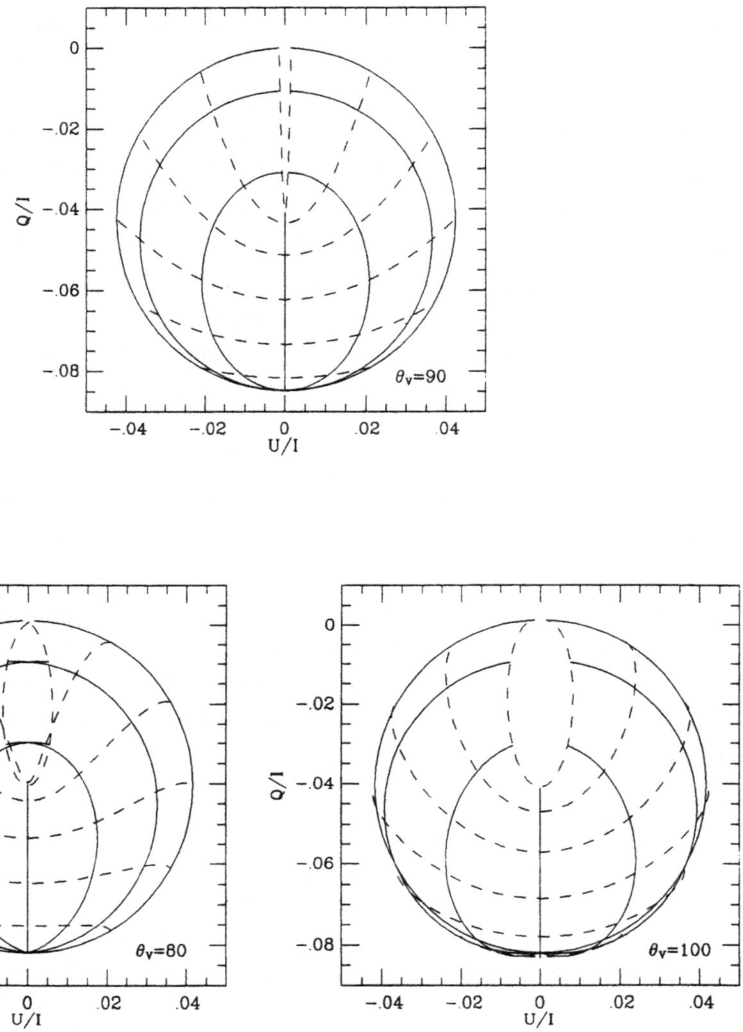

Transition J=1 J'=2

θ_B=90 χ_V=0 h=70" $A=7.06\times10^7$

χ_B values: 0, 30, 60, 90, 120, 150, 180

B values: 1.07, 2.3, 4.0, 6.9, 15., 230.

Figure 4

Same as figure 4 for a prominence seen at θ_V angles different from 90°. The θ_B value is 90° for all the three panels. The top panel is the same as the upper left panel of Figure 3. The meaning of the lines is the same as in Figure 3.

curves at χ_B = const, with χ_B ranging from 0° (at the left end of the figure to 180° (at the right end); the dashed lines are curves at B = const, with B increasing from the bottom to the top of the diagram. Figure 3 shows that for zero magnetic field the polarization is parallel to the solar limb (Q<0, U=0) and that a depolarization of a factor of the order of 1/2 is reached for $B \cong 5$ Gauss. When the inclination of the magnetic field decreases, however, its depolarizing effect also decreases and it reduces to zero in the limiting case Θ_B = 0°. Figure 4, on the contrary, is obtained for Θ_B = 90° and for various values of the angle Θ_V specifying the line-of-sight direction. These diagrams show the importance of knowing correctly the position of the prominence for an accurate deduction of the magnetic field vector.

5b). Two-level atom with polarized ground level

We can now start generalizing Equation (1) by releasing the approximation of considering the ground level as unpolarized. In general, if the polarization of the ground level is taken into account, it is impossible to express the emission coefficient in a closed analytical form. However, in the limit of weak anisotropy of the exciting radiation field, or, in other words in the limiting case $d_2 \ll d_0$ (an approximation that is usually well satisfied for prominences), Equation (1) still holds but the depolarizing factor W(k,q) has to be substituted by the following expression (see Landi Degl'Innocenti, 1985) :

$$W(0,0) = 1$$

$$W(k,q) = 3(2J_u+1)\left[\begin{Bmatrix} 1 & 1 & k \\ J_u & J_u & J_l \end{Bmatrix}^2 (1+i\,\Gamma_l q) + \right.$$

$$\left. + (-1)^{J_u+J_l} (2J_l+1) \begin{Bmatrix} 1 & 1 & k \\ J_u & J_u & J_l \end{Bmatrix} \begin{Bmatrix} 1 & 1 & k \\ J_l & J_l & J_u \end{Bmatrix} \begin{Bmatrix} J_u & J_u & k \\ J_l & J_l & 1 \end{Bmatrix}\right] \times$$

$$\times \left[(1+i\,\Gamma_l q)\,(1+i\,\Gamma_u q) - (2J_l+1)\,(2J_u+1) \begin{Bmatrix} J_u & J_u & k \\ J_l & J_l & 1 \end{Bmatrix}^2\right]^{-1} \qquad (k,q \neq 0)$$

where :

$$\Gamma_u = 2\pi\nu_L g_u B / A$$

$$\Gamma_l = 2\pi\nu_L g_l B / (\mathcal{B}\, \bar{J}^0_0 + A_l)$$

where A_ℓ is the Einstein coefficient for spontaneous deexcitation of the lower level.

Figure 5 shows the effect on the Hanle-diagrams of the presence of atomic polarization in the ground level. The various diagrams have been obtained by writing the central intensity I_{ph} of the photospheric radiation field in the form :

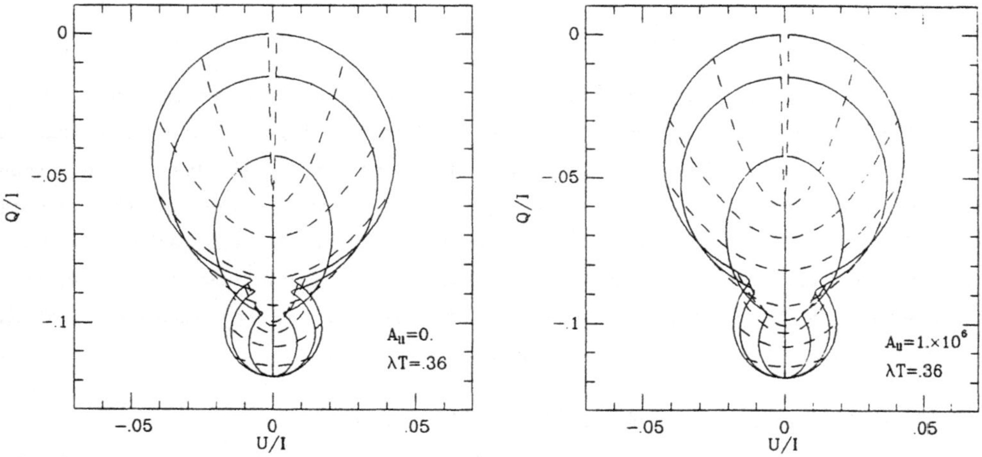

B values: .01, .02, .04, .1, .49, 1.6, 2.9, 5.5, 12., 230.   B values: .03, .06, .12, .25, .69, 1.6, 3.3, 5.5, 12., 230

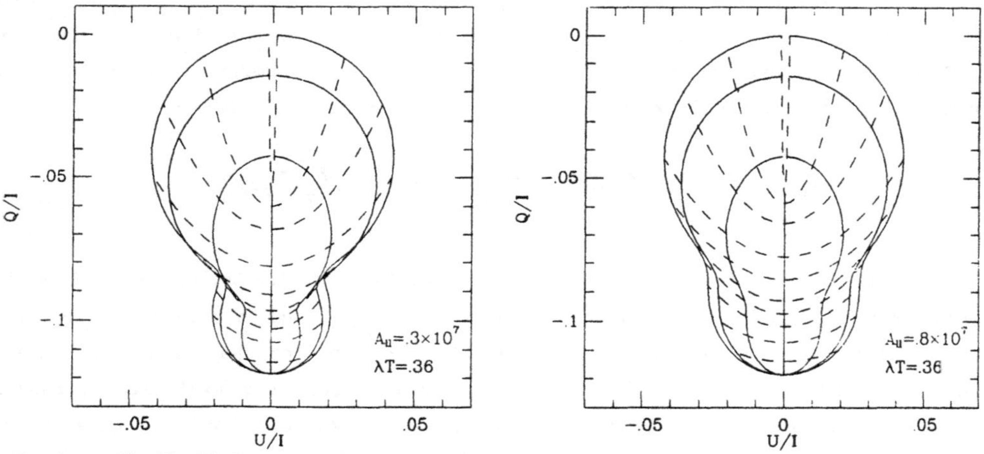

B values: .07, .15, .25, .51, 1., 1.9, 3.3, 6.3, 12., 230.   B values: .16, .31, .52, .88, 1.5, 2.5, 3.8, 7.3, 16., 230.

Transition J=1 J'=2 (with ground level polarized)

θ_B=90 θ_V=90 χ_V=0 h=70" A=7.06×10^7

χ_B values: 0, 30, 60, 90, 120, 150, 180

Figure 5

Polarization diagrams for a two-level atom with polarized ground level. The diagrams refer to a prominence seen in the plane of the sky with θ_B = 90°. The four panels differ for the Einstein coefficient of the lower level. The meaning of the lines is the same as in Figure 3.

$$I_{ph} = \frac{2hc}{\lambda^3}\left[\exp\left(\frac{hc}{\lambda kT}\right) - 1\right]^{-1}$$

and assuming for the parameter λT the value of 0.36 cm °K, which again mimics the D_3 line of He I. The most stricking features of Figure 5 are : the increase in linear polarization for B = 0 with respect to the case of the unpolarized ground level and the appearence of a secondary blob for small values of the magnetic field. In the case considered here, the secondary blob is exterior to the usual Hanle diagram; both these features are connected to the J-values of the transition considered ($J_\ell = 1$, $J_u = 2$); for different values of J, the secondary blob may well be found inside the usual Hanle diagram.

5c) Two-terms atom with unpolarized ground term

Equation (1) can also be generalized by considering the effect of the presence of fine structure (or hyperfine structure) in the line considered. In the case of an unresolved fine-structured line originating from the transition between a lower term, having spin S and orbital angular momentum L_ℓ, and an upper term having orbital angular momentum L_u, neglecting stimulated emission and supposing the ground term unpolarized, we have, for the Stokes parameter scattered in the direction $\vec{\Omega}_0$, an analytical expression similar to Equation (1), namely :

$$\varepsilon_i(\nu,\Omega_0) = \frac{h\nu_0}{4\pi}\,\phi(\nu_0-\nu)\,\mathcal{B}\,\mathcal{N}\sum_{Kkq} W_{fs}(K,k,q)\,(-1)^q\,\bar{J}^K_{-q}\,\mathcal{T}^k_q(i,\Omega_0)$$

where $\mathcal{B}$ is the Einstein coefficient for absorption in the multiplet, N is the overall population of the ground term and where $W_{fs}(K,k,q)$ is a depolarizing factor which accounts for the effect of the magnetic field and for the presence of interferences between different J-levels of the upper term; the explicit expression for the depolarizing factor is the following (the formal proof will be given elsewhere):

$$\begin{aligned} W_{fs}(K,k,q) &= \frac{3(2L_u+1)}{2S+1}\begin{Bmatrix} 1 & 1 & K \\ L_u & L_u & L_l \end{Bmatrix}\begin{Bmatrix} 1 & 1 & k \\ L_u & L_u & L_l \end{Bmatrix} \\ &\times \sum_{JJ'J''J'''MM'} (-1)^{J+J'-J''-J'''+k-K}\sqrt{(2K+1)(2k+1)(2J+1)(2J'+1)(2J''+1)(2J'''+1)} \\ &\times \begin{Bmatrix} L_u & L_u & k \\ J & J' & S \end{Bmatrix}\begin{Bmatrix} L_u & L_u & K \\ J'' & J''' & S \end{Bmatrix}\begin{pmatrix} J & J' & k \\ M & -M' & -q \end{pmatrix}\begin{pmatrix} J'' & J''' & K \\ M & -M' & -q \end{pmatrix} \\ &\times \sum_{jj'} C^j_J(M)\,C^{j*}_{J''}(M)\,C^{j'}_{J'}(M')\,C^{j'*}_{J'''}(M')\left[1+2\pi i\nu_{jM,j'M'}/A\right]^{-1} \end{aligned} \qquad (4)$$

where $C^j_J(M)$ are the coefficients of the energy eigenvectors of the upper term expressed as a linear combination of the standard angular momentum states :

$$|\alpha j M\rangle = \sum_J C^j_J(M)\,|\alpha L S J M\rangle$$

with

$$H_0\,|\alpha j M\rangle = E_{jM}\,|\alpha j M\rangle$$

H_0 being the atomic hamiltonian in the Paschen-Back regime; finally, the quantities $\nu_{jM,j'M'}$ appearing in Equation (4) are given by :

$$\nu_{jM,j'M'} = (E_{jM} - E_{j'M'})\,/\,h$$

and represent the frequency difference between different eigenstates of the upper term.

To understand the importance of fine structure on the scattered polarization, Equation (4) has to be compared with the simplified expressions that can be obtained either neglecting the effect of fine structure "tout-court" or by neglecting the effect of crossing level interferences; neglecting fine structure is equivalent to let S = 0 in Equation (4); this brings (after some Racah algebra) to the expression :

$$\left[W_{fs}(K,k,q)\right]_{\text{no fine structure}} = \delta_{K,k}\, 3(2L_u+1) \begin{Bmatrix} 1 & 1 & k \\ L_u & L_u & L_l \end{Bmatrix}^2 \times$$
$$\times \left[1 + 2\pi i q \nu_L \,/\, A\right]^{-1}$$

which is strictly similar to the expression contained in Equation (2) as a consequence of the principle of spectroscopic stability. The corresponding Hanle diagram is shown in the first panel of Figure 6; this diagram is exactly the same as the first diagrams of Figures 3 and 4.

On the contrary, neglecting crossing-level interferences or, in other words, considering all the J-levels as independent, one obtains the following expression

$$\left[W_{fs}(K,k,q)\right]_{\text{no } J-J' \text{ interference}} = \delta_{K,k} \frac{3(2L_u+1)}{2S+1} \begin{Bmatrix} 1 & 1 & k \\ L_u & L_u & L_l \end{Bmatrix}^2 \times$$
$$\times \sum_J (2J+1)^2 \begin{Bmatrix} L_u & L_u & k \\ J & J & S \end{Bmatrix}^2 \left[1 + 2\pi i q \nu_L g_J q \,/\, A\right]^{-1} \quad ;$$

the corresponding Hanle diagram is shown in the second panel of Figure 6, while the third panel shows the diagram as computed according to Equation (4).

A direct comparison among the three diagrams of Figure 6 shows that for the actual computation of the polarization scattered in the D_3 line of He, the influence of fine structure is extremely important as it reduces the scattered polarization of approximately a factor 1/2; moreover, also the influence of crossing level interferences is important, especially for magnetic fields of the order or larger

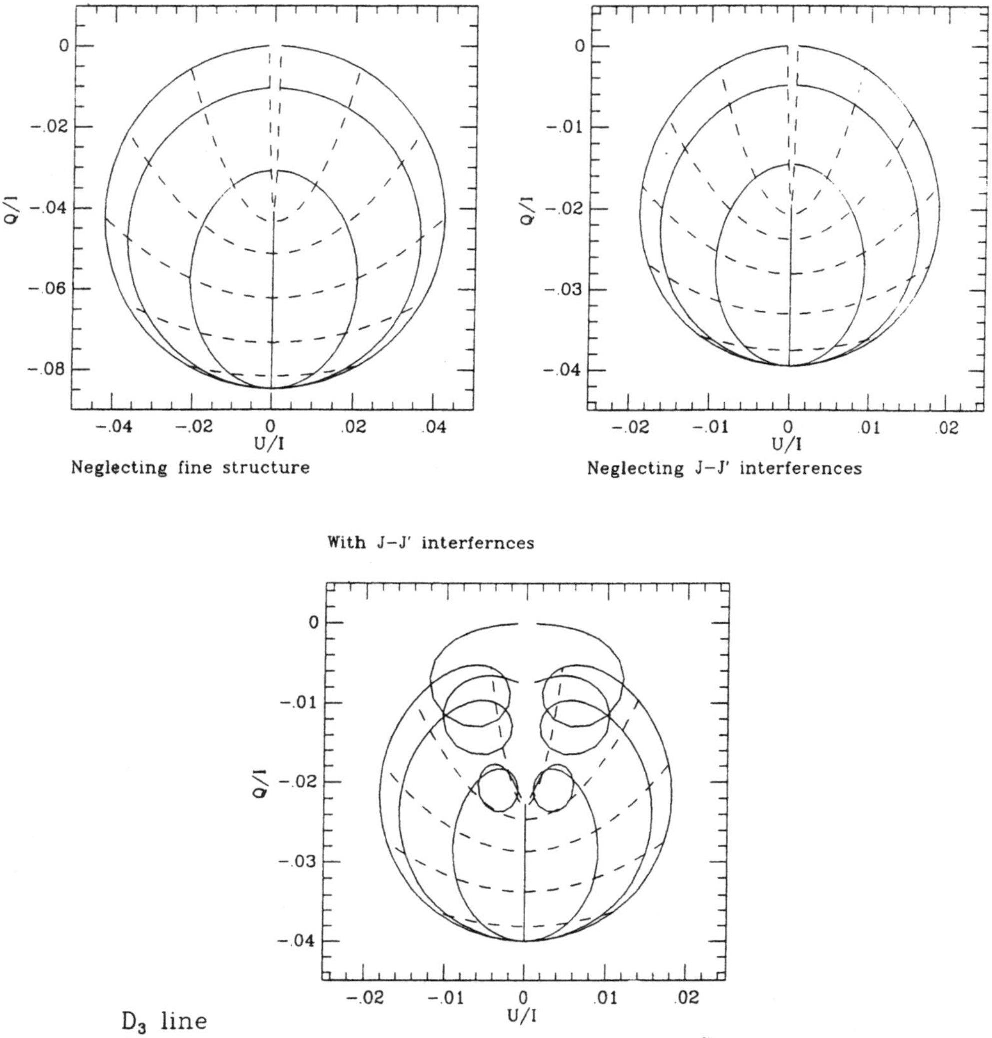

D_3 line

θ_B=90 θ_V=90 χ_V=0 h=70" A=7.06×10^7

χ_B values: 0, 30, 60, 90, 120, 150, 180

B values: 1.07, 2.3, 4.0, 6.9, 15., 230.

Figure 6

Polarization diagrams for the D_3 line of He with unpolarized ground level. The diagrams refer to a prominence seen in the plane of the sky with θ_B = 90°. The upper left diagram is obtained neglecting fine structure and it is the same as the upper left diagram of Figure 3. Taking into account the effects of fine structure, but neglecting interferences between different J-levels, the upper right panel is obtained. Finally, when interferences are fully taken into account, the bottom panel is obtained. The meaning of the lines is the same as in Figure 3.

than 20 G, as it is responsible for the appearence of the characteristic "loops" that are shown in the third panel of Figure 6.

Fine structure and crossing-level interferences are also responsible for the appearence of another important phenomenon that is described by the term $K = 2$, $k = 1$ in Equation (4). The presence of this term shows that the anisotropy of the photospheric radiation field is capable of introducing a net amount of circular polarization in a fine-structured line, a phenomenon that will be called in the following the "alignment-orientation transfer effect". This circular polarization signal adds to the usual one due to the Zeeman effect and results in a complicated profile which is neither symmetrical nor antisymmetrical with respect to line center. Great care has to be taken for the presence of this phenomenon when trying to measure the magnetic field in prominences through the usual Zeeman effect.

The aligment-orientation transfer effect is not very conspicuous for the He I D_3 line where the scattered circular polarization is approximately one order of magnitude less than the scattered linear polarization. It has however to be remarked that a similar phenomenon is also present in hyperfine structure multiplets, like for instance in the Na I D-lines.

6. DIAGOSTIC FROM OPTICALLY THICK LINES

The interpretation of the polarization observed in optically thick lines is much more complicated than for the case of optically thin lines. The problem here is that the self-consistency loop of Figure 1 has in principle to be solved in its generality and this requires also that a kind of geometrical model has to be assumed for the prominence plasma. The geometrical models that have been proposed up to now are the "infinitely sharp" model and the "elliptical" model.

In the first case the prominence is described as a slab of infinite optical thickness (but of negligible geometrical thickness) standing vertically over the solar surface and extending indefinitely along the y and z directions defined in Figure 2. In the second case the prominence is modeled as a cylinder extending indefinitely along the y direction and having an elliptical cross-section in the x-z plane. As a first approach to the actual solution of the self-consistency loop, a perturbative method has been proposed (based on the fact that the polarization of the radiation propagating inside the prominence is very weak, being of the order or less than 1%). According to this method, one starts from a theoretical (or empirical) non-LTE model of a prominence (like for instance the models developed by Heasley and Milkey, 1976, 1978). From these models one can calculate the "zero-order", unpolarized radiation field propagating inside the prominence and repeat steps a) to d) described in Section 3 for several points inside the prominence. The difference with the case of the optically thin lines is twofold : i) in the radiative rates two contributions are present, one from the photospheric radiation field which is attenuated during its propagation inside the prominence, and one from the internal radiation field that is calculated from the zero-order model; ii) the final values of the Stokes parameters to be compared with observations result from an integration along the line-of-sight.

For the simple case of a two-level atom with an unpolarized ground level, Equation (1) still gives the local emission coefficient in the four Stokes parameters, with the difference that the polarization tensor $\bar{J}^k_q$ assumes a more complicated expression. For the sake of completeness we write down explicitely these expressions in the geometry of the "infinitely sharp" model shown in Figure 7. In the reference system xyz of Figure 7, the non-zero components of the polarization tensor in a particular point P are given by :

$$\bar{J}^0_0 = \frac{S_0}{2}\left[2 - E_2(\tau_+) - E_2(\tau_-)\right] + $$

$$+ \frac{I_{ph}}{2\pi}\int_0^\gamma d\theta\,(1-u+u\cos\alpha)\left[\mathcal{J}_0(\tau_+/\sin\theta) + \mathcal{J}_0(\tau_-/\sin\theta)\right]\sin\theta$$

$$\bar{J}^2_0 = \frac{S_0}{8\sqrt{2}}\left[-E_2(\tau_+) - E_2(\tau_-) + 3E_4(\tau_+) + 3E_4(\tau_-)\right] + $$

$$+ \frac{I_{ph}}{4\pi\sqrt{2}}\int_0^\gamma d\theta\,(1-u+u\cos\alpha)\left[\mathcal{J}_0(\tau_+/\sin\theta) + \mathcal{J}_0(\tau_-/\sin\theta)\right](3\cos^2\theta - 1)$$

$$\bar{J}^2_{\pm1} = \pm\frac{\sqrt{3}\,I_{ph}}{4\pi}\int_0^\gamma d\theta\,(1-u+u\cos\alpha)\left[\mathcal{J}_1(\tau_+/\sin\theta) - \mathcal{J}_1(\tau_-/\sin\theta)\right]\cos\theta\,\sin^2\theta$$

$$\bar{J}^2_{\pm2} = \frac{\sqrt{3}\,S_0}{16}\left[E_2(\tau_+) + E_2(\tau_-) - 3E_4(\tau_+) - 3E_4(\tau_-)\right] + $$

$$+ \frac{\sqrt{3}\,I_{ph}}{8\pi}\int_0^\gamma d\theta\,(1-u+u\cos\alpha)\left[2\mathcal{J}_2(\tau_+/\sin\theta) + 2\mathcal{J}_2(\tau_-/\sin\theta)\right.$$

$$\left. - \mathcal{J}_0(\tau_+/\sin\theta) - \mathcal{J}_0(\tau_-/\sin\theta)\right]\sin^3\theta$$

where S_0 is the (constant) line source function inside the prominence, $E_n(x)$ is the usual exponential-integral of order n, $\mathcal{J}_n(x)$ is a function defined by : $\mathcal{J}_n(x) = \int_0^{\pi/2} e^{-x/\cos\beta}\cos^n\beta\, d\beta$ that can be related to the repeated integral of the modified Bessel function K_0, τ_+ and τ_- (see Figure 7) are the line optical depths specifying the position of the point P inside the prominence, θ is the polar angle characterizing the single ray through P and, finally : $\cos\alpha = (1-\sin^2\theta/\sin^2\gamma)^{1/2}$, with $\sin\gamma = (1+h/_{R_\odot})^{-1}$. (*)

(*) Note : In the expressions given here the frequency dependence of the various quantities (I_{ph}, u, τ_+, τ_-) has been neglected to shorten the notations. More correctly one should consider frequency integrals over the line profile of the quantities written above.

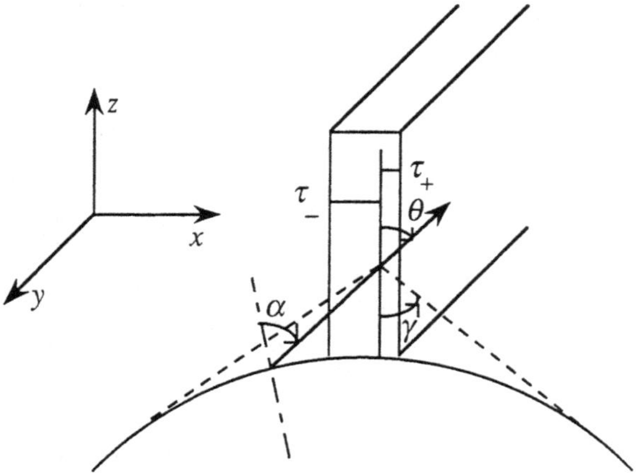

Figure 7

Schematic representation of the "infinitely sharp" model.

From these expresions, the polarization tensor in the magnetic field reference system (the tensor that has to be substituted in Equation (1)) can be easily recovered through the simple transformation :

$$\left[\bar{J}^K_Q\right]_{\text{mag. field ref. system}} = \sum_P \left[\bar{J}^K_P\right]_{xyz} \mathcal{D}^K_{PQ}(\chi_B, \theta_B, 0) \quad .$$

With this substitution the emission coefficient in the four Stokes parameters can be calculated for any point P inside the prominence and the emerging Stokes parameters are then recovered by means of a direct integration. As a final results of these calculations one gets the relative values for the Stokes parameters (Q/I, U/I, V/I) as complicated functions of the form : f $(h, \theta_V, \chi_V, B, \Theta_B, \chi_B, \tau, S_o, \Gamma)$ where τ is the total optical thickness of the prominence and where Γ is a parameter (or in some cases a set of parameters) specifying the type and geometry of the assumed prominence model (infinite slab, elliptical cylinder, etc ...).

Figure 8 shows various Hanle diagrams relative to a simple transition with unpolarized ground level for a prominence observed in the plane of the sky and described by the "infinitely sharp" model. The various panels (that are drawn on the same scale) refer to different values of τ and are obtained supposing $S_o = 0.5\ I_{ph}$. The effect of increasing optical thickness is a marked decrease of the scattered polarization and a less obvious modification of the signature of the diagram itself that shows up more clearly when the diagrams are plotted on different scales (Figure 9).

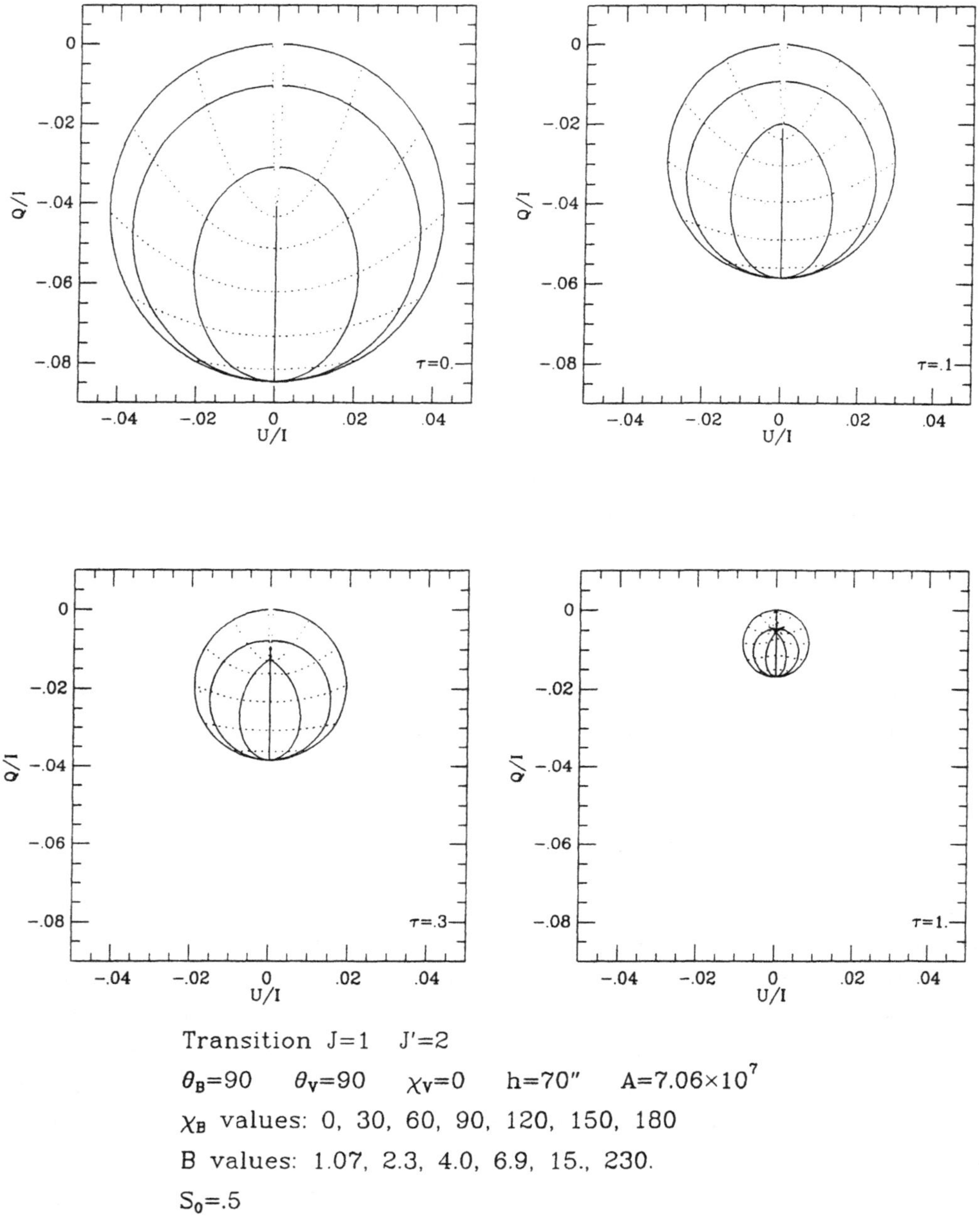

Transition J=1 J'=2

θ_B=90 θ_V=90 χ_V=0 h=70" A=7.06×10^7

χ_B values: 0, 30, 60, 90, 120, 150, 180

B values: 1.07, 2.3, 4.0, 6.9, 15., 230.

S_0=.5

Figure 8

Polarization diagrams for a two-level atom with unpolarized ground level. The diagrams refer to an optically thick prominence seen in the plane of the sky with θ_B = 90°. The upper left diagram is the same as the corresponding diagram of Figure 3. The meaning of the lines is the same as in Figure 3. Note the decrease of polarization for increasing τ.

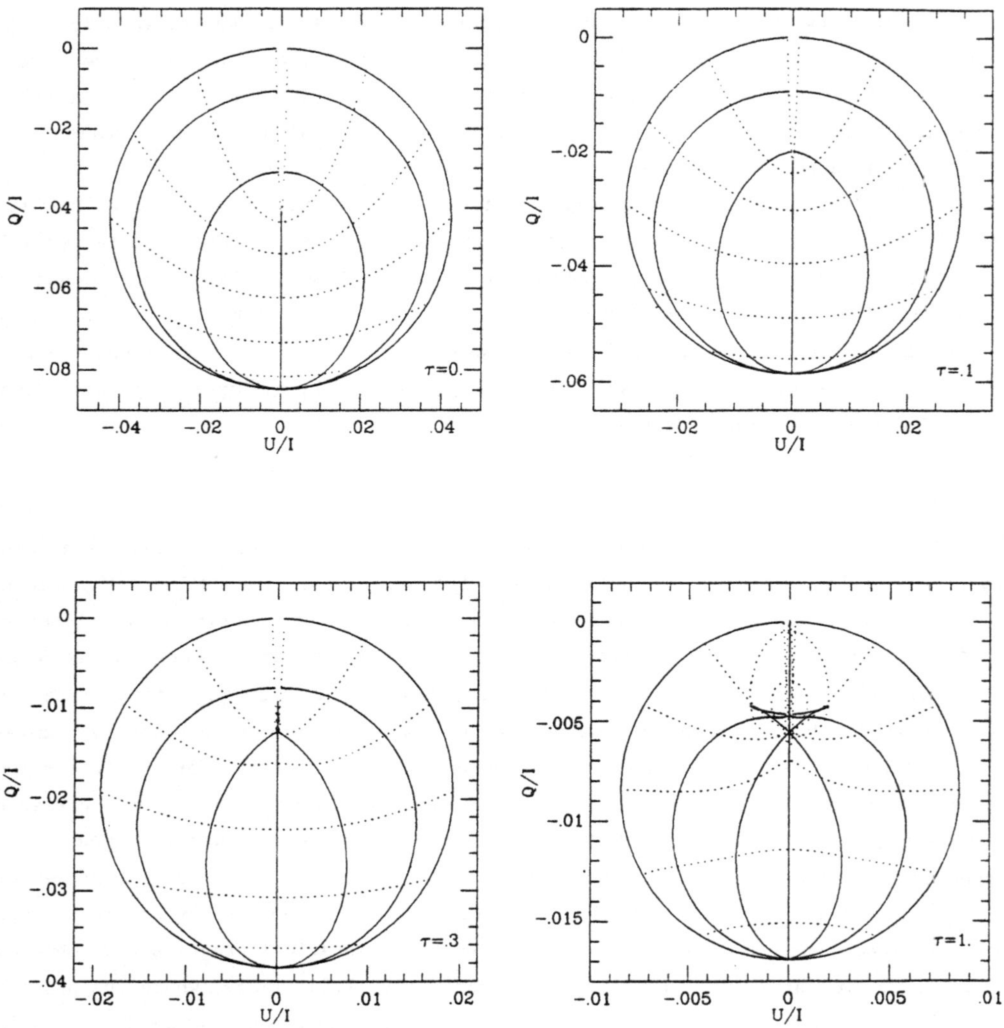

Transition J=1 J'=2

θ_B=90 θ_V=90 χ_V=0 h=70" A=7.06×10^7

χ_B values: 0, 30, 60, 90, 120, 150, 180

B values: 1.07, 2.3, 4.0, 6.9, 15., 230.

S_0=.5

Figure 9

Same as Figure 8 with the diagrams plotted on different scales.

The most important consequence of optical thickness for the diagnostic of magnetic field is however the fact that the elongated structure of a prominence introduces into the scattering geometry a further physical direction which is capable of breaking the typical symmetry of the optically thin case. Although the same value for the scattered radiation is obtained, also in the optically thick case, for two different determinations of the magnetic field vector, this ambiguity does not have the simple behavior of the 180° symmetry around the line-of-sight, characteristic of the optically thin case and already discussed in Section 3. This fact suggests a method for solving the 180° ambiguity that is present in field vector determinations based on the Hanle effect analysis of optically thin lines; to this aim observations of linear polarization in optically thin and optically thick lines should be performed simultaneously.

7. HISTORICAL REVIEW

After the first suggestion of employing the Hanle effect for diagnosing magnetic field vectors in prominences (Hyder, 1965) the first significant contributions to the establishment of a theory capable of describing the phenomena of scattering polarization in the presence of a magnetic field were due to House (1970a,b; 1971). Only in the late seventies, however, a rigorous theory based on the density-matrix formalism was developed by Bommier (1977) and applied by Bommier and Sahal-Bréchot (1978) to derive the theoretical expectations for the integrated linear polarization of the HeI D_3 line in optically thin lines in the presence of weak magnetic fields ($B < 10$ G). The formalism presented in these papers was subsequently generalized by Bommier (1980) to allow for crossing-level interferences and by Landi Degl'Innocenti (1982) to interpret the fine structure of the D_3 line both in linear and circular polarization. Further theoretical progress was achieved in a series of papers by Landi Degl'Innocenti (1983,1984, 1985), where the problem of the generation and transfer of polarized radiation was attacked in full generality, by Landolfi and Landi Degl'Innocenti (1985) who computed the expected polarization of the NaI D lines in optically thin prominences, and by Bommier et al. (1986a,b) who performed analogous computations on H_β taking properly into account the effect of depolarizing collisions with electrons and protons.

More recently, the more involved problem of $H\alpha$ polarization in optically thick prominences has been attacked, by means of a perturbative approach, by Landi Degl'Innocenti et al. (1987) and by Bommier et al. (1989a) for two different geometries of the prominence model ("infinitely sharp" model and "elliptical cylinder" model, respectively). Some further developments for solving the coupled set of radiative transfer and statistical equilibrium equations in a fully consistent way are now in progress (Bommier et al., 1989b,c).

The theory developed in the papers previously quoted has been applied to the interpretation of a large set of data obtained through the Pic-du-Midi coronograph polarimeter (Leroy, 1977, 1978; Bommier et al., 1981; Leroy et al., 1983, 1984; Bommier et al., 1986a,b) and through the HAO Stokes parameters (Athay et al., 1983). The results, that are summarized in several reviews (Leroy, 1987; Zirker, 1989; Kim, 1989), have raised some controversies especially with respect to two points that is worthwhile to discuss here in some detail.

The first point concerns the fact that in many cases, and especially for high latitude quiescent prominences with vertical threads, the magnetic field is found to be rather homogeneous and practically parallel to the solar surface. This fact is rather puzzling because, quoting Zirker (1981): "the vector field is apparently horizontal, uniform and static over an object that is riddled with transient, vertical, fine structures". Although this apparent contradiction has been thoroughly discussed by Leroy (1987) we just want here to stress the fact that the rotation of the plane of linear polarization with respect to the solar limb, which is observed in D_3 in the large majority of cases, is totally incompatible with a vertical field. For a vertical field, in fact, the linear polarization should be parallel to the solar limb, this conclusion being a basic result of the physics of the Hanle effect, independent of any detailed modeling of the D_3 line in prominences. We can then conclude that the observations exclude the possibility of the magnetic field being aligned with the vertical threads observed in prominences.

The second point concerns a result that has been obtained by Bommier et al. (1986b) in their analysis of joint observations of D_3 and H_β lines in quiescent prominences. According to Bommier et al., typical electron densities in prominences range from 10^9 to 4×10^{10} cm^{-3}, while the values that are usually deduced through different diagnostic methods are approximately one order of magnitude larger (Hirayama, 1978, 1989). It has to be remarked that hydrogen lines (differently from the lines of other elements) are quite sensitive to collisions with charged perturbers due to the typical ℓ-degeneracy of the hydrogen eigenstates. The effect of collisions with an isotropical distribution of perturbers is the one of reducing the polarization of hydrogen lines by a factor that is, roughly speaking, proportional to the density. As, on the contrary, observations show that the depolarizing effect of collisions on H_β is rather small, Bommier et al. (1986b) are forced from their data to deduce a relatively low value for the density. In other words, at electron densities of the order of 10^{11} cm^{-3} the linear polarization in H_β should be practically destroyed by the effect of collisions, contrarily to what is observed.

As there is no reason to think that the depolarizing collisional cross-sections computed by Bommier et al. may be wrong by one order of magnitude, and excluding selection effects in the observations, the only way out to reconcile the discrepancy found on the determinations of Ne is to invoke the presence in the prominence plasma of an additional polarizing mechanism that has been up to now neglected. Impact polarization due to vertical motions along the prominence threads, or to macroscopic electrical currents is a possibility that may be worth investigating in the future.

8. CONCLUSIONS AND RECOMMENDATIONS

From the discussion presented in the previous sections we can draw the following basic conclusions :

1) The theory of the Hanle effect has reached a sufficient degree of sophistication such as to provide a reliable method for measuring the vector magnetic field in prominences with a high degree of confidence.

2) For optically thin lines, insensitive to depolarizing collisions, the theory provides a diagnostic method that is "model independent" in as far as the photospheric radiation field and its center-to-limb variation are known for all those spectral lines that are involved in populating or depopulating the levels from which the line is originating.

3) The He I D_3 line is particularly suitable for the diagnostic of magnetic fields in prominences because, in typical cases, its optical thickness is negligible and moreover, it is practically insensitive to depolarizing collisions. Another advantage of the He I lines is the fact that they are absent from the photospheric spectrum which makes their diagnostic insensitive to effects of Doppler dimming. The only disadvantage of the D_3 line is the fact that its lower level is connected to the (metastable) ground level of the Helium triplet system by the λ 10830 line line, which, in some cases, may reach non negligible values of optical thickness.

4) In order to measure all the three components of the magnetic field vector it is necessary either to measure the linear polarization in two different lines (or in two components of the same line - like the two components of D_3 -) or, alternatively, to measure the linear <u>and</u> circular polarization in a single line, although in typical case the circular polarization signal is rather weak and hence more affected by noise. However, if the lines are optically thin, the determination of the vector field remains ambigous for the symmetry property (outlined in Section 3) typical of the Hanle effect and of the Zeeman effect.

5) Optically thick lines (like e.g. Hα) lead to a diagnostic that is more involved and, moreover, somewhat "model-dependent". However, observations in these lines can be efficiently used to remove the ambiguity (between "true solutions" and "spurious solutions") typical of optically thin lines. In most cases even a rough geometrical model of the prominence is sufficient to discriminate between the two alternatives.

6) Although their interpration may be rather involved, optically thick lines provide the unique posibility of determining the magnetic field vector from disk observations. Linear polarization observations in Hα filaments will probably

become possible in the near future with the development of new spectropolarimetric instrumentation (THEMIS Project). Such observations should provide the exciting possibility of getting important informations on the magnetic configuration in the higher layers of prominences.

7) The measurement of the longitudinal component of the magnetic field vector from circular polarization profiles suffers, for many lines, from the inconveniences that have been outlined in Section 5c). All the lines that are either fine-structured or hyperfine structured (Hα, Hβ, HeI D_3, NaI D) are expected to show in prominences a complicated circular polarization profile that results from the combination of an antisymmetrical component (due to the Zeeman effect) and a symmetrical component (due to the effect of crossing-level interferences). The determination of $B_{//}$ from such lines requires a careful calibration of the observations for removing the contribution, due to the symmetric component, to the circular polarization signal.

ACKNOWLEDGEMENTS

This work has been carried out while the author was holding a sabbatical leave at the Instituto de Astrofisica de Canarias, Tenerife, Spain. The author wishes to express his sincere thanks to the Director of the IAC, Prof. Francisco Sanchez, and to all the members of the solar physics division for their kind hospitality.

REFERENCES

Athay, R.G., Querfeld, C.W., Smartt, R.N., Landi Degl'Innocenti, E., Bommier V., : 1983, Solar Phys. 89, 3.
Bommier, V. : 1977, Thèse de 3ème cycle, Université de Paris VI.
Bommier, V. : 1980, Astron. Astrophys. 87, 109.
Bommier, V., Sahal-Bréchot, S. : 1978, Astron. Astrophys. 69, 57.
Bommier, V., Leroy, J.L., Sahal-Bréchot, S.: 1981, Astron. Astrophys. 100,-231.
Bommier, V., Leroy, J.L., Sahal-Bréchot, S.: 1986a, Astron. Astrophys. 156, 79.
Bommier, V., Leroy, J.L., Sahal-Bréchot, S.: 1986b, Astron. Astrophys. 156, 90.
Bommier, V., Landi Degl'Innocenti, E., Sahal-Bréchot, S.: 1989a, Astron. Astrophys. 211, 230.
Bommier, V., Landi Degl'Innocenti, E., Sahal-Bréchot, S.: 1989b, in H. Frisch and N. Mein (eds.) "Second Atelier Transfert du Rayonnement", p. 44.
Bommier, V., Landi Degl'Innocenti, E., Sahal-Bréchot, S.: 1989c, These proceedings.
Heasley, J.N., Milkey, R.W.: 1976, Astrophys. J. 210, 827.
Heasley, J.N., Milkey, R.W.: 1978, Astrophys. J. 221, 677.
House, L.L.: 1970a, J. Quant. Spectrosc. Radiat. Transfer 10, 909.
House, L.L.: 1970b, J. Quant. Spectrosc. Radiat. Transfer 10, 1171.
House, L.L.: 1971, J. Quant. Spectrosc. Radiat. Transfer 11, 367.
Hirayama, T.: 1978, in E. Jensen, P. Maltby, F.Q. Orrall (eds.), IAU Coll. 44, 4.
Hirayama, T.: 1989, These proceedings.
Hyder, C.L.: 1965, Astrophys. J, 141, 1374.
Kim, I.C.: 1989, These Proceedings.
Landi Degl'Innocenti, E.: 1982, Solar Phys. 79, 291.
Landi Degl'Innocenti, E.: 1983, Solar Phys. 85, 3.
Landi Degl'Innocenti, E.: 1984, Solar Phys. 91, 1.
Landi Degl'Innocenti, E.: 1985, Solar Phys. 102, 1.
Landi Degl'Innocenti, E., Bommier, V., Sahal-Bréchot, S.: 1987, Astron. Astrophys. 186, 335.
Landolfi, M., Landi Degl'Innocenti, E.: 1985, Solar Phys. 98, 53.
Leroy, J.L.: 1977, Astron. Astrophys. 60, 79.
Leroy, J.L.: 1978, Astron. Astrophys. 64, 247.
Leroy, J.L.: 1987, in J.L. Ballester and E. Priest (eds.), "Dynamics and Structure of Solar Prominences", Universitat des Illes Balears, pag. 33.
Leroy, J.L., Bommier, V., Sahal-Bréchot, S.: 1983, Solar Phys. 83, 135.
Leroy, J.L., Bommier, V., Sahal-Bréchot, S.: 1984, Astron. Astrophys. 131, 33.
Zirker, J.B.: 1989, Solar Phys. 119, 341.

Discussion

Wiehr: Considering your warning to measure the Zeeman effect in Balmer lines (and perhaps also Helium lines?) would you then recommend to use Ca^+ lines? The infrared line 8542 has large splitting, is fairly bright and almost optically thin.

Landi Degl'Innocenti: Ca^+ lines fall in the category of recommended lines because they are not affected by either fine or hyperfine structure.

Heinzel: I think that the effect of an enhanced irradiation by plages is not so negligibile. The central intensity of the H-alpha line may be a factor 2 - 3 higher as compared to the quite Sun radiation and the line profile of $H\alpha$ in plages is much more flat. Moreover, incident $L\alpha$ and $L\beta$ are also enhanced significantly so that the hydrogen exitation and ionization within the filament can be modified.

Engvold: In your talk you mentioned the possible effect on the polarization by Doppler dimming. How large would the vertical velocity be in prominences to affect the interpretation of the Hanle effect.

Landi Degl'Innocenti: The vertical velocity would have to be such to produce a Doppler shift comparable to the typical width of the photospheric line. However, the D_3 line does not suffer from the problem as there is no HeI photospheric spectrum.

FORMATION OF A FILAMENT AROUND A MAGNETIC REGION

B.Schmieder[*], P.Démoulin[*], J.Ferreira[*], C.E.Alissandrakis[**]

[*] Observatoire de Paris Meudon, F-92195 Meudon Principal Cedex, France,
[**] Section of Astrophysics, Department of Physics, University of Athens, GR-15783 Athens, Greece

ABSTRACT

The evolution of the active region AR4682 observed in 1985 during six rotations was dominated by three different phenomena:

. the large scale pattern activity: relationship between two active regions, formation of a quiescent filament during the decay phase of the active region,

. the presence of two pivot points along the filament surrounding the sunspot-with the long term one is associated the existence of the filament , with the short term one the activity with partial disappearance,

. the magnetic shear during one rotation.

The magnetic field lines have been extrapolated from photospheric data using Alissandrakis code (1981). The magnetic configuration with the existence of a dip favors the formation of a filament. We note that the shearing of the sunspot region and of the filament are both well described by force-free magnetic fields with the same constant α. This suggests that they are both a consequent of the same shear process.

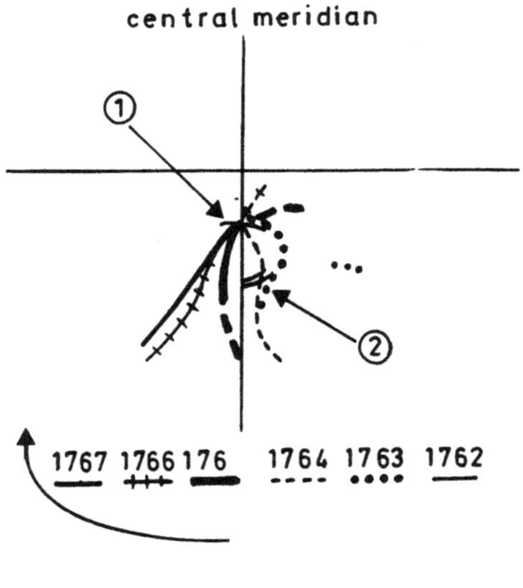

Pivot points found from synoptic map study during six rotations. Pivot 1 (respectively pivot 2) is a long-term (short) pivot (courtesy of M.J.Martres).

VECTOR MAGNETIC FIELD AND CURRENTS AT THE FOOTPOINT OF A LOOP PROMINENCE

A.Hofmann[1], V.Ruždjak[2] and B.Vršnak[2]

[1]Central Institute for Astrophysics of the GDR Academy of Sciences, Solar Observatory "Einsteinturm", Telegrafenberg, DDR-1561 Potsdam

[2]Hvar Observatory, 58 450 Hvar, Yugoslavia

Abstract. Using H_α -filtergrams and vector magnetograms we study the structure of the magnetic field at the footpoint of a loop prominence rooting deep in the penumbral photosphere of a sunspot. In the region investigated the footpoint -field is well marked in the transversal field map. The field has a predominantly transverse character and is directed parallel to the axis of the prominence. The flux bundle forming the prominence left the photosphere by an angle of about 26°, i.e. close to the horizontal. In the maps of current densities inferred from the vector magnetic field we find a pair of up- and downflowing currents, being situated symmetrically to the axis of the prominence. This indicates on a current ($\approx 3{,}8 \cdot 10^{11}$ A) flowing round the flux bundle and generating the Lorentz forces causing the concentration of flux at the footpoint region.The vertical gradients of the longitudinal field hint on an increase of the field strength with height, i.e. toward the axis of the prominence.

PHOTOSPHERIC FIELD GRADIENT IN THE NEIGHBOURHOOD OF QUIESCENT PROMINENCES *

B.S.NAGABHUSHANA AND M.H.GOKHALE
Indian Institute of Astrophysics, Bangalore 560034

ABSTRACT

We have determined statistically the horizontal gradiant of the vertical magnetic field in the neighbourhood of filaments inside and outside the active regions during a few months in 1981 and in 1984. The results show that there are meaningful upper and lower limits on the gradiant of the surrounding large scale photospheric magnetic field for the existance of a filament. These limits represent a necessary but not sufficient condition.

TABLE I. Mean values and root mean square deviations of dB_r/ds across the filaments

OUTSIDE ACTIVE REGIONS:

1981 : 4.77 ± 1.80 (10^{-5} G/km) (sample size : 96)

1984 : 2.02 ± 0.84 (10^{-5} G/km) (sample size : 293)

IN ACTIVE REGIONS:

1981 : 11.89 ± 4.30 (10^{-5} G/km) (sample size : 59)

TABLE II

Mean values and root mean square deviations of dB_r/ds across neutral lines without filaments were found to be as given below

OUTSIDE ACTIVE REGIONS:

1981 : 4.93 ± 2.52 (10^{-5} G/km) (sample size : 68)

1984 : 1.92 ± 1.13 (10^{-5} G/km) (sample size : 282)

IN ACTIVE REGIONS:

1981 : 13.98 ± 7.10 (10^{-5} G/km) (sample size : 111)

---

*Detailed version to be published in Hvar Obs. Bull.

EVOLUTION OF FINE STRUCTURES IN A FILAMENT

B. Schmieder*, P.Mein *

* Observatoire de Paris, Section de Meudon, Dasop, F 92195 Meudon Principal Cedex, France.

ABSTRACT

A quiescent filament observed in June 1986 underwent a slow Disparition Brusque which lasted 4 days. Here, we focus our study on the dynamical behaviour of the fine structures (Full-Width Half-Max $\sim$ 350 km) in this filament which were observed at Pic du Midi with the Multi-Channel Subtractive Double Pass (MSDP) spectrograph during a period of 30 minutes. We observed no changes in intensity during this period, but we did observe changes in the velocity field with no correlation from one minute to the next. High velocities were detected at the footpoints where the filament is anchored in the photosphere , of the same order than those observed at the boundaries of the supergranules (between ± 10 km s^{-1}). To explain these observations we suggest a spicule-like model which supplies material to the prominence.

I_C

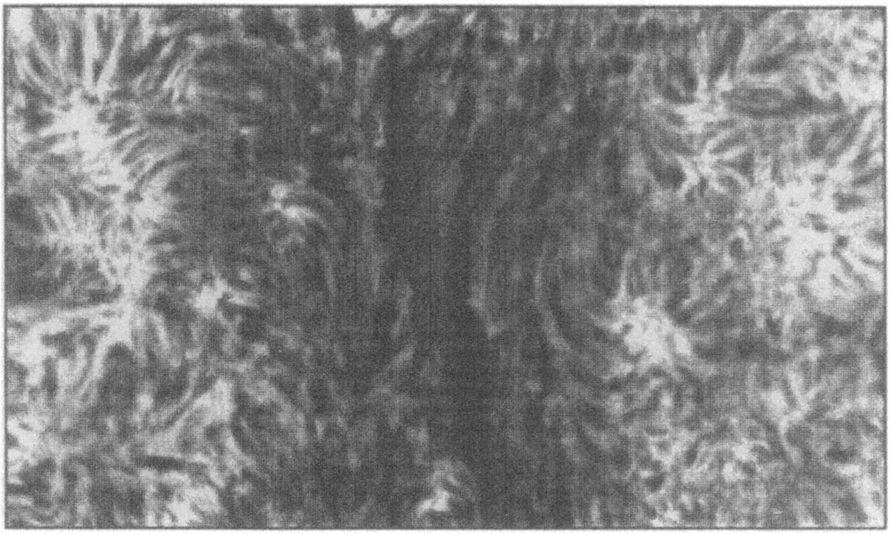

(a)

MSDP observations of a filament at Pic du Midi (209"x122"- 1"= 6 pixels).

MAGNETIC 'POLE-ANTIPOLE' CONFIGURATION AS AN ALTERNATIVE MODEL FOR SOLAR PLASMA LOOPS AND SUNSPOTS' NATURE

Metod Saniga
Astronomical Institute, Slovak Academy of Sciences
059 60 Tatranská Lomnica, Czechoslovakia

EXTENDED ABSTRACT: We propose here an alternative model of a solar plasma loop where the configuration of a loop matches that of the lines of force of a magnetic monopole-antimonopole system instead of a fictitious magnetic dipole. It is shown that although identical in large distances these two configurations crucially differ in a short distance from the place of location. Some light is shed on a close connection between loop prominences and sunspots; a conclusion is arrived at that sunspots could be viewed as 3-dimensional topological solitons, i. e., as regions in the Sun's atmosphere where the magnetic charge is smoothed out and where, hence, the classical Maxwell theory is rather unapplicable.

Spherical symmetry and double size-scale of sunspots naturally occurs in this model. It should also be stressed that since the Higgs fields and gauge fields fall exponentially to their vacuum values there must be a sharp boundary between the photosphere and the penumbra as well as between the penumbra and the umbra.

Finally, since the magnetic charge is of a topological origin it is conserved independently of dynamics; so in our model a sunspot does not need any other forces to be present to keep it compact-hence also a relative "long-livedness" of sunspots.

ON THE SPATIAL DISTRIBUTION OF PROMINENCE THREADS

J. B. Zirker, National Solar Observatory, Sunspot, NM, USA
S. Koutchmy, Institut d'Astrophysique, CNRS, Paris,France

Abstract of Oral Contribution

A quiescent prominence with pronounced vertical fine structure was observed on November 24, 1988 with the Echelle spectrograph and universal filter at the Vacuum Tower Telescope at Sunspot. The observations were obtained under good seeing conditions; the spatial resolution of the Hα spectra was 1.6 arcseconds.

A model was proposed to interpret the observed contrast of threads in the Hα spectra. The model assumes that each thread is composed of a large number of sub–arcsecond elements, each optically thin, which are randomly distributed in space. Simulations were compared with the observations, and an average spatial density of 2 elements/arcseconds^2 in the prominence cross–section was deduced. A typical observable fine–structure in the spectrum is found to consist of a cluster of 7 to 20 sub–arcsecond elements.

MICRO- AND MACROINHOMOGENEITIES OF DENSITY IN A QUIESCENT PROMINENCE

L.N.Kurochka[1], A.I.Kiryukhina[2]

1. Astronomical Observatory of the Kiev University, Kiev, USSR
2. Sternberg Astronomical Institute, 119899 Moscow, USSR

UDC 523.987.2-355.7
conference paper

Abstract. The analysis of a bright prominence of August 13, 1972 was carried out. Continuous spectrum due to Thomson scattering of solar radiation on free electrons and the Balmer precontinuum stipulated by free bounded transitions in the volumes with different electron concentration n_e (Kurochka, Kiryukhina, 1989) were observed in the spectrum of this prominence. Heir intensities can be explained only by assuming that there are volumes with different values of electron concentration ($10^{10} \leqslant n_e \leqslant 10^{12}$ cm^{-3}) and extension in the prominence ($10^5 \leqslant \ell \leqslant 10^9$).

In addition, there are extended volumes (with ℓ of the order of the active region size and even larger) with relatively low temperature ($T_e < 10^4$ K) and electron concentration intermediate between the concentration in the prominence and that of the corona ($10^9 \leqslant n_e \leqslant 10^{10}$ cm^{-3}).

The problem of existence of micro- and macroinhomogeneities of density in prominence and in the volumes around the prominence originated at bringing to the agrement the observed intensity of the continuous spectrum of the prominence $\mathcal{J}_{sc}$ in the region λ 3700 Å with the intensity of Balmer continuum $\mathcal{J}_{c,2}$ which increased from λ 3685 Å up to λ 3650 Å in the prominence studied.

The spectrum of the prominence was obtained on August 13, 1972 ($\lambda = +90°, \varphi = +12°$) under highland conditions (altitude 3000 m) and at high transparency of the earth atmosphere.

HIGH RESOLUTION ANALYSIS OF QUIESCENT PROMINENCES AT NSO / SACRAMENTO PEAK OBSERVATORY

Tron A. Darvann[1,2] , *Serge Koutchmy*[1,3] , *Fritz Stauffer*[1] and *Jack B. Zirker*[1]

1) NSO / Sacramento Peak Observatory, Sunspot, NM 88349, USA
2) Inst. of Theor. Astrophysics, Univ. Oslo, 0315 OSLO 3, Norway
3) Paris Inst. d'Astrophysique, CNRS, 75014, France

ABSTRACT.

We present preliminary results of several experiments carried out at the National Solar Observatory / Sacramento Peak (NSO/SP) Vacuum Tower Telescope (VTT) with the aim to resolve velocities and magnetic fields of the fine scale structures of filaments and prominences.

1. Velocity field.

Filtergrams recorded at different positions along the Hα and Hβ line profiles permit the deduction of the small scale motions of prominence features. In Hα, the Universal Birefringent Filter (UBF) with a 220mA passband gives images for analysis of the proper motions at 400mA ($\simeq$20 km/s) and up to 700mA ($\simeq$35 km/s) Doppler velocities. Images are digitized with a video-CCD camera and processed with a VICOM image processing system. Large velocities ($\pm$30 km/s) are observed in extended parts of the prominence, especially in the faintest parts, including the edges. Hα spectra with a 0.5 arcsec slitwidth and a 7mm/A dispersion were obtained at video rate using an image tube attached to the Echelle Spectrograph of the VTT. Thread-like features are observed at sub-arcsec scales, showing large Doppler shifts; up to 3 structures with different velocities have been observed at the same location of the slit. Typical FWHM is not less than 400mA and a large dispersion is observed on central brightnesses of the line profiles. Features are mostly unresolved and much overlapping is clearly present.

2. Magnetic field.

Attempts to observe the magnetic field underlying filaments have been made on the disc with a new video-CCD longitudinal-field magnetograph working with the UBF. Photospheric magnetograms were first obtained with high resolution (integration time 1 sec) in the wing of the CaI 6103A line around plage filaments. Further, chromospheric magnetograms in the wing of Hα have been obtained for the first time with encouraging results. When a sufficient integration time is used ($\simeq$20sec) at + or - 0.2 A from the line center, a polarization signal well above the noise is observed at the location of the filament, giving a first evidence of a uni-polar vertical field at the level of the filament channel, this signal is completely absent on the photospheric magnetogram.

All observations have been made with the NSO/SP facilities which are operated by the Association of Universities for Research in Astronomy, Inc., under contract with the National Science Foundation. We would like to thank the staff of the VTT for their skillful work and continuous attention to our program.

A MOVIE OF SMALL-SCALE DOPPLER VELOCITIES IN A QUIESCENT PROMINENCE

Serge Koutchmy[1,2], *Jack Zirker*[2], *Lou B. Gilliam*[2], *Roy Coulter*[2], *Stephen Hegwer*[2], *Richard Mann*[2] *and Fritz Stauffer*[2]

[1] Institut d'Astrophysique, CNRS, F-75014 Paris, France
[2] NSO – Sac Peak, Sunspot NM 88349, USA

Abstract. A movie made of selected $H\alpha$ off-band images of a typical Quiescent Prominences has been produced with the optical printer of NSO–S.P. High speed pictures were obtained with the UBF of the VTT on June 21, 1987 during 30 min of very good seeing, at a 20 sec cycling rate. Blue and red wings images are made at plus and minus .040 nm from line center, with a .022 nm passband. Original Pictures were enlarged to give an effective field of view of 100 x 80 arcsec^2. Negative to positive superposition allows the mapping of strongly Doppler–shifted features ($\pm$ 20 km sec^{-1}) on a grey scale. The prominence threads are mostly discrete, allowing accurate measurement of proper motions. Typical transverse velocities of proper motions of small knots moving vertically downward are about 10 km sec^{-1}. The movie also demonstrates the turbulent behaviour of the prominence plasma. Large–scale motions at lower transverse velocities are also clearly present.

$V_{\parallel}$

Mapping of strongly Doppler-shifted features ($\pm$ 32 km.sec^{-1} typical amplitude) on a gray scale of the lower part of a quiescent prominence observed on Aug.3, 1989 (14^H45 UT) at the SE-limb. Note the spatially uncorrelated distribution of the line of sight velocities of large amplitude.

Fibril structure of solar prominences

J.L.Ballester. Departament de Fisica.Universitat de les Illes Balears.Spain.
E.R.Priest. Mathematical Sciences Department.University of St Andrews.Scotland

Limb observations of quiescent prominences have revealed it to be composed of many fine structures.Different observations in Hα and UV lines also suggest that quiescent filaments are made up of many clusters of small scale loops inclined to the filament axis ones.This suggests that quiescent filaments are composed of fine structures at different temperatures.In Hα ,the dimensions of these structures are about 7000 Km long and 1000 Km thick,evolving over a typical time scale of about 8 min.Active-region prominences have been modelled as cool loops along prominence.However,other observations suggest that they could be interpreted in term of loops of plasma inclined to the filament.Taking into account the observational background, our aim has been to construct a model for the fibril structure of solar prominences in terms of slender magnetic flux tubes. We consider the fibril structure of the prominence as composed of flux tubes containing hot plasma ($T_i \sim T_c$) over most of their lengths and cool parts ($T_i \ll T_c$) near their summit representing the cool region of the prominence.We start with a hot flux tube and assume that a cool condensation appears near the top of the flux tube.This produces a downward anti-buoyancy force that must be balanced by other forces acting on the flux tube and the general shape of the structure therefore be as shown in Figure 1.We solve the equations for the hot and the cool part of the flux tube,matching the solutions at the connection point.We have used different values of internal and external densities,internal and external temperatures of hot flux tube and corona,and looked for the values of the depression, the width and the mass in the cool region.On the other hand ,in the case with external field,we have performed the calculations for values of external field typical of quiet and active regions.
The most realistic results are obtained by including the effect of an external magnetic field in the corona,then,we are able to reproduce realistic values for the width of the cool region and the mass contained in quiescent and active region prominences.

Ballester,J.L.,Priest,E.R.:1989,Astronomy&Astrophysics (in press)

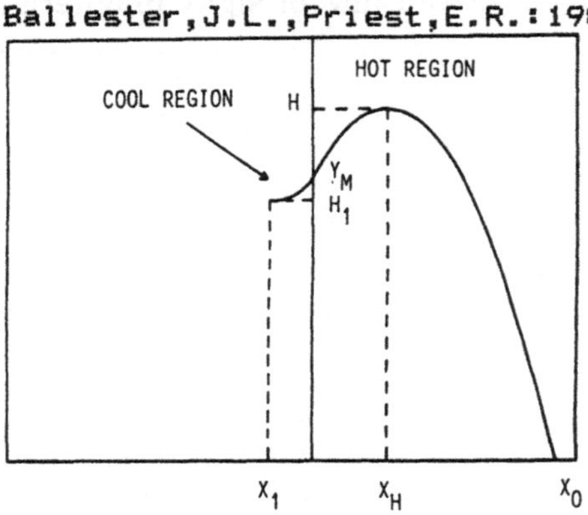

FIGURE 1

ESTIMATION OF THE LINE OF SIGHT AMPLITUDE OF THE MAGNETIC FIELD ON THREADS OF AN ACTIVE REGION PROMINENCE

S. Koutchmy[1,2]
J. B. Zirker[2]
[1]Paris Institut d'Astrophysique, CNRS, Paris, France
[2]National Solar Observatory, Sacramento Peak, Sunspot, NM, USA

Active Prominence magnetic fields :

A preliminary analysis of an active prominence observed at the E-limb in October 2, 1988, has shown high amplitudes (up to 1000 gauss) of the magnetic field in low lying thin horizontal threads extending between two parts of the prominence. The line of sight amplitude of the magnetic field is determined from simultaneous pictures obtained in circularly polarized light in the wing (plus 0.2 or minus 0.2 Å) of Hα line, using the UBF and a Wollaston prism at the VTT of SPO. The noise level at a typical 0.6 arcsec spatial resolution is $\pm$ 250 gauss for each couple of frames. Integration in space (along the threads, for ex.) and in time (use of successive frames) has been also used.

Large fields can be seen only in the finest parts of an active prominence (see the figure), at a typical distance from the chromospheric limb of 5 to 20 arcsec. Outside the field is always smaller than 150–200 gauss, although the region is typically situated between two sunspot groups. Note, also, the absence of large signal in the chromospheric fringe. The picture in negative (lower part) displays the distribution of the line of sight magnetic field in gray scale.

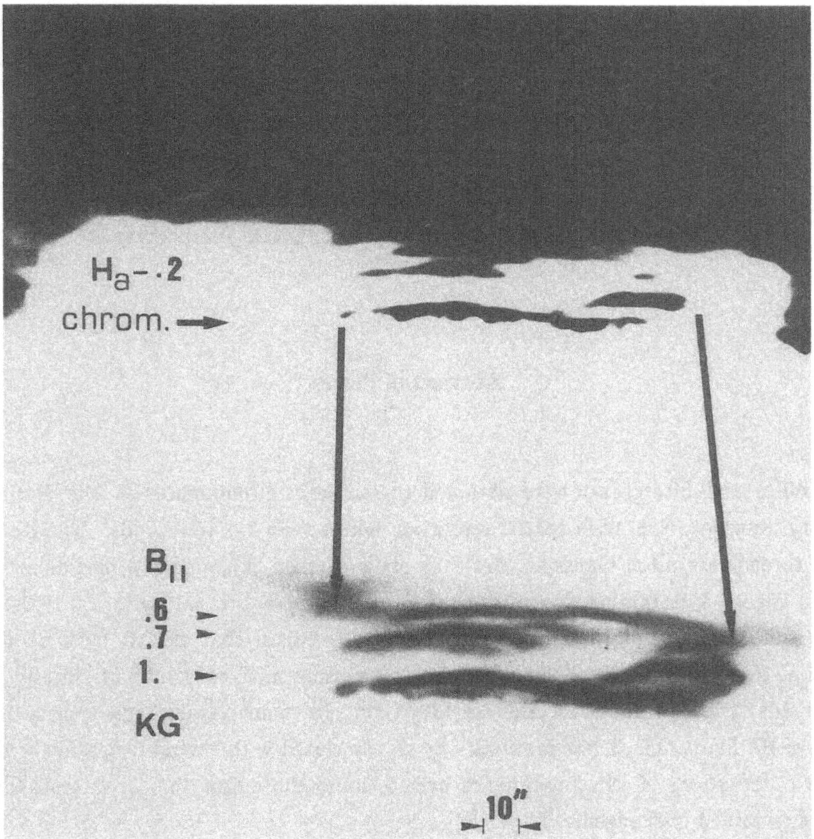

Figure 1: Top and middle: Hα intensity image; Bottom: Line of sight magnetic field in threads.

HIGH RESOLUTION OBSERVATIONS OF MOTIONS AND STRUCTURE OF PROMINENCE THREADS

J. B. Zirker, National Solar Observatory, Sunspot, NM, USA
S. Koutchmy, Institut d'Astrophysique, CNRS, Paris, France

Abstract of Poster

Hα profiles and filtergrams were obtained of quiescent prominences at the National Solar Observatory, Sunspot, NM, with spatial resolution better than 1.6 arcseconds. The Hα profiles of individual threads are often Gaussian, but may show marked asymmetries, particularly near the prominence edges. Filtergrams, taken at ± 0.7 Å (± 30 km/sec) in Hα (with a 0.18 Å passband) show high speed knots and threads at the prominence edges, that persist with no perceptible change for at least 10 minutes. This result implies plasma motions (along essentially horizontal magnetic fields) over distances as large as 18000 km, i.e. much larger than typical thicknesses (5000 km) of Hα filaments. These motions may be associated with thread formation or decay, but continuous observations of this prominence over 6 hours show that the *large–scale* form of the prominence remained unchanged.

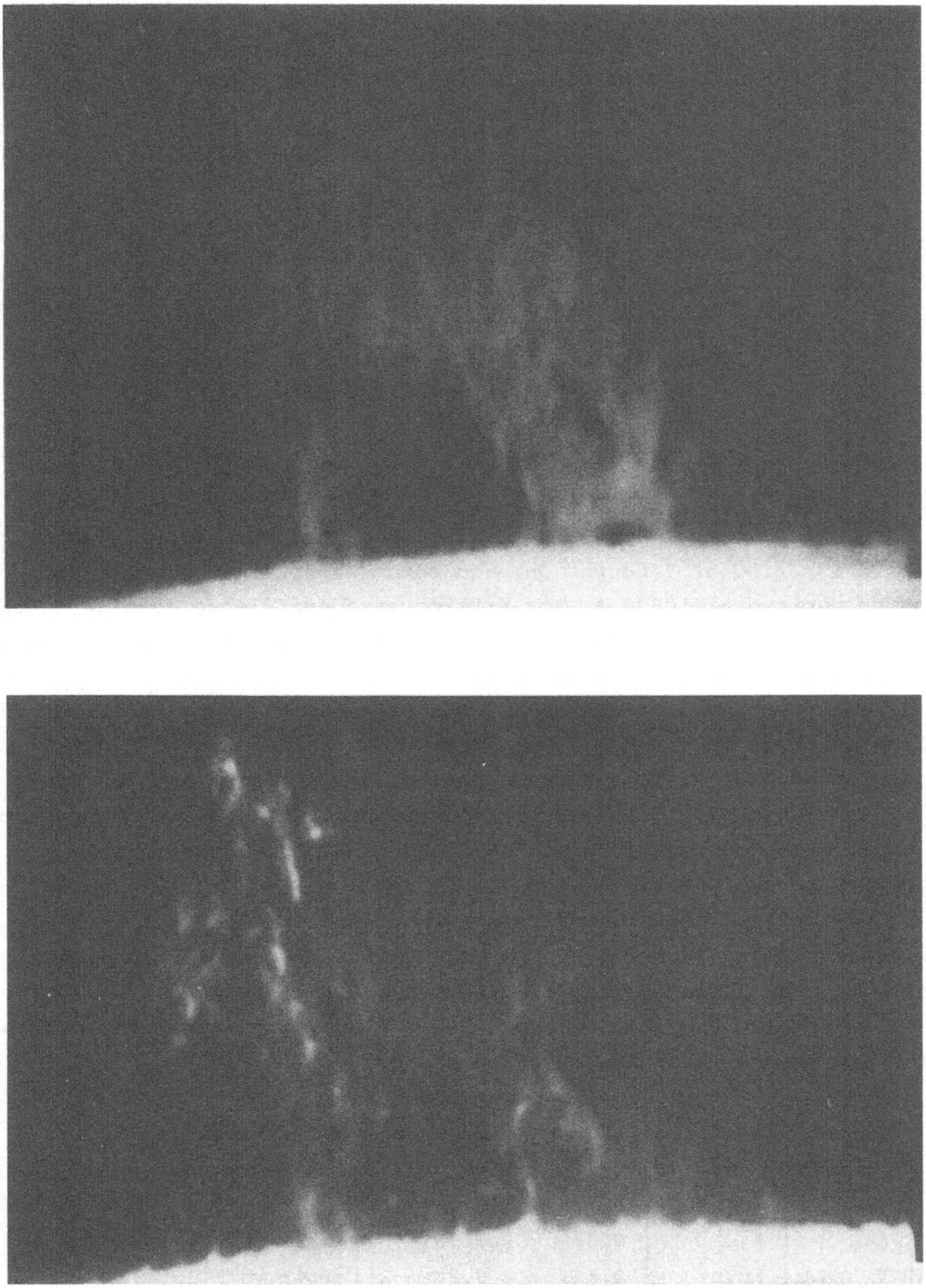

Fig. 1 Hα filtergrams of quiescent prominence of November 24, 1988. Top (Hα + 0.7 Å), middle (Hα + 0.0 Å), bottom (Hα − 0.7 Å).

OBSERVATIONAL ASPECTS OF A PROMINENCE FROM HeI 10830 DATA ANALYSIS

J.Deliyannis
Section of Astrophysics, Astronomy and Mechanics
University of Athens, 15783 Zografos, Greece

Z.Mouradian
Observatoire de Paris-Meudon, URA326
92195 Meudon Pr. CEDEX, France

The whole image of a quiescent prominence had been scanned three times in a period of 7 hours. On each point we obtained a 4 A-wide spectrum centred at 10830 A. For fitting each profile we considered the two most intense components of this He triplet line as one with intensity I_{12} at λ_{12}=10830.31 A. We noticed that:

1) There was an increase of I_{12} from the edges to the centre of the prominence, but, in general, I_{12} decreased with time.

2) The Doppler width remained constant against time and position.

3) At first the radial velocities were positives in the lower part of the prominence and negatives in the higher one, but later on all velocities became positives.

4) The optical depth was always ≤ 1 in the lower part of the prominence. In the higher part the values of τ_o decreased with time and finally, in the whole prominence, the optical depth was ≤ 1.

Below we give the mean values of I_{12}, $\Delta\lambda_D$, V and τ_o for the main part of the prominence which had a column shape.

Param.	Unit	I image	II image	III image
I_{12}	$I_{o,c}$=1.	0.131	0.104	0.085
$\Delta\lambda_D$	A	0.35	0.33	0.34
V	$Km.s^{-1}$	0.15	0.88	1.42
τ_o		0.83	0.56	0.78

($I_{o,c}$ is the intensity of the continuum for λ=10830 A at the centre of the solar disc).

SOME OBSERVATIONS OF THE CORONAL ENVIRONMENT OF PROMINENCES

Jacques-Clair Noëns (1), Zadig Mouradian (2)

(1): Observatoire du Pic-du-Midi, LA 285 du CNRS, 65200 Bagnères-de-Bigorre, France
(2): Observatoire de Paris-Meudon, DASOP, UA 326, 92195 Meudon, France

As it as been shown in eclipse photographs, quiescent prominences are frequently located at the bottom of streamers and are surrounded by arch systems with a dark cavity. It may be assumed that there is a mass depletion in the cavity because the mass was condensed to form the prominence.

At the present time we have no a real proof of the existence of a cavity around quiescent prominences. This problem is a fundamental one concerning the formation and evolution of prominences. Our purpose is to study the relations between the prominences and the coronal material surrounding it. In this paper we give some preliminary observational results. The multichannel 20 cm. spectro-coronagraph at the PIC-DU-MIDI is used to scan the coronal environment of prominences. Simultaneous measurements of the total flux in the Fe XIII, 10747 and 10798 A coronal emission lines, the He I, 10830 A cool emission line and the continuum at 10700 A are obtained.

The distribution of the coronal Fe XIII abundance can be found from the data by the intensities of the iron lines, and the electron density at Te=1.8 million degres in the observed regions by the ratio of the intensity of the two Fe XIII lines.

The general conclusion of this study is that the coronal cavities above a prominence do not always exist, or if they do they form a complex system with time variation.

CORONA-PROMINENCE INTERFACE AS SEEN IN H-ALPHA

Vojtech Rušin, Milan Rybanský
Astronomical Institute, Slovak Academy of Sciences,
059 60 Tatranská Lomnica, Czechoslovakia

Vlado Dermendjiev, Georgi Buyukliev
Department of Astronomy and National Observatory
Lenin Bld 72, 1174 Sofia, Bulgaria

It is generally assumed that the mass balance between corona and prominences will be exist, but unambiguously observational proofs for a such exchange of material are rare. Especially, there is a little data that can be used to adress the problem of flow mass from prominence to the corona. We relate prominence-corona interface as was obtained in short-term variation of special observations in H-alpha with Lomnický Štít coronograph. The two different prominence types (quiescent, and highly structured, stable, but not quiescent, like as a surge) were performed with a 20-cm lens coronograph over 24, respectively 17 minutes on 1988 August 28 and 18 (FWHM = 0.8 nm, a 1 minute record). A detailed photometry was made with Joyce-Labell microdensitometer at Rožen Observatory (a slit 40 x 40 μm, a step 20 μm).

There were no changes in the outer shape in the quiescent prominence on 1988 August 28 over 24 minutes (isolines, calibrated to the Sun's disk were used).

A remarkable, but exceedingly faint changes should be seen in the outer shape on August 18, 1988 prominence. The prominence mass at the top of two curved columns continuously disappeared during 10 minutes. One may suppose that this cool prominence plasma was continuously heated to the coronal temperature via some mechanism of heating (fast-mode waves, current dissipation etc.). The total disappeared prominence mass was estimated of 8×10^{11}g. It is necessary to stress that prominence-corona interface outside of disappeared area, including the first one, did not change over these sequences. It seems that surge-like type of prominence could be a next candidate for a transport of mass from dense underlying layers to the corona (isolated knots od disparition brusque or slowly ascending material connected with flares are the first ones). Prominence areas change over cycle in the same course as the total brightness of the white-light corona (both display their mass) i.e. they are of two times higher (minimally) in the maximum as in the minimum of cycle.

DIAGNOSTIC STUDY OF PROMINENCE-CORONA INTERFACE[+]

P.K.Raju and B.N.Dwivedi[**]

[*]Indian Institute of Astrophysics
Bangalore, India

[**]Department of Applied Physics
Institute of Technology
Banaras Hindu University
Varanasi 221005, India

Theoretical EUV line intensity ratios from Ne V, Ne VI, Mg VI, Mg VII, and Mg VIII are useful for electron density determinations within prominence-corona interface (PCI). Skylab observations of an eruptive prominence [1] have been used to infer electron density within PCI. The physical parameters thus derived are given in Table 1. The 'a' values are from [2]. The 'b' row values and the values for Mg VI and Ne VI are from [1]. The Mg VIII values are from [3]. The new values for the pressure parameter are given in set 'B' of Table 1.

Table 1. Physical parameters for a prominence-corona interface

Ion	A:	T_e	N_e	N_eT_e	B:	T_e	N_e	N_eT_e
Ne V	a	2.5+5[x]	1.85+8	4.63+13		2.5+5	4.63+9	1.16+15
	b	2.5+5	5.44+9	1.36+15				
Mg VI Ne VI		4.0+5	1.10+9	4.40+ 14		4.0+5	1.50+9	6.00+14
Mg VII	a	5.0+5	3.69+9	1.85+15		5.0+5	1.00+9	5.00+14
	b	5.0+5	3.41+10	1.71+16				
Mg VIII		8.0+5	5.80+8	4.64+14		8.0+5	5.80+8	4.64+14

[x]2.5+5 means 2.5 x 10^5

It would be necessary to obtain accurate line intensities for many more lines in order to model the P-C interface.

References

1. K.G.Widing, U.Feldman and A.K.Bhatia: (1986) Astrophys. J. 308, 982.
2. P.K.Raju and B.N.Dwivedi: (1979), Pramana, 13, 319.
3. B.N.Dwivedi: (1988), Solar Phys., 116, 405.

[+]Detailed version to be published in Hvar Obs. Bull.

The Prominence-Corona Transition Region Analyzed from SL-2 HRTS

O. Engvold, V. Hansteen, O. Kjeldseth-Moe
Institute of Theoretical Astrophysics
University of Oslo
Norway

G. E. Brueckner
US Naval Research Laboratory
Washington DC
USA

ABSTRACT

The ultraviolet spectrum of a large prominence has been observed with the *High Resolution Telescope and Spectrograph (HRTS)* on *Spacelab 2* August 5, 1985. The spectrum covers the wavelength range λλ1335-1670Å and shows numerous emission lines from gas at chromospheric and transition region temperatures. A spectral atlas of these data is available.

The data reveals a variation with height of the line intensities. The prominence becomes "hotter" with height. A value of ~0.12 dyn cm^{-2} for the gas pressure in the prominence-corona transition region is obtained from line ratios. The resolved fine structure of the He II λ1640.400Å line indicates that a major part of this emission comes from "cold" gas. A broad Fe XI λ1467.080Å suggests high velocities in the coronal cavity region. The Fe XI line in the cavity region is a factor ~5 less bright in the normal corona at the same height. Assuming that the temperature is the same in the two regions the present obervations suggest that the pressure in the cavity region is lower by a similar factor.

RADIO EMISSION FROM QUIESCENT FILAMENTS

Kenneth R. Lang

Department of Physics, Tufts University
Medford, MA 02155 U.S.A.

ABSTRACT.

Full-disk VLA synthesis maps of the quiet Sun indicate that filaments can be seen in emission at 91.6-cm wavelength; they are detected in absorption at shorter microwave wavelengths. The 91.6-cm emission has a brightness temperature of $T_B = 3 \times 10^5$ K. It is hotter, wider and longer than the underlying filament detected at Hα wavelengths, but the similarity between the shape, position, elongation and orientation of the radio and optical features suggests their close association. The 91.6-cm emission is attributed to the thermal bremsstrahlung of a hot transition sheath that envelopes the Hα filament and acts as an interface between the cool, dense Hα filament and the hotter, rarefied corona. The transition sheath is seen in emission because of the lower optical depth of the corona at 90-cm wavelength, and the width of this sheet is 10^9 cm. A power law gradient in pressure provides a better match to the observations than a constant pressure model; definitive tests of theoretical models await simultaneous multi-wavelength studies of filaments at different observing angles. When the thermal bremsstrahlung is optically thin, the magnetic field strength in the transition sheath can be inferred from the observed circular polarization. Variable physical parameters of the sheath, such as width, electron density, and electron temperature, can explain controversial reports of the detection of, or the failure to detect, the meter-wavelength counterpart of Hα filaments.

DYNAMICAL STRUCTURE OF A QUIESCENT PROMINENCE

P. Mein*, N. Mein*, B. Schmieder*, J.C. Noëns**

* Observatoire de Paris Section de Meudon, DASOP (URA 326) F 92195 Meudon P[al].

* Observatoire Midi-Pyrénées F 65200 Bagnères de Bigorre

Abstract.

A statistic analysis of H_α profiles in a quiescent prominence is consistent with the superposition of individual velocity structures (typically 10), with standard deviation $\sim$ 15 km s^{-1}.

A quiet prominence was observed simultaneously by the M.S.D.P. spectrograph of the turret dome and the 20 cm coronagraph at the Pic du Midi, on June 7,1988. No coronal cavity can be detected in the vicinity of the prominence. The M.S.D.P. H_α profiles are approximated by gaussian functions in each point of the 2D-field:

$$I(\lambda) = I_0.exp - [\frac{\lambda-\lambda_0-V.\lambda_0/c}{W}]^2$$

In the optically-thin range, we assume that each profile is due to the superposition of N profiles emitted by individual velocity cells (or threads). For a given I_0-value, the scattering of V (Doppler velocity) and W (line width) can be related to a stochastic distribution of velocities. The figure shows a comparison of the observations with various models corresponding to different individual opacities. The best fit should be intermediate between the two broken lines (T=5000 K)

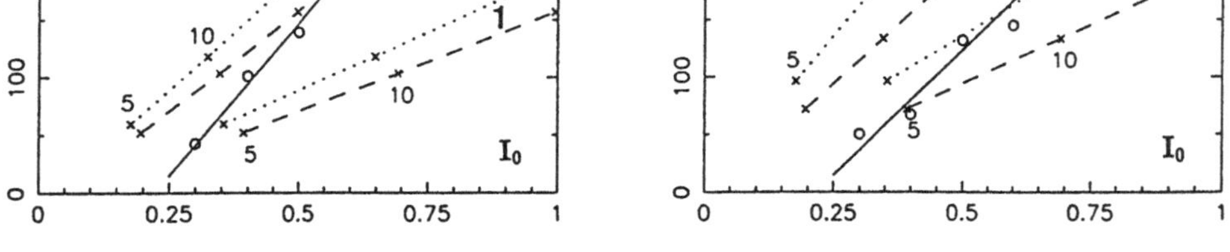

Standard deviations of V and W.
$< (V - \overline{V})^2 >^{-1}$ and $< (W - \overline{W})^2 >^{-1}$ are plotted versus I_0. The observations fit approximately a straight line(——).Two sets of calculations are made with two values of cell opacities. The assumed temperatures are 5000 K (- - - -) and 8000 K (· · · · ·).

DYNAMICS OF SOLAR PROMINENCE ON DECEMBER 7, 1978

Vojtech Rušin
Astronomical Institute, Slovak Academy of Sciences,
059 60 Tatranská Lomnica, Czechoslovakia

ABSTRACT. The 1978 December 7 eruptive prominence with a helical structure is analysed. Attention is given to the distribution of magnetic helicity, a measure of helical structure, and the morphology and evolution of this event. The prominence was an edge-on case. The southern leg of the prominence consisted of two (four) helical threads which suggested a tube. A disruption probably occurred in the upper part of this leg. The threads in the northern leg were nearly radially oriented to the solar surface, and consisted from several subfilaments, tangled in a small-scale. The central region unwound after disruption, similarly as at the top (these threads were during next evolution connected with the northern leg), showing a much the helicity, both, large- and small-scale. Moreover, there tangled kinks probably displayed in some parts of the prominence. Observed height: $2 - 7 \times 10^5$ km, the projected speed of the prominence head of 180 km s^{-1}, estimated of magnetic field strength of 1×10^{-4} T.

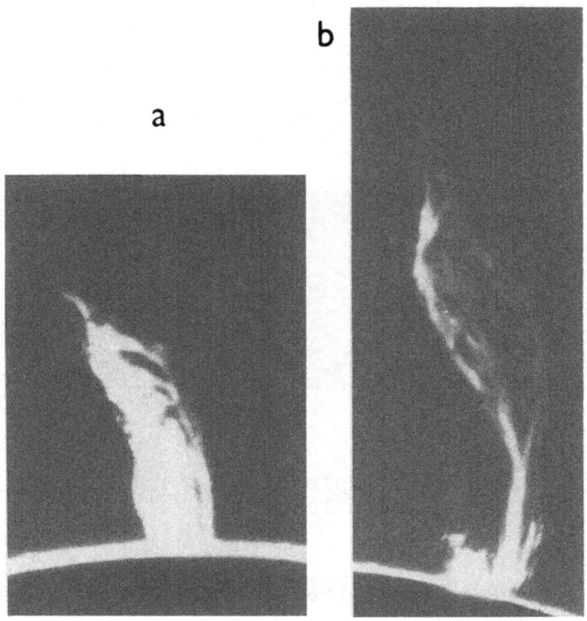

FIGURE 1

A sample of the prominence at (a) 08:17 UT (the first its observation) at P.A. 55 - 59^0 and (b) at 08:32 UT.

POST - FLARE LOOPS ON AUGUST 15 - 16, 1989

Milan Kamenický
Observatory and Planetarium, 080 01 Prešov, Czechoslovakia

Vojtech Rušin
Astronomical Institute, Slovak Academy of Sciences,
059 60 Tatranská Lomnica, Czechoslovakia

Several flares have arised during the west limb passage of sunspot group (NOAA/USAF Region 5629: S 16, W 64 - 84) on August 15-16, 1989. Sequences of post-flare loops in H-alpha, connected with them, were taken by coronographs at Lomnický Štít coronal station and Prešov Observatory and Planetarium (see Figs. 1 and 2). We present here some preliminary results.

(1) Dynamical system of thin (2-4 arcsecs) post-flare loops (7-9) with a stable height above the solar surface (max. height: 100 000 km) was observed between 08:48-12:18 UT. They formed loop channel,slightly declined with respect to the line of sight, and prominence plasma in their individual loops was seen to 15:36 UT, when probably definitely decayed. There a flow of matter was seen in the legs. It follows that this system had its existence of about 11 hours (in individual legs may be more or less) because they arised after the flare occurence at 01:02 UT. A new system of low-lying loops was observed at 15:32-16:23 UT in the same day (max. height: 26 000 km).Development of them started from the very bright chromospheric mass ejection. One may suppose it could be connected with flares which began at 15:18 UT.

(2) New dynamical systems of post-flare loops, probably connected with flares originated at 01:00 UT at the same active region,have been observed on August 16, 1989 between 06:41-16:46 UT. Maximum height varied between 75 000-130 000 km (it growed-up slowly). Existence of lower, very bright loops lasted more as 11 hours

(3) It is remarkable that the enhancement in the 1-8 x-rays (SGD, 541), connected with these active features in the photosphere, appeared nearly with a periodicity of 24 hours over 14-17 August 1989 at the beginning between 01:00-02:00 UT.

1

2

Figs. 1 and 2: A sample of 1989 August 15 (left) and 16 (right) post-flare loops.

REFERENCES

Solar Geophys. Data,1989,No.541,Part 1,(U.S.Dept. of Comm.,Boulder,USA.

DOPPLER VELOCITY OSCILLATIONS IN QUIESCENT PROMINENCES

E.Wiehr, H.Balthasar
Universitäts-Sternwarte, D-3400 Göttingen
G.Stellmacher
Institute d'Astrophysique, F-75014 Paris

From a series of first observations using a 4 arcsec measuring aperture centred at one location within the prominence, the authors found periods near one hour (Sol.Phys.94, 285), and 5 min (Astr.Astrophys.163,343). From 'one-dimensional' observations, i.e. along an extended spectrograph slit crossing the prominence, the spatial behaviour of these periods could be investigated (Astr. Astrophys. 204, 286). Table 1 summarizes our various observations:

date	site	line	time	prom.	3	5	12	20	60 min	remark
28.09.86	TF	H_α	5.0	F	+	+	+	+	+	(e)
30.09.86	TF	H_α	7.2	E	(+)	(+)	+	+	+	(e)
3.10.86	TF	H_α	3.4	I	(+)	+	+	+	+	
5.10.86	TF	H_α	3.4	I	-	(+)	+	+	+	
10.11.87	TF	K	2.2	I	-	(+)	+	+	+	
14.11.87	TF	H/H_ε	5.4	F	?	-	+	+	+	e
10.11.87	SP	8542	1.6	I	+	+	+	+	+	
22.08.88	SP	8542	0.5	F	(+)	+	-	?	?	
23.08.88	SP	HeD_3	0.7	E	+	+	+	?	?	
24.08.88	SP	HeD_3	1.7	I	+	+	+	+	+	
17.09.88	TF	Hβ	3.4	F	-	+	+	-	+	
22.09.88	TF	Hβ	5.8	E	-	-	-	+	-	(e)
4.07.89	TF	H_α	0.7	?	-	(+)	+	(+)	(+)	2 tel.
6.11.88	SP	Hβ	1.0	F	-	+	+	+	+	UBF

TF:Tenerife, SP:Sac.Peak; E end-on, F face-on, I inclined
+ present, - absent, (+) close to noise, ? undefined, e Eigenmodes

It can be seen that the long periods near one hour occur in all prominences (in agreement with Bashkirtsev et al.,Sol.Phys.82,443, and 91, 93). The photospheric 5 min are also found in most cases, whereas the chromospheric 3 min occur only occasionally. In addition, there is a tendency for periods near 12 and 20 min, resp. (see also Tsubaki et al., PASJ 40,121). The different periods occur at different times and at different locations within the prominences (cf.Wiehr et al. in 'Seismology of the Sun and Sunlike Stars'; p.269; E.J.Rolfe ed.)

We checked whether these periods are of solar origin or may possibly be due to imperfect guiding, which moves different parts of the spatially highly structured velocity field of the prominence (Engvold, Sol.Phys.70,315, or Kubota and Uesugi,PASJ 38,903) over the slit. Observations using two telescopes simultaneously (Gregory and VTT at Tenerife) yield uncorrelated Doppler shifts except for general trends. This is due to the highly uncorrelated local seeing in both telescopes (separated by about 150m) as was clearly seen by a TV link. Hence, some of the quasi-oscillatory Doppler variations might well be due to imperfect guiding of the prominence on the spectrograph slit.

As a consequence,'two-dimensional' Doppler observations are required as were done with the universal filter at the Sacramento Peak observatory on Nov 6,1988, as well as from May 27 to June 4, 1989.

OSCILLATORY RELAXATION OF AN ERUPTIVE PROMINENCE

B. Vršnak[1], V. Ruždjak[1], R. Brajša[1], F. Zloch[2]

[1] Hvar Observatory, 58 450 Hvar, Yugoslavia

[2] Astronomical Institute, Czechoslovak Academy of Sciences, 251 65 Ondrejov Czechoslovakia

ABSTRACT

Different types of oscillatory motions were detected in the late phases of eruption of a prominence. We found oscillations of the prominence axis and diameter with periods of 4.3 and 9.1 minutes corresponding to the eigenmodes m=4 and m=8 with a damping factor $4.6\ 10^{-3}\ s^{-1}$. A period of about 4.5 minutes was found for oscillations of the pitch angle of the helically twisted filaments. The m=2 and m=3 eigenmodes could be also identified and they led to the final relaxation of the prominence axis. The observations are interpreted in analogy with damped oscillations of an elastic string. The lowest eigenmode was not excited due to $\delta > \omega$ while the m=2 and m=3 eigenmodes were highly damped. The frequency of free oscillations due to restoring forces and the decay constant were inferred using the dispersion relation for oscillation of the elastic string and the observed frequences in the m=4 and m=8 modes to $\omega = 3.1\times10^{-3-1}$, corresponding to a period of T=34 min. and $\delta = 4.6\times10^{-3} s^{-1}$.

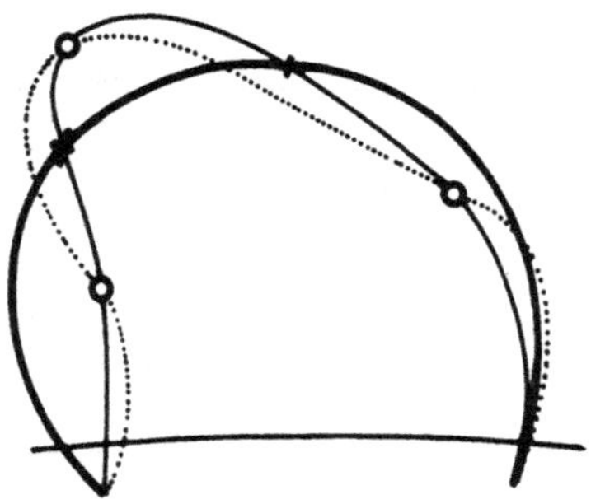

In the Figure we present the prominence axis at different moments: 07 43 35 UT (thin line) 07 50 26 (dotted line) and 08 08 30 UT (thick line). Note the knots of eigenmodes m=3 (crosses) and m=4 (circles).

The contribution in full length will be published in Hvar Obs. Bull 13 (1989).

ON OSCILLATIONS IN PROMINENCES

L. A. Gheonjian
Abastumani Astrophisycal Observatory,
Academy of Sciences of the Georgian SSR

V. Yu. Klepikov and A. I. Stepanov
Institute of Terrestrial Magnetism, Ionosphere and
Radio Wave Propogation, Academy of Sciences of the USSR

In this work the problems of the observation of longitudinal magnetic field, Doppler velocity, Hα-line intensity and half-width oscillations in the prominences with periods of up to 40 minutes are being considered. Into this class of oscillations one may put the short period resonance oscillations with periods of up to 10 minutes and also the long-period eigen-modes with the periods ranging from tens of minutes up to one hour and more. It is known that in the terrestrial atmosphere wave phenomena with similar periods are being observed which appear as the variations of transparensy and of the amplitude of refractive index fluctuations - the latter parameter determining the Solar image quality. Simultaneous observations by the Nikolsky's magnetograph and by the register of the Solar image quality have been carried out in order to determine the influence of the terrestrial atmosphere on the observations of the oscillations in prominences. The amplitude of high frequency trembling of solar disk edge was registered which appears as blurring of the image. The correlation analysis has shown that the temporal series of parameters measured by the magnetograph don't correlate The line profiles are reliably correlating with the amplitude of the high frequency image trembling. The magnetic field correlates with the derivative of this amplitude and the variation of the Doppler velocity slightly correlates both with the amplitude and with its derivative. This means that the variations of the Doppler velocity in time appears to be real. The variations of all other parameters most likely are connected with the changing observation conditions. Maybe the waves in the terrestrial atmosphere are accompained not only by the transparency and the amplitude of the refractive index fluctuations' variations, but also by the variations of the polarizing (or depolarizing) properties of the terrestrial atmosphere. The preliminary results of the polarimetrical observations of the Solar aureole undertaken by us with ad hoc constructed device are pointing at that.

MATTER FLOW VELOCITIES IN ACTIVE REGION EMISSION LOOP OBSERVED IN H- ALPHA

Delone A., Makarova E., Porfiréva G., Roschina E., Yukunina G.
Sternberg Astronomical Institute, 119899 Moscow, USSR

At 11:03 - 11:53 UT on 9 July 1982, a system of emission loops was observed in a long-lived flare active region (12N, 313L) with a complex morphology. Observations were fulfilled on the H_α Opton filter at the wavelength set: H_α, $H_\alpha \pm 0.25$, $H_\alpha \pm 0.62$, $H_\alpha \pm 1.0$, $H_\alpha \pm 1.5$ Å.

The loop chosen for investigations was associated with the flare 2B, max 11-31 UT. The loop footpoints ended in emission kernels on the solar disc and its main part was visible above the limb. The mathematical method for reconstructing the true solar loop geometry, proposed by Loughhead R. E. et al. (Solar Phys., 1984, 92, 53), has been used. Our own calculation algorithm is briefly described.

In its own plane the loop proved to be symmetrical with respect to the axis inclined at the angle 88°.2 to the line joining its footpoints P_1 and P_2. The loop plane has been found to be only slightly inclined to the solar surface vertical at the angle 5°.4. The azimuthal angle between P_1P_2 and the tangent to the heliocentric latitude circle is 34°.7. The loop base and height have been evaluated to be about 65900 and 53700 km respectively.

Matter flow velocities along the loop length have been deduced from the observed line-of-sight velocities and maximum intensity contrastes relatively to the loop neighbourhood measured at a number of points. Loop matter proved to rise along one leg from the solar surface with the velocities 145-35 km/s and to descend along the other one with the velocities 14-65 km/s, the least ones being near the loop top. Mean flow velocities turned out to be greater than the sound one. Mean accelerations or decelerations were smaller than the gravitational one. A possible mechanisam of such a matter motion is briefly discussed.

The loop shape has been shown to have a close correspondence to the magnetic line of a point dipole, i. e., in the first approximation, the loop field can be considered as a potential one.

QUANTITATIVE RESEARCH ON THE VELOCITY FIELD OF A LOOP PROMINENCE SYSTEM

Gu Xiao-ma, Lin Jun and Li Qiu-sha
Yunnan Observatory, Kunming, China

ABSTRACT

The falling motions of the matter within the loop system are studied under the united actions of solar gravity, magnetic stress of dipole and gradient force of atmospheric pressure and the two-dimensional velocity field of the loop system is calculated by use of numerical method in the conditions of isotherm and quasi-closed loop system. The results calculated theoretically are in a good agreement with that of observations made by Gu, et al. (1988). The calculations indicate that the density and magnetic field in the loop system have big influence on the falling motion of matter, but the influence of temperature on the falling motion is relative small.

ON THE PROBABLE DOUBLE-LOOP STRUCTURE OF THE FLARE-LIKE DISC OBJECT.

T.P.Nikiforova, A.M.Sobolev
Astronomy department,Urals State University,Lenin str.51,Sverdlovsk 620083,USSR

Spectral observations of faint emission disc object (may be faint flare) were carried out in H_δ , H_ε and CaII H and K. Variations of H and K CaII profiles of the object along spectrographic slit were analysed. It is shown that double-loop flare model with opposite directions of plasma movements along the loops and the point of contact (and more intensive energy release) in their tops may qualitatively explain some observed structural features of H and K lines : a) increase of wavelength separation between I_{2v} and I_{2r} peaks with distance from the site of brightest profile; b) approximately linear dependence of the emission peaks radial velocities on the distance across dispersion; c) opposite signs of asymmetry $A=I_{2v}/I_{2r}-1$ on the different sides of the brightest profile site.

On the basis of this model estimations of the lower limits of the loops' radii $R_1>7800$ km, $R_2>8700$ km and of velocities along the loops $v_1>13.6$ km/s, $v_2>15.3$ km/s are obtained. The angle between the loops' planes is less than $1^o32'$.

This explanation is not unique and one can use the approach from the point of view of source function variations in the presence of velocity gradients or of the other geometries.

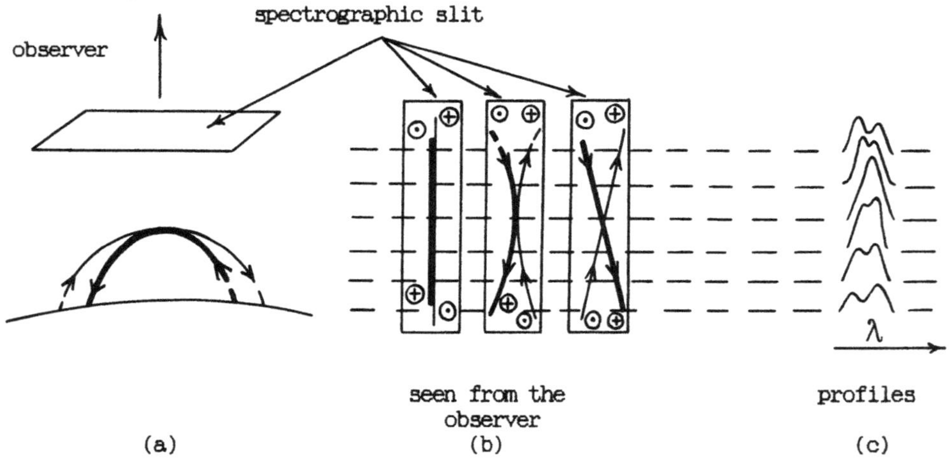

Fig.1 a) Double-loop geometrical model of the flare-like object on June 24, 1971; b) 3 possible orientations of the loops; ⊙ and ⊕ denote movement of plasma toward us and away; c) scheme of the observed photometrical scans.

AN ANALYSIS OF SURGES TRIGGERED BY A SMALL FLARE

Ágnes Kovács and L. Dezső
Heliophysical Observatory, Hungarian Academy of Sciences
H-4010 Debrecen, Hungary

(Extended Abstract)*

On 22 October 1980, near the solar central meridian, in the western vicinity of the large spot group, a subflare of the two-ribbon type was observed. Three surges were associated with this flare. Their starting points were situated close to the principal flare patches on both sides of a short filament that was visible for only a few hours. True flow velocities and decelerations along the arch-shaped surge trajectories have been determined for two of the surges.

The event was studied from a series of Hα on- and off-band filtergrams taken through the Debrecen coronagraph of 53 cm aperture.

The following principal results are concluded:

- The highest velocities of surges are measured at their visible onset in agreement with Webb *et al.*(1980); this causes us to deduce that the plasma streams of surges are accelerated below the visible chromospheric layers.
- It is demonstrated that these surges which traverse complete arches have velocity curves similar to surges that do not reach the apex of an arch as shown in Webb *et al.* (1980).
- Even a small subflare can reveal the characteristic signs of a two-ribbon flare (if the circumstances of observation are fairly favourable).

Reference:

Webb,D.F., Cheng,C.-C., Dulk,G.A., Edberg,S.J., Martin,S.F., McKenna Lawlor,S., McLean,D.E.: 1980, in P.A.Sturrock (ed.), *Solar Flares* (Skylab Workshop II),Colorado Assoc.Univ.Press,Boulder,p.471.

* The full paper is submitted to the journal *Solar Physics*.

Vertical Flows in a Quiescent Filament

by

You Jian-qi
Purple Mountain Observatory
Nanjing
China

Oddbjørn Engvold
Institute of Theoretical Astrophysics
University of Oslo
Norway

ABSTRACT

High resolution spectra of the $H\alpha$, and Ca II H and K lines were observed simultaneously for a quiescent filament on the solar disk on 27 July, 1974. The narrow absorption line profiles of the disk filament show asymmetries which give rise to differences in flow velocities derived from measurements of line center positions. The bi-sector at a higher intensity level of the three lines gives consistent values for vertical flow velocities. The velcities range from -1.7 to +2.7 $km\ s^{-1}$, and mean value 0.5 $km\ s^{-1}$, for $H\alpha$, and -1.9 to 2.5 $km\ s^{-1}$, and mean value 0.3 $km\ s^{-1}$, for the Ca II H and K lines.

Distribution of Velocities in the Pre-Eruptive Phase of a Quiescent Prominence

by

Oddbjørn Engvold, Eberhart Jensen
Yi Zhang and Nils Brynildsen
Institute of Theoretical Astrophysics
University of Oslo
Norway

ABSTRACT

High resolution Ca II K line spectra of a large quiescent prominence were obtained about one hour before a prominence eruption May 2nd 1974. The observations were made with the main spectrograph of the vacuum tower of Sacramento Peak Observatory. The observed velocities in a wide range from -50 to +25 km/s suggest that the very initial stages of destabilization of the prominence were in fact recorded.

The distribution of line intensities and line widths versus line shift suggest that one observes ≥ 3 "threads" in the line-of-sight for low velocities. For larger shifts one is evidently able to observe individual threads.

A high-velocity "tail" in the velocity distribution indicating a substantial deviation from a Maxwellian distribution may be attributed to energy and mass being fed into the prominence plasma. Alternatively, it could be an effect of local contraction (*pinching*) resulting in a temperature increase.

MASS MOTIONS IN A QUIESCENT PROMINENCE AND AN ACTIVE ONE

Vazha I.Kulidzanishvili
Abastumani Astrophysical Observatory,Georgian SSR
Academy of Sciences,383762 Abastumani,Mt.Kanobili,
Georgia,USSR

Abstract

1.Simultaneous spectral and light filter observations of a quiescent prominence were made with the large coronagraph of Abastumani Astrophysical Observatory equipped with a special attachment.

About 500 H_α line profiles,belonging to 70 different details of the prominence were drawn.

The histogram of radial velocities has a considerable asymmetry, its maximum corresponding to the velocity of $\sim$10 km/sec.This indicates the existence of an oriented flow of matter having a velocity component along the line of sight,without totally changing the structure of the formation.That is,the prominence does not seem to be stationary,but it represents a formation in a relative dynamic equilibrium.

2.Motions of knots in an active prominence are investigated by the analysis of the H_α line filtrograms obtained at Sacramento Peak Observatory.

It is found that in the upper part of the trajectory,i.e.at velocities less than that of the sound,separate fragments of the prominence practically move when affected by the gravity,while gas-dynamic forces turn out to be a principal reason of decelerating of the knots moving downward.

An indirect conclusion is drawn on a short-lived action of the mechanism directly resulting in eruption of the prominence.

RADIAL VELOCITIES OF ACTIVE AND QUIESCENT PROMINENCES

A. I. Kiryukhina
Sternberg State Astronomical Institute, Moscow, USSR

Results of determination of radial velocities ξ_r of some active and quiescent prominences with bright metalic lines are presented. A spectral method used for determination of prominence radial velocities |1|.It is based on the observed central intensities ratio of optically thin emission lines of the same metal. It is known that the main mechanism for emission of optically thin metal lines in prominences is exitation of matter of prominence by the photospheric radiation of the appropriate frequency |2|. Due to the motion of the prominence with respect to the Sun the metal lines are exited not by the central residual intensity but by intensity of the line wing shifted from the centre of the absoption line in the solar spectrum by $\Delta\lambda = \frac{\xi_r}{c}\lambda$

In order to determine ξ_r we have selected several pairs of TiII lines belong to the same multiplet. They are close to each other in the spectrum and have equal or nearly equal theoretical intensities. The dependence of the ratio of residual intensities of corresponding absorption lines on the values of radial velocities was plotted for the selected line pairs. Using on such dependences and ratios of the observed central intensities of emission lines in a prominence the value of ξ_r is found. It satisfies the ratios of central intensities of all the selected pairs of TiII lines.

Following the spectral method, we have determined the radial velocities for 50 bright prominences. The values of radial velocities are between 1 and 5 km/s for different prominences or different parts of the same prominence. In the investigated active prominences these velocities reach 60 km/s. The motion is directed mainly away from the Sun. Large difference of metal lines intensity ratios in prominences is naturally explaned by their different radial velocities.

References:

1. Kiryukhina A.I.: Astron. Circ. N 1389, 4,1985.
2. Yakovkin N.A., Zel'dina M.YU., Rakhobovsky A.S.: Astron.Zh. 52,332, 1975.

SPECTRAL LINES STRUCTURAL FEATURES OF THE ACTIVE PROMINENCE

T.P.Nikiforova
Astronomy department,Urals State University,Lenin str.51,Sverdlovsk 620083,USSR

Spectral observations of low and long prominence in extremely flare-active region were made on July 8, 1982 ($\varphi = 10^{\circ}$, $l = -90^{\circ}$). 44 cm horizontal solar telescope with 7 m grating spectrograph of the Urals University Astronomical Observatory were used . The spectrographic slit was fitted nearly parallel to the limb. Structural features of H_{α} , H_{δ} , CaII H and K lines were analysed in details.

Radiation in H_{α} and CaII H and K was concentrated in a number of discrete emission features. These structural features had very wide range of intensities. The weakest features showed intensities of 0.002-0.007 in units of disc centre continuum; the brightest ones showed subflare intensities of 0.4-0.6.

The features radial velocities were horizontal to the solar surface and vary in the range (-29 - +35) km/s.

Spectral features in H_{α} showed the fast temporal evolution during 1.5 min in their number, luminousity and shapes.

The growth of the H_{α} line luminousity was caused by appearance of additional discrete bright features . Some of them showed the inclination to the direction of dispersion.

A number of features were double-structured and consisted of two parallel threads with different radial velocities.

Several bright features seemed to be places of intersection of two faint inclined threads . The most thin H_{α} thread had the halfwidth at half intensity of 0.196 Å , which corresponds to T_{kin}= 6500 K with nonthermal velocities ξ = 2.8 km/s.

The features of small dimensions (5") appeared simultaneously inside of (or in projection on) the two brightest H_{α} features. Under the assumption that their inclinations are caused by rotational motions, the values of v_{rot}= 144 km/s and 44 km/s were determined.

AN AUTOMATED PROCEDURE FOR MEASUREMENT OF PROMINENCE TRANSVERSE VELOCITIES

Tron Andre Darvann[1,3], *Serge Koutchmy*[2,3] and *Jack B. Zirker*[3]

1) Inst. of Theor. Astrophysics, Univ. Oslo, 0315 OSLO 3, Norway
2) Paris Inst. d'Astrophysique, CNRS, 75014, France
3) NSO / Sacramento Peak Observatory, Sunspot, NM 88349, USA

ABSTRACT.

A computer algorithm for measurment of transverse velocities (proper motion) in prominences has been developed. We present the method and examples of computed proper motion maps. The method is a modified version of the local cross correlation technique previously applied to granulation images (November 1986, Title et al 1987, November and Simon 1988, November 1988, Darvann 1988, Brandt et al 1988).

Prominence images show much steeper intensity gradients and a wider range of spatial scales of fine structure than granulation images. Due to this we find it necessary to replace the prominence images by an image showing the intensity gradients (derivative of the intensity image). Furthermore, in our algorithm we compute absolute differences instead of correlation coefficients in order to reduce the influence of large scale intensity gradients across a local window (Karud 1988). We have tested the method on datasets obtained at the Vacuum Tower Telescope of NSO/SP. The accuracy of the algorithm is seen to be ±0.3 pixels which, in our data, corresponds to about 1/10 arcsec. Seeing effects are effectively reduced by averaging N cross correlation functions formed from images sampled Δt apart. We find that $\Delta t = 120s$ gives the highest accuracy in the proper motion measurement when applied to our data consisting of quiescent prominences. The correlation coefficient between two interlaced, independent proper motion maps is as high as 0.92 when N=50. The size of the smallest structure for which a proper motion velocity can be measured is limited by the size of the smallest local window that can successfully be applied in the measurement. It needs to be large enough to contain some high contrast structures, typically 4x4 arcsec in our data. Our algorithm is "self-adaptive" to the data in the sense that the window size is changed automatically depending on the presence of local high contrast structures. We conclude that the method successfully produces prominence proper motion maps in addition to being able to correlation track prominence images. Furthermore the algorithm will be useful for destretching of prominence images before producing Dopplergrams or carrying out oscillation studies at high spatial resolution.

Brandt, P.N., Scharmer, G.B., Ferguson, S.H., Shine, R.A, Tarbell, T.D, Title, A.M.: 1988, In *Solar and Stellar Granulation (R.Rutten, R.J.Severino,(eds.))*, 305.
Darvann, T.A.: 1988, *Proc. of the 10th SPO Workshop, Sunspot.*
Karud, J.: 1988, *Proc. of the NOBIM-conference, Oslo 1988,* 93, (in Norwegian).
November, L,J.: 1986, *Appl. Opt.*, **25**, 392.
November, L.J.: 1988, *Proc. of the 10.th SPO Workshop, Sunspot.*
November, L.J., Simon, G.W.: 1988, *Ap.J.* **333**, 427.
Title, A.M., Tarbell, T.D., Topka, K.P.: 1987, *Ap.J.* **317**, 892.

A THREE-DIMENSIONAL MODEL FOR SOLAR PROMINENCES

P. Démoulin[1], E.R. Priest[2], U. Anzer[3]

[1] Observatoire de Paris, Section de Meudon, DASOP (URA 326) F 92195 Meudon Principal Cedex, France

[2] Mathematical Sciences Department, University of St. Andrews, KY 16 9SS St. Andrews - Scotland

[3] Max-Plank-Institut fur Physik und Astrophysik, D-8046 Garching F.R.G.

Abstract: We suggest here a model for the 3D structure of quiescent prominences by a superposition of two fields. A 3D force-free field with constant α is assumed to exist in the corona prior to the prominence formation. The prominence itself is represented by a line current which interacts with the coronal field. The three-dimensional field is represented by analytical functions and concentration of the magnetic field at the photospheric level by convection cells is taken into account. When the field created by the photospheric pattern supports the prominence, the prominence feet are found to be located at supergranule centres otherwise; they are located at cell boundaries.

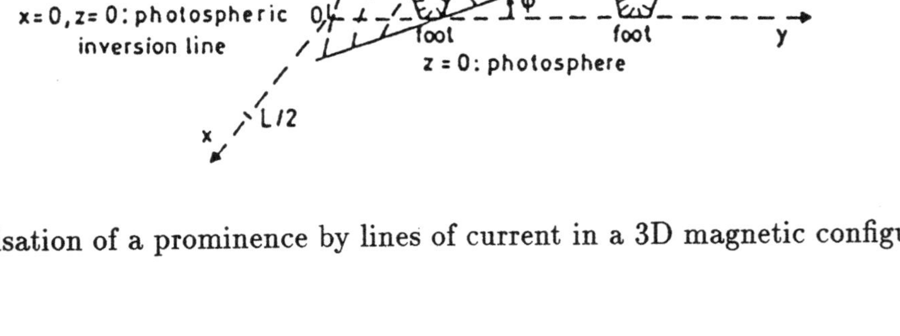

Modelisation of a prominence by lines of current in a 3D magnetic configuration.

HOW TO FORM A DIP IN A MAGNETIC FIELD

BEFORE THE FORMATION OF A SOLAR PROMINENCE

P. Démoulin[1], E.R. Priest[2]

[1] Observatoire de Paris, Section de Meudon, DASOP (URA 326) F 92195 Meudon Principal Cedex, France

[2] Mathematical Sciences Department, University of St. Andrews KY 16 9SS St. Andrews - Scotland

Abstract : Magnetic fields with downward curvature are not favourable for prominence formation since the presence of a small quantity of dense material at the summit of a low-beta arcade cannot deform sufficiently the magnetic field lines to remain there in a stable manner. Thus a dip at the field line summit is needed before a prominence can form. We investigate different ways of forming such an upward curvature. Results with a twisted flux tube or a sheared arcade are reviewed, and a third possibility, namely a quadrupolar region is proposed.

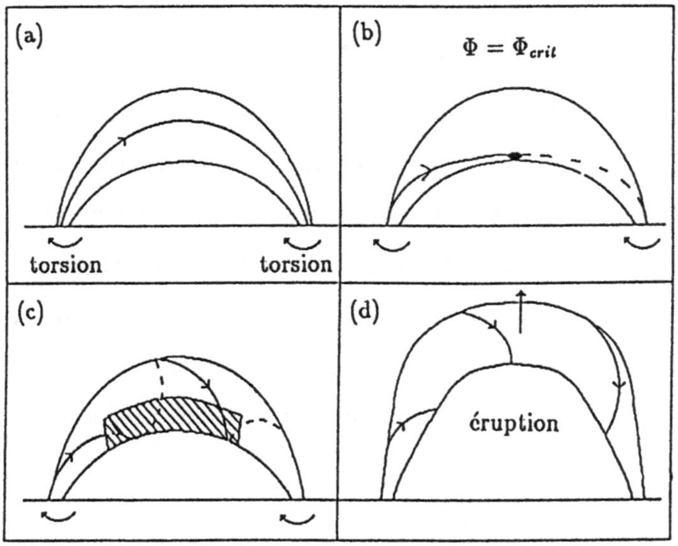

Formation of a prominence in a twisted flux tube. a: an initial untwisted magnetic flux tube, b:the beginning of dip formation, c: the prominence grow as the twist increase, d: eruption of the prominence when the twist is to important.

MHD STABILITY OF LINE-TIED PROMINENCE MAGNETIC FIELDS

P. De Bruyne and A. W. Hood
Department of Mathematical Sciences, University of St. Andrews, St. Andrews, Fife, KY16 9SS, Scotland, U.K.

Magnetostatic equilibria are widely used as a first approximation to modelling a rich variety of large-scale long-lived structures in the solar atmosphere. Under the frozen-in assumption, the dense photosphere provides an effective stabilising mechanism by anchoring the magnetic field lines and thus restricting the wide range of destabilising plasma displacements. Stability theory should account for both the longevity and the sudden eruption of prominences.

The stability properties of a prominence model by Low [1] are investigated using a procedure presented by De Bruyne and Hood [2] based on the energy principle of Bernstein *et al.* [3]. In Low's model the dimensionless flux function A and plasma pressure p may be written as

$$A(U,z) = \ln \frac{U}{[2(z_1+\delta)z+U]^2} ,$$

$$p(U,z) = \frac{2a^2}{(z_1+\delta)^2 z^2}\left[\frac{(z_1+\delta)^2 z^2}{U^2} - \frac{2(z_1+\delta)z}{U} + \ln\left(1+\frac{2(z_1+\delta)z}{U}\right)\right] + \frac{\beta}{2}\exp(z_1 - z) - \frac{1}{2}B_y^2(A),$$

where y is the longitudinal direction, z is height, $U = x^2 + (z-z_1)^2 + a^2$, $a^2 = \delta^2 - z_1^2$, and β is a characteristic ratio of plasma to magnetic pressure. δ fixes the length scale of the bipolar field and z_1 varies between $-\delta$ and δ. The requirement that p must be positive severely restricts the ranges of δ, z_1, β and B_y. It is found that for parameter values $\gamma = 5/3, \delta = 1, \beta = 1, 0.7 < z_1 \leq 0.95$, the shearless field (i.e. $B_y = 0$) is Rayleigh-Taylor unstable. Introducing shear (i.e. $B_y \neq 0$) does not suppress this instability, mainly because it decreases the plasma pressure and hence the stabilising effect of compression. Further details can be found in [4].

References

[1] Low, B. C.: 1981, *Astrophys. J.* **246**, 538.
[2] De Bruyne, P. and Hood, A. W.: 1989, *Solar Phys.* **123**, 241.
[3] Bernstein, I. B., Frieman, E. A., Kruskal, M. D., and Kulsrud, R. M.: 1958, *Proc. Roy. Soc. London* **A244**, 17.
[4] De Bruyne, P. and Hood, A. W.: 1990, *Hvar Obs. Bull.* , submitted.

A Model for Quiescent Solar Prominences
with Normal Polarity

A.W. Hood* and U. Anzer
Max-Planck-Institut für Astrophysik
Karl-Schwarzschild-Str. 1
8046 Garching, FRG

* permanent address:
Department of Mathematical Sciences
University of St. Andrews
Fife, KY16 9SS, U.K.

We have developed 2-d models of quiescent prominences with normal magnetic polarity. Here normal polarity means that the magnetic field traverses the prominence in the same sense as in a potential field configuration. The classical model of this type is the Kippenhahn-Schlüter configuration. In this investigation we have extended their model to allow for several additional effects: We match the internal prominence structure to a global coronal configuration. We allow for a shear in the magnetic field. This then implies that the fields in the corona is now force-free instead of being potential. We also take the finite height of the prominence into account by assuming an exponential density fall-off with height.

We make the following simplifications: we take separable functions for all physical quantities We use a 2 temperature configuration with a sharp boundary between the low prominence temperature and the hot corona. We then obtain simple solutions for both regimes. These solutions are then matched in such a way that the pressure and magnetic field are continuous. The field in the corona is a force-free constant α field. Our calculations show that global equilibria of this type are possible. We find that the internal structure of the prominence differs only slightly from the Kippenhahn-Schlüter model. When one increases the shear (by increasing α) the vertical field component in the prominence becomes smaller but the density remains almost unchanged.

We have extended this model further by allowing for a lower boundary of the prominence. So far we have only preliminary results on this aspect because we have not yet obtained models with a complete force balance at the interface between prominence and underlying corona. But we believe that a relaxation procedure could yield the desired results.

The details of this investigation can be found in a paper which will appear in Solar Physics.

Reference:

Hood, A.W., and Anzer, U.: Solar Phys. (accepted Oct 1989)

THE NONLINEAR EVOLUTION OF MAGNETIZED FILAMENTS

G. Van Hoven, L. Sparks, and D.D. Schnack
Department of Physics, University of California, Irvine
Irvine, California 92717, USA

The anisotropic influence of magnetic shear on the linear dynamics of the filament-condensation instability has been examined extensively in a series of studies [Chiuderi and Van Hoven, 1979; Van Hoven and Mok, 1984; Van Hoven et al., 1986, 1987; and Sparks and Van Hoven, 1988].

To understand the character of the excitations that lead to significant cooling and condensation in a sheared field, we consider a simplified model of the small-amplitude behavior of this plasma system. To do so, we first linearize the MHD equations around an isothermal, isobaric equilibrium with the force-free field $\mathbf{B}_0 = B_0$ [sech (y/a) $\hat{\mathbf{e}}_x$ + tanh (y/a) $\hat{\mathbf{e}}_z$] by using $T = T_0 + T_1(y)\exp(\nu t + ikz)$ where $T_1 << T_0$. In two parallel-wavenumber limits one can simplify the mass-conservation law

$$\nu\rho_1 = -\rho_0\nabla\cdot\mathbf{v}_1 \rightarrow \begin{cases} -ik_{\|}\rho_0 v_{1\|} & k_{\|}v_A >> \nu \\ 0 & k_{\|}v_A >> \nu >> k_{\|}v_s \end{cases} .$$

In both of these $k_{\|}(=\mathbf{k}\cdot\mathbf{B}/B)$ limits, the perpendicular (to $\mathbf{B}$) flow perturbations disappear and we need only consider the parallel component of the force equation [Drake et al., 1988]

$$\nu\rho_0 v_{1\|} = -ik_{\|}p_1 = -ik_{\|}p_0(\hat{\rho}_1 + \hat{T}_1)$$

where $\hat{T}_1 = T_1/T_0$, for example. Finally, we linearize the energy equation to obtain

$$\nu\hat{p}_1 = \gamma\nu\hat{\rho}_1 - (\gamma-1)p_0^{-1}[k_{\|}^2\kappa_{\|}T_1 + \frac{\partial C}{\partial\rho}\rho_1 + \frac{\partial C}{\partial p}p_1 + \left(k^2 - k_{\|}^2\right)\kappa_{\perp}T_1 - \kappa_{\perp}T_1'']$$

where $T_1'' = d^2T_1/dy^2$. By eliminating $\hat{\rho}_1$ and $\hat{p}_1$, we can obtain a 1-D eigenvalue form of the energy equation

$$0 = \Omega_{\perp}\hat{T}_1'' - [k_{\|}^2a^2\Omega_{\|} + (k^2 - k_{\|}^2)a^2\Omega_{\perp} + (\nu - \Omega_\rho) + \frac{k_{\|}^2v_s^2[(\gamma-1)\nu - \Omega_T]}{\nu^2 + k_{\|}^2v_s^2}]\hat{T}_1$$

where typical rates [Sparks and Van Hoven, 1988] include $\Omega_{\|} \equiv (\gamma-1)\,\kappa_{\|}T_0/a^2p_0$ and $\Omega_\rho \equiv -(\gamma-1)[\partial C/\partial p]_\rho$. The last (fractional) term in the square bracket disappears when $\rho_1 \approx 0$.

One can have an idea of the transverse (shear-direction) character of these modes by looking at $\hat{T}_1''/\hat{T}_1$ which is shown in Fig.1(a), along with a schematic solution $\hat{T}_1$ (and -$\hat{\rho}_1$) for the (third) eigenvalue ν. The critical wavenumbers $k_1^2 \approx \nu^2(\nu - \Omega_\rho)/v_s^2(\Omega_p - \nu)$ and $k_2^2 \approx \gamma(\Omega_p - \nu)/a^2\Omega_{\|}$ give the approximate location of the zeros of the square bracket [equivalent to α_1 and α_p of Chiuderi and Van Hoven, 1979].

Fig. 1(b) interprets the turning points (approximately) in terms of the essential local rates of the problem. At large y the parallel thermal conduction suppresses the temperature perturbation. For $k_{\|}(y) < k_2$, radiative losses lead to cooling, which can then cause pressure-driven parallel-to-$\mathbf{B}$ flow and significant condensation in those locations $k_{\|}(y) > k_1$ where $k_{\|}v_s$ is the fastest rate in the problem.

The growth rates and structure of these *kinematic* modes have been discussed by Drake et al. [1988], and numerically detailed and interpreted by Van Hoven et al.[1986; Sparks and Van Hoven, 1988]. Excitations composed of these short-wavelength modes provide the key to the attainment of significant nonlinear cooling and condensation in a sheared magnetic field.

We initiate a computer simulation of these filamentation processes [Van Hoven et al., 1987] by specifying equilibrium values appropriate to the solar atmosphere. We then add a small-amplitude *noise* excitation T_1 to the equilibrium T_0. Two aspects of the final nonlinear state of this noise (multi-mode) excitation are shown in the following figures. The surfaces (and contours below) plotted in Fig. 2(a) represent the magnitude of the temperature $T(y, z)$ at $t = 290$ sec. Here the z axis is linear, but the y axis utilizes a coordinate transformation $y \rightarrow \sinh y$. We show this variation implicitly by plotting the direction of the equilibrium magnetic field as a function of y. Corresponding surfaces displaying the magnitude of the mass density $\rho(y, z)$ are plotted in Fig. 2(b). Vectors have been superimposed on the density contours to indicate the magnitude and direction of flows in the $y - z$ plane.

By $t \sim 210$ sec, an irregular temperature well had formed near the *center* of the shear layer where the plasma had cooled from its equilibrium value by almost an order of magnitude, nearly to the point at which local nonlinear saturation occurs when the radiative cooling rate comes back into balance with the heating rate. As shown in Fig. 3, this leads to a pressure drop and subsequently to a large peak in the density that arises off-center as a consequence of plasma flow that is directed primarily parallel to the magnetic field. At t = 290 sec the maximum number density has increased to about $10^{11.5} cm^{-3}$, and the local temperature has fallen to nearly 10^4K. At the position of the density peak, the magnetic field forms an angle of approximately 36° with respect to the filament (x) axis, a value within the range of those observed [Leroy, 1989]. Nonlinear saturation of the condensation process should occur when temperature gradients become sufficiently large that thermal conduction can restore energy balance.

By performing nonlinear magnetohydrodynamic simulations, we have demonstrated that the radiative-condensation instability in a sheared magnetic field is capable of generating small-scale, filamentary, plasma structures with densities and temperatures, and angular orientations with respect to the local field, characteristic of solar prominences. We have found that the most rapid growth of the condensation results from mass flow parallel to the magnetic field in regions of the plasma where the field orientation inhibits thermal conduction. Like the linear kinematic modes, nonlinear condensations possess density maxima located away from the center of the shear layer where the parallel sound-speed rate is finite.

A fuller description of the physical mechanisms of the linear thermodynamic excitations is given in Sparks and Van Hoven (1988) and Van Hoven (1990). A complete report of the nonlinear simulations will appear in the ***Astrophysical Journal*** [Sparks, Van Hoven and Schnack, 1990].

This research was supported by NASA and NSF, and computations by NSF and DOE.

Chiuderi, C., and Van Hoven G.: 1979, ***Astrophys. J. (Letts.)*** **232**, L69.

Drake, J.F., Sparks, L., and Van Hoven, G.: 1988, ***Phys. Fluids*** **31**, 813.

Hildner, E.: 1974, ***Solar Phys.*** **35**, 123.

Leroy, J.L.: 1989, in ***Dynamics and Structure of Quiescent Solar Prominences***, ed. E.R. Priest (Dordrecht: Kluwer), pp. 77-113.

Sparks, L., and Van Hoven, G.: 1988, ***Astrophys. J.*** **333**, 953.

Sparks, L., Van Hoven, G., and Schnack, D.D.: 1990, ***Astrophys. J.*** **353**, 000.

Van Hoven, G.: 1990, in ***Physics of Magnetic Flux Ropes***, eds. C.T. Russell, E.R. Priest and C. Lee (Wash., D.C.: Amer. Geophys. Union), in press.

Van Hoven, G., and Mok, Y.: 1984, ***Astrophys. J.*** **282**, 267.

Van Hoven, G., Sparks, L., and Tachi, T.: 1986, ***Astrophys. J.*** **300**, 249.

Van Hoven, G., Sparks, L., and Schnack, D.D.: 1987. ***Astrophys. J. (Letts)*** **317**, L91.

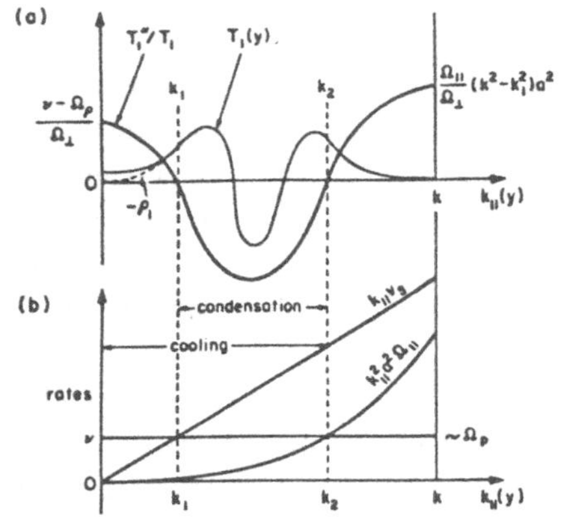

Fig. 1. (a) The effective binding potential, for $\Omega_p < \nu < \Omega_p$ and $ka > (\gamma\Omega_p/\Omega_\parallel)^{1/2}$, along with a typical T_1 (and $-\rho_1$) eigenfunction; (b) an interpretation of the turning points in terms of the fundamental local rates of the radiative-condensation problem.

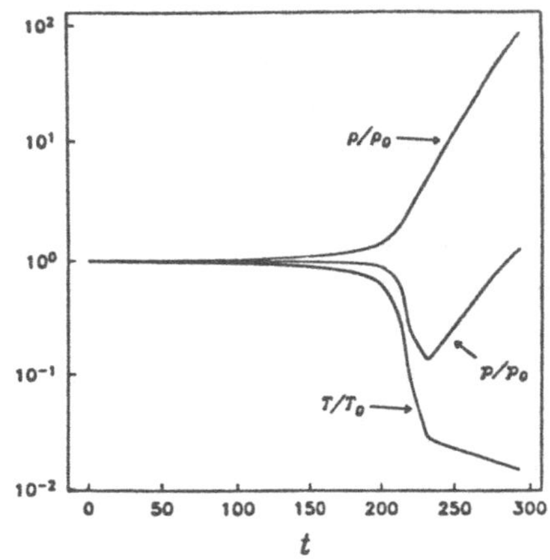

Fig. 3. Typical values of the thermodynamic variables *vs* t/τ_A taken from a diagnostic probe located at the position of the density peak, in a case where the angle between the filament axis and the local magnetic field is 14°.

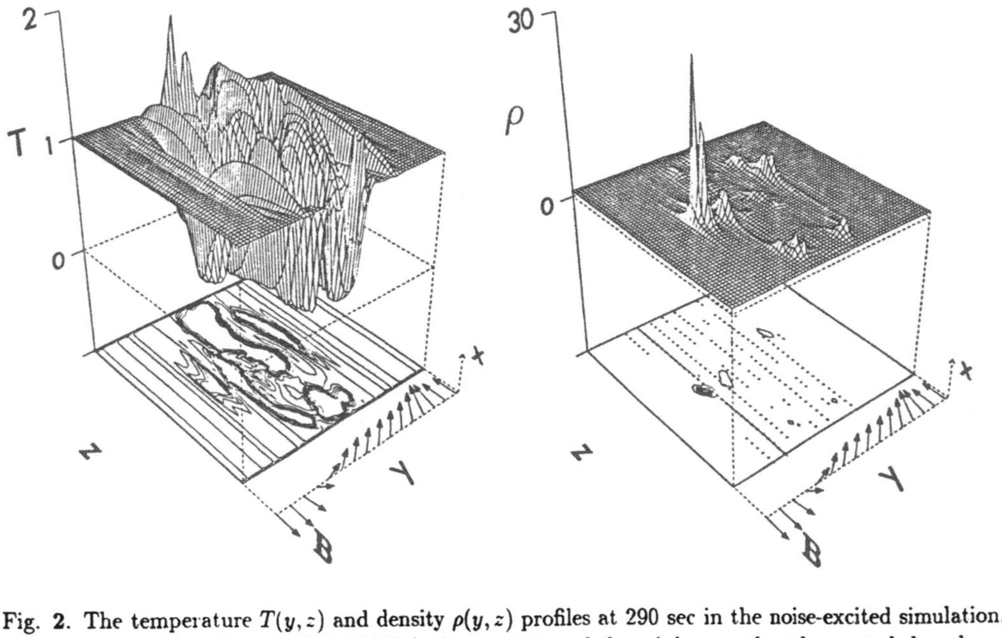

Fig. 2. The temperature $T(y,z)$ and density $\rho(y,z)$ profiles at 290 sec in the noise-excited simulation, along with the sheared magnetic field $\mathbf{B}(y)$; the y axis is scaled as sinh y, so that the central shear layer is expanded while the range $(-\infty,\infty)$ is covered. The density at the peak is $\sim 30\rho_0$, and the temperature is $\sim 0.02T_0$; the angle between the field and the filament axis is 36°.

THERMAL EQUILIBRIUM OF CORONAL LOOPS AND PROMINENCE FORMATION

C. D. C. Steele and E. R. Priest
Dept. of Mathematical Sciences
University of St. Andrews
Scotland

The equations for thermal equilibrium of a coronal loop are solved. Boundary conditions are that the footpoints are at temperature 20 000 K and, at the top of the loop, the temperature gradient is zero. Two parameters, which basically represent the length of the loop and the amount of coronal heating give the properties of the solutions. As these two parameters are varied, different types of loops occur. **Hot loops** have summits at much higher temperatures than the footpoints. For **cool loops** the variation in temperature along the loop is much less. **Hot-cool** loops have cool footpoints and cool summits but hot regions elsewhere. One particular class of hot loops, namely **warm loops**, have cooler summits than most hot loops and a region where the temperature is almost constant. A single well-developed cool loop may appear as an active region prominence, whereas the summits of an ensemble of hot-cool loops in an arcade may show up in the solar atmosphere as a quiescent prominence. An assembly of loops may be fitted together to form an arcade; such an arcade may contain several areas where the different classes of loops may occur. For some arcades there are areas where more than one class of loop may apply while for certain other arcades there are areas where no equilibrium solutions for the temperature may be found.

Thermal Instability in Planar Solar Coronal Structures

R.A.M. Van der Linden* and M. Goossens,
Astronomisch Instituut, KU Leuven
Celestijnenlaan 200B, 3030 Heverlee, Belgium.

Abstract.

Prominences and filaments are thought to arise as a consequence of a magnetized plasma undergoing thermal instability. Therefore the thermal stability of a magnetized plasma is investigated under coronal conditions. The equilibrium structure of the plasma is approximated by a 1–D slab configuration. This is investigated on thermal instability taking into account optically thin plasma radiation and anisotropical thermal conduction. The thermal conduction perpendicular to the magnetic field is taken to be small but non–zero.

The classical rigid wall boundary conditions which are often applied in the litterature, either directly on the plasma or indirectly through some other medium, are replaced by a more physical situation in which the plasma column is placed in a low–density background stretching towards infinity. Results for a uniform equilibrium structure indicate the major effect of this change is on the eigenfunctions rather than on the growthrate. Essentially, perpendicular thermal conduction introduces field–aligned fine–structure. It is also shown that in the presence of perpendicular thermal conduction, thermal instability in a slab model is only possible if the inner plasma has the shortest thermal instability time–scale.

* Research Assistant of the National Fund for Scientific Research (Belgium).

TO THE PROBLEM OF INSTABILITY OF THE SOLAR ATMOSPHERE CAUSED BY ABSORPTION OF RADIATION ENERGY

Yurij N. Redcoborody
Astronomical Observatory of Kiev University,
Observatornaya St. 3, Kiev, 252053, USSR

The instability of internal gravity waves (IGW) in the solar atmosphere is investigated. Departures from adiabatic variations are taken into account to consider the presence of heat sources in a medium (the heat releasing caused by absorbtion of radiation). Adiabaticity relation is replaced by the general energy balance equation

$$\frac{dp}{dt} = C^2 \frac{d\rho}{dt} + (\gamma - 1)\rho G$$

(we consider $T \ll 10^5$ K and this implies that the radiative loss function $L_{rad} \cong o$; $G = G_{rad}$ is absorbed radiation energy). The reverse influence of disturbances $\vec{V}, \delta\rho$, ... on the gain function G is taken into account. This implies that in addition to usual equations we have one more equation

$$\frac{\partial G}{\partial t} + \vec{v}\,\nabla G = 0.$$

The boussinesq approximation is used. The linearised simultaneous equations have unstable solutions. Roots of the dispersion equation are complex and this fact corresponds to oscillating and monotone instability.

Considered instability, in contrast to the thermal instability (Field, 1965), takes place for $T \ll 10^5$ K. The obtained results may be applied to interpretation of the observed activation of quiescent prominences (the eruption, "a winking filament" etc.) that may be initiated by a disturbance from a flare, and to formation of BN-type prominences.

R E F E R E N C E S

Field G.B., 1965, Ap.J., 142, N 2,531.

Radiative transfer in cylindrical prominence threads

Pierre Gouttebroze

Institut d'Astrophysique Spatiale

B. P. 10, 91371 Verrieres-le-Buisson

France

Abstract : Whereas radiative transfer in Astrophysics generally deals with plane-parallel or spherical objects, the use of other geometries is desirable in some particular cases. We propose to treat infinitely long cylinders, a geometry which is relevant to, e. g., prominence threads or coronal loops. We study two different situations : the first one corresponds to an incident radiation field which is symmetrical by rotation around the axis of the cylinder, and may be reduced to a 1-dimension formalism. This problem is usually treated in an approximate way, replacing the cylinder by an equivalent slab, but a real solution in cylindrical coordinates can give a better precision. The other case is that of anisotropic incident radiation, which results in a 2-dimension problem. We review the different available techniques to solve these two kinds of problems, and discuss their range of applicability and their utility with respect to the diagnostic of prominence threads. Prospects for new methods that could be developed are also examined.

HYDROGEN LINE FORMATION IN FILAMENTARY PROMINENCES

Petr Heinzel
Astronomical Institute of the Czechoslovak Academy of Sciences
251 65 Ondrejov, Czechoslovakia

Most of the non-LTE prominence work has been confined to a 1D slab geometry which approximately describes the radiative transport in quiescent prominences, at least within some line and continuum transitions. Only recently, a few authors have considered schematically the inhomogeneous nature of prominence structures. Here we briefly review this effort and propose some improvements in this direction. In fact, it is rather difficult to reproduce the observed profiles of resonance hydrogen lines without taking into account a prominence porosity. This is particularly evident in the case of Lyman α and Lyman β lines detected on OSO-8 satellite. We start with models of Fontenla and Rovira (Solar Phys. 96, 1985, 53), where the non-LTE problem is solved for one representative fine-structure element, taking into account an interface between its cold core and the surrounding interfilar medium. Our modification consists in an iterative improvement of the boundary conditions for the radiative transfer in this element. We call this rather heuristic approach as an Iterative Boundary Conditions (IBC) method, in which the irradiation of each fine-structure element inside the prominence is expressed as a linear combination of the direct diluted solar radiation and the iteratively computed radiation coming from nearby elements. The branching ratio has a stochastic nature and characterizes the prominence porosity. As a special case we get the models of Fontenla and Rovira. On base of extended numerical simulations, we discuss here mutual effects of the multilevel interlocking, partial redistribution, prominence porosity and the prominence-corona interface. In general, both Lyman α and Lyman β line profiles (and particularly their peaks) differ from those computed for an "equivalent" homogeneous model or for the limiting case of Fontenla and Rovira´s models.

TOWARD HYDROGEN EMISSION IN STRUCTURALLY INHOMOGENEOUS PROMINENCES

Valtentina V. Zharkova
Physical Department, State University, Vladimirskaja, 64
252017, Kiev, USSR

It is considered the model prominence with a filamentary structure as proposed by Morozhenko (1978) to be described by a filamentary degree $\gamma = l^*/(l+l^*)$ where l is a filament thickness, l^* is a distance betwen filaments. A prominence is supposed to be irradiated by photospheric, chromospheric and coronal radiations which penetrate as in the faces of prominence so in the intervals between filaments. Material of a prominence consists of 5-level plus continuum atoms and ions of hydrogen and free electrons. In the assumption of a complete redistribution the radiative transfer equations together with statistical equilibrium ones for H_α, L_α and L_c - frequencies and steady state equations for other transitions are written as in paper Zharkova (1983). The equations are solved by iterations with varying of physical and structural conditions in prominences.

In result the relative emission measures are obtained to be close connected with the second level populations so them dependency of a filamentary structure and physical conditions is similar. The appearance of a ingomogeneous structure influences on hydrogen atom excitation and ionization mainly in the central parts of a prominence and almost does not vary the values near the edges. At low electron temperatures the relative emission measure $\frac{n_e n^+}{n_1}(\tau)$ and second level populations $\frac{n_2}{n_1}(\tau)$ increase with rising of a filamentary degree and absolute value of increasing is independent of electron density in the feature. The relative third level populations decrease with γ-growth. If electron temperatures become higher the functions $\frac{n_e n^+}{n_1}(\tau)$ and $\frac{n_2}{n_1}(\tau)$ fall off with rising of a filamentary degree and third level populations in contrary increase.

Computed H_α and L_α line profiles are shown a filamentary structure to be appeared mainly in the line wings. Observed full intensity H_α to L_α ratios for faint prominences are fitted by computed ones for the features with filamentary degrees $\gamma = 0.3$-0.7.

Morozhenko N.N. Solar phys., 1978, 58, P.47-56.
Zharkova V.v. Preprint TPI Ac.Sci.UkSSR, 1983, N83-141P, 20P.

LINEAR POLARIZATION OF HYDROGEN Hα LINE IN FILAMENTS: METHOD AND RESULTS OF COMPUTATION

BOMMIER, V.[(1)], LANDI DEGL'INNOCENTI, E.[(2)], SAHAL-BRÉCHOT, S.[(1)]

[(1)]Laboratoire *Astrophysique, Atomes et Molécules*, URA 0812 du CNRS, DAMAp
Observatoire de Paris, Section de Meudon, F-92195 Meudon Cedex, France

[(2)]Dipartimento di Astronomia e Scienza dello Spazio
Università di Firenze, Largo E. Fermi, 5, I-50125 Firenze, Italia

A method for solving on individual cases the ambiguity encountered in field vector determination using the Hanle effect consists in performing simultaneous observations of one optically thin and one optically thick line. Such observations have been performed at the Pic-du-Midi, in the Helium D_3 line (optically thin) and the Hydrogen Hα line (optically thick). The interpretation of Hα observations requires the computation of the linear polarization of this line.

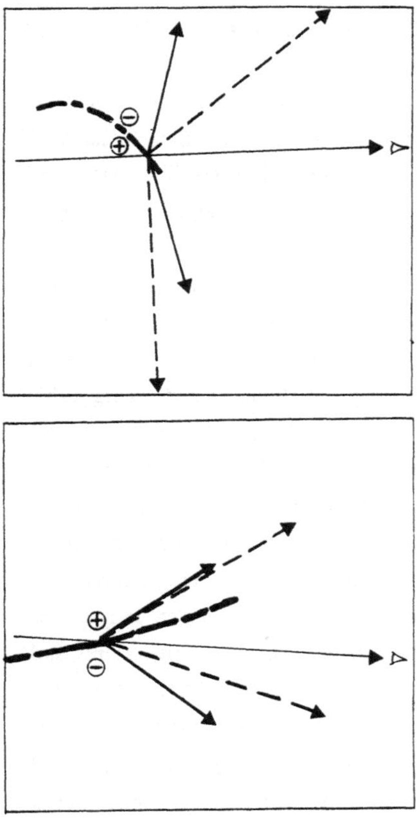

Fig. 1: The 2 solutions obtained with the optically thin line (HeI D_3) are in bold full lines, symmetrical with respect to the line-of-sight. The 2 solutions obtained with the optically thick line (Hα) are in bold dashed lines, not symmetrical with respect to the line-of-sight. The most different solutions are the parasitic ones; the closest solutions are the true ones. In these 2 prominences, the true solution is of inverse polarity.

A computation has been done, using an iterative method restricted to the first iteration, and results have been provided for a prominence seen at the limb (Landi Degl'Innocenti *et al.*, 1987) and for a filament seen on the disk (Bommier *et al.*, 1989a). However, the validity of the restriction to one iteration has to be established.

This can be done by comparing the results given by the iterative method and a global method (Bommier *et al.*, 1989b), which is of the integral type, in the case of an optically thick plane-parallel atmosphere for a 2-level atom. The comparison show that the iterative method restricted to the first iteration can be used when the line *is not* a normal Zeeman triplet; the previous results for the Hα line from prominences are thus validated.

These results can be used for interpreting the polarization data obtained at Pic-du-Midi; preliminary results for 2 prominences are given on fig. 1, showing that the parasitic solution is of normal polarity and that the true solution is of inverse polarity with respect to the underlying photospheric field.

References

Bommier, V., Landi Degl'Innocenti, E., Sahal-Bréchot, S.: 1989a, *Astron. Astrophys.* **211**, 230

Bommier, V., Landi Degl'Innocenti, E., Sahal-Bréchot, S.: 1989b, *"Transfert du rayonnement II"* CNRS workshop, Paris, February 9-10, H. Frisch and N. Mein ed.

Landi Degl'Innocenti, E., Bommier, V., Sahal-Bréchot, S.: 1987, *Astron. Astrophys.* **186**, 335

MULTITHREAD STRUCTURE AS A POSSIBLE SOLUTION FOR THE Lβ PROBLEM IN SOLAR PROMINENCES

J.-C. VIAL(1), M. ROVIRA(2), J. FONTENLA(3), and P. GOUTTEBROZE(1)

(1) Institut d'Astrophysique Spatiale, B. P. 10, 91371 Verrières-le-Buisson Cédex, France

(2) I.A.F.E., cc 67 Suc. 28, Buenos-Aires 1428, Argentina

(3) Space Science Laboratory, NASA/ Marshall Space Flight Center, Alabama 35812, USA

ABSTRACT

Following the pioneering works of Heasley, Mihalas, Milkey and Poland (see e.g. Heasley and Milkey, 1983) who built non LTE onedimensional models of solar prominence, much attention has been paid to the spectral signatures of the Lyman lines as observed with OSO 8 (Vial, 1982_a). In spite of a better treatment of the frequency redistribution and boundary conditions, one-dimensional low-pressure models lead to Lyman β intensities much lower than observed ones (Heinzel, Gouttebroze and Vial, 1987). Different atomic processes of formation of hydrogen lines (Cooper, Ballagh and Hubeny, 1988) or the inclusion of a Prominence Corona Transition Region or PCTR (Heinzel, Gouttebroze and Vial, 1988) have been proposed to explain this discrepancy. We present here a different approach where the filamentary nature of prominences which provides the hydrogen lines with different opacities, offers their photons different escaping possibilities. The thread models we use derive from an energy equation where radiative losses are balanced by conductive flux (Fontenla and Rovira, 1983, 1985). We show that no superposition of threads gives good values of Lyman α, β and H α intensities for too high and too low pressures. Solutions are found for pressure around 0.05-0.1 dyn/cm^2 and a number of threads between 100 and 400. Two improvements have been performed : first, the inclusion of Partial Redistribution leads to a decrease of Lα (and Lβ) intensity and models now require a higher number of threads; second, the inclusion of the ambipolar diffusion along the steep temperature gradient which changes the hydrogen ionization in the lower regions (Fontenla, Avrett and Loeser, 1990). The new run of temperature and density implies more material at low temperatures and hydrogen lines intensities increase. A solution for the Lβ problem can be found for a pressure of about 0.1 dyn cm^{-2}. However the Hα intensity appears to be rather high. Moreover, the number of threads required (about 200) is far larger than the number derived by Zirker and Koutchmy (this issue) and Mein (this issue) from observed Hα profiles. Our neglect of the radiative interaction between threads may explain our results (Heinzel, this issue). To conclude, these computations of non-lte radiative transfer in realistic geometrical and physical models, appear to be a promising path for the investigation of solar prominences.

ON THE BALMER AND PASCHEN ENERGY DECREMENTS IN DIFFERENT BRIGHTNESS PROMINENCES

E. G. Rudnikova
Astronomical Observatory of Kiev University
Observatornaya St., 3, Kiev, 252053, U.S.S.R.

The dependence of the Balmer and Paschen energy decrements on the physical conditions in prominences is illustrated by means of computed energy curves $[\lg E_{2i}, \lg \tau_{2i}]$, $[\lg E_{3i}, \lg \tau_{3i}]$ using statistical equilibrium equations for the 10-level hydrogen atom at the volume-average parameters with different values n_e, T_e, v_t, length and height. The calculated Balmer and Paschen decrements change from ratios 12:1:0.31:0.12 ... and 3.1:1:0.42:0.20 ... at $n_e \simeq 10^{10}$ cm^{-3} (τ(H)<1) to 2:1:0.66:0.50 ... and 0.73:1:0.77:0.52 ... at $n_e \simeq 10^{12}$ cm^{-3} ($\tau(H\alpha) \simeq 300$).

The method is considered to determine the porosity on the picture-plane for prominences which are optically thick in the $H\alpha$ line basing on the observed energy curves of the Balmer lines. The prominence porosity is investigated for its influence on the determination of the matter parameters from the measured strengths of emission lines. It is found that the ignoring of the porosity leads to that the level populations, optical thickness, electron density become lowered, and the line-of-sight length becomes overstated, sometimes a few times.

SEMI-EMPIRICAL MODELS AT DIFFERENT HEIGHTS OF A QUIESCENT PROMINENCE

Fang Cheng, Zhang Qizhou, Yin Suying
(Astronomy Depertment, Nanjing University)
and W. Livingston
(National Solar Observatory, U.S.A.)

ABSTRACT

Semi-empirical models of a quiescent prominence observed on Dec.12, 1972 with the McMath Telescope at Kitt Peak have been deduced for different heights. The transfer, statistical equilibrium equations as coupled with the hydrostatic equilibrium, and the partical conservtion equations have been solved. The models reproduce well the simultaneously observed H , H , H_r, CaII K,H and infrared triplet line profiles. The study indicates that the pressure near the edge of prominences and the microturbulence velocity in the prominences basically do not vary with height, but the temperature decreases monotonously from the edge toward the center. It is found that the temperature near the edge of prominences increases only slightly with height, but the central temperature decreases significantly. The results also indicate that near the edge of prominences there is no radiative equilibrium and the total radiative loss has a maxium which is mainly due to L . The radiative loss due to CaII is negligible in comparison with that due to hydrogen.

The ionization problem of Calcium has also been discussed.

ANALYSIS OF HeI 10830 A LINE IN A QUIESCENT PROMINENCE

P. Kotrč and P. Heinzel
Astronomical Institute of the Czechoslovak Academy of Sciences
251 65 Ondřejov, Czechoslovakia

In order to explain an enhanced emission in the wings of the photoelectrically observed HeI 10830 A line, Landman et al. (Astrophys. J. 218, 1977, 888) have introduced the concept of a multilayer prominence structure with different temperatures and/or turbulent velocities characterizing the prominence-corona transition region (PCTR). As previously reported (Heinzel et al., Contr.Astron.Obs. Skalnate Pleso 15, 1986, 171), our photographically measured profiles of this line also exhibit substantial surplus radiation (about 10 %) in the line wings, which can hardly be explained in terms of simple isothermal models. However, isothermal models of cool structures can be used to fit the line cores of observed profiles(both peaks in our case), assuming that the line-core emission comes predominantly from cool parts of the prominence. Theoretical profiles corresponding to these models are typically lower in the wings, namely in the gap between both peaks. Intensity differences in the gap serve then us as an indicator of the amount of a hot plasma radiation, emergent from optically-thin PCTR structures. Since the amount of this surplus radiation was found to increase almost linearly with the total optical thickness of cool structures (which we relate to their number) , we deduce that cool prominence threads are surrounded by hot plasma sheets rather than that the prominence periphery as a whole is hot. In the latter case, no simple relation between the surplus radiation and cool structures optical thickness was expected. Finally, we have also found an increase of PCTR emission with geometrical height in the prominence which may be connected with the coronal temperature rise.

QUIESCENT FILAMENT "APPEARANCES AND DISAPPEARANCES"

Z. Mouradian and I. Soru-Escaut
Observatoire de Paris, Meudon, DASOP, URA 326
92195 Meudon Principal Cedex FRANCE

The intention of the present paper is to outline the evolution of quiescent filaments, from their genesis to their evanescence. One or more "accidents" may occur during this period, namely sudden disappearances, or DB ("disparitions brusques") of the filaments. Very often the DB's are followed by reappearances. These events are classified in one of two categories, depending on the physical nature of their cause : dynamic sudden disappearance (DBd) and thermal ones (DBt).

The genesis of a filament begins with the alignment, growth and settling of the vortices before the quiescent filament itself actually appears. The evanescence, or destruction, of the filament occurs by a gradual decline in its size and intensity. It seems that the evanescence is due to the gradual submergence of the magnetic support.

A small percentage of filaments (prominences) undergo sudden changes in Hα visibility.

i) The dynamic DB consists of an ejection of matter and of the magnetic field into the corona, due to a local reorganization of the magnetic field. Two types of dynamic DBs were detected : DBd caused by the emergence of an Active Region and DBd caused by a modification in the magnetic support of the prominence. These examples illustrate the classical idea of prominence eruption. Let us call that DBd affects filaments without pivot points.

ii) The thermal DB consists in a heating-up of the plasma, while filaments that have pivot points are likely to give rise to thermal DB. This occurs subsequent to the heating of the plasma, which leads to on-site ionization of HI. This type of phenomenon is followed by the reappearance of the filament, due to cooling, when the heating process stops. After disappearance in Hα by heating, the filament becomes visible in the EUV lines indicating that the temperature reaches 10^5 to 10^6 K.

References :

Mouradian, Z., and Soru-Escaut, I. : Proceedings IAU Colloquium N°117, HVAR, September 25-29, 1989 (V. Ruzdjak and E. Tandberg-Hanssen eds.) Public. of HVAR OBSERVATORY

NUMERICAL SIMULATION OF A
CATASTROPHE MODEL FOR PROMINENCE ERUPTIONS

T.G. Forbes
Institute for the Study of Earth, Oceans, and Space
University of New Hampshire, Durham, NH 03824, USA

In 1978 W. Van Tend and M. Kuperus proposed a simple catastrophe model which suggests that a prominence containing a current filament will lose equilibrium when the filament current exceeds a critical value. Here I use a two-dimensional numerical simulation to test how the Van Tend - Kuperus model works in ideal MHD. The simulation exhibits the expected loss of mechanical equilibrium near the predicted critical value, but the current filament only moves a short distance upward before coming to rest at a new equilibrium. However, this new equilibrium contains a current sheet which is resistively unstable to magnetic reconnection, and if magnetic reconnection occurs rapidly, the filament can continue to move upwards at Alfvénic speeds.

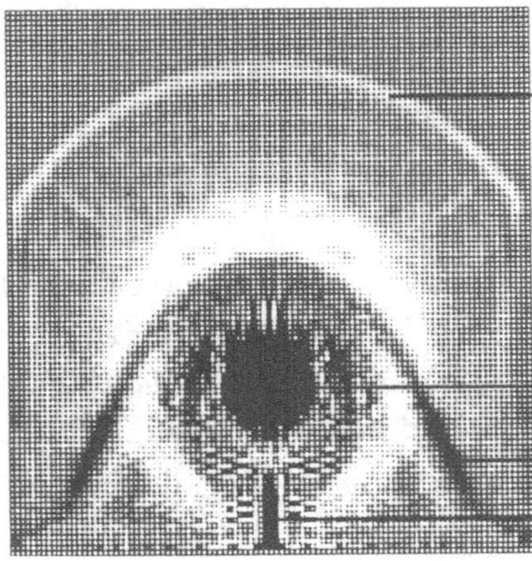

Fig. 1. Current density in the simulation at 0.8 Alfvén scale-times when the filament is moving rapidly upwards. Dark and light regions have opposite signs.

The initial magnetic field configuration for the numerical simulation is equivalent to the combined field from a line-current of strength I at $y = h$, an image line-current at $y = -h$, and a two-dimensional dipole of strength m at $y = -d$. That is

$$B_y + iB_x = i(\mu/2\pi)\left[\frac{2hI}{z^2 + h^2} - \frac{m}{(z + id)^2}\right]$$

where $z = x + iy$, and μ is the magnetic permeability of free space. The image line-current repels the current filament, but the dipole attracts it, and repulsion and attraction are balanced when

$$h/d = M - 1 \pm \sqrt{M^2 - 2M}$$

where $M = m/Id$ is the relative strength between the dipole and the filament current. When M becomes less than 2, equilibria no longer exist, and the filament erupts. However, in the absence of reconnection, the filament travels only a short distance upwards before coming to rest.

Reference: Van Tend, W., and M. Kuperus, *Sol. Phys.*, *59*, 115, 1978.

LAWS OF EVOLUTION AND DESTRUCTION OF SOLAR PROMINENCES

G.P.Apushkinskij
Astronomical Observatory, State University
198904 Leningrad USSR

Abstract. Many prominences after some time of quiet existence undergo disparition brusque. These cases were observed on 22 meter radiotelescope at wavelength 0.8 and 1.35 cm and in H_α line. Figure 1 shows prominence parametres during its disparition brusque.

Figure 1:

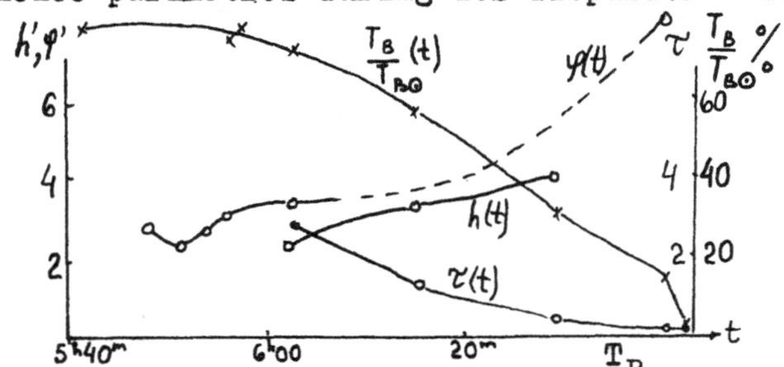

They are relative brightness temperature $\frac{T_B}{T_{B\odot}}$ (t), size $\varphi(t)$, height h(t) and optical thickness $\tau(t)$. The brightness temperature decreases in several times because of the decrease of $\tau(t)$, the cynetic temperature is constant during the evolution process. The second case under observation is the active region prominence evolution preceeding the flare on July, 28 1983 (flare 13^h04^mUT) and July, 29 1983 (flare 2^h17^m). Figure 2 shows the scans, radiomap and H_α one.

Figure 2:

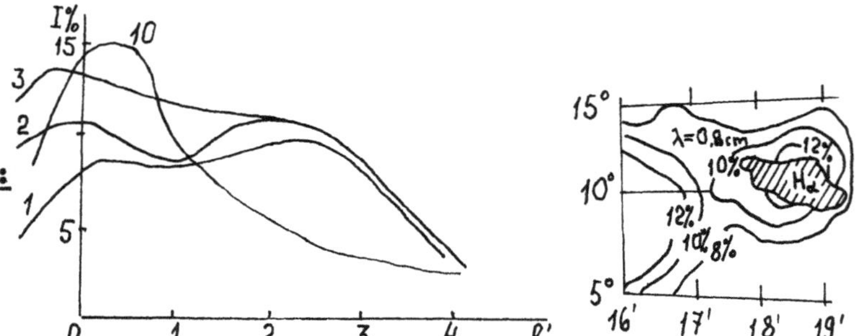

Scan moments are following: 1-8^h30^mUT, 2-8^h47^m, 3-9^h02^m all on July, 28 and 10-15^h45^m on July, 29. Several hours before the flare the brightening on solar limb appears. Then the overlying prominence becomes active and 10 minutes before the flare brightens in H_α line. The prominence turns out to be partly (on July, 28) or fully (on July,29) destructed. The paper will be published in the Hvar Obs. Bull.

PROTO-ELEMENTS OF DARK SOLAR FILAMENTS

A. A. GALAL
National Research Institute of Astronomy
and Geophysics, Helwan, Cairo, ARE

ABSTRACT

Rapid sequences of H-alpha filtergrams (2-6 seconds interval) have been secured with the aim to study the vertical structures of solar dark filaments. The registerd changes of the filamentry features with the displacement in H-alpha line clearly point to the inhomogeneous pattern of the velocity field at various physical levels in filaments.

Proto-type features of a quiescent filament have been recorded at certain displacements from the center of the line. The dark protofeatures at the base of the quiescent filament are notched by an elongated plage area having exactly the same shape of the filament. At this level in the solar atmosphere, one may hardly distinguish between proto-elements of filaments and the neighbouring dark intergranular spaces. The general pattern of plage elements adjacent to the filament implies that it is built over a fishbone-like magnetic region.

Some sections of the filament are formed at the boundaries of supergranular cells. It is obvious in some of the filtergrams obtained that filaments may arise from the alignment of dark fibrils existing at the boundaries of supergranules. The footopoints of filaments are also detected in the dark intergranular spaces.

ON THE BEHAVIOUR OF THE LONG-LIVING SOLAR FILAMENTS

V. Dermendjiev, P. Dukhlev and K. Velkov
Department of Astronomy and NAO
72, Lenin Blvd., BG-1784 Sofia

The statistical studies of the solar filaments distribution by the latitude and the phase of the solar cycle have shown some not so obvious regularities which, to our opinion, are related to the organization of the large-scale subphotospheric circulation. The surface distribution of the filaments shows the tendency of forming active longitudes and non-random configurations with typical dimensions of 10° and 25°. The main zone of solar prominences shows fine structure. The polar zone shows complex structure and two maxima in the prominence distribution by latitude. If trace the filament activity in 5 latitude zones using the index computed in Meudon, we shall see that the high latitutde phase of the activity pass 3-4 years ahead of the lower latitude phase and in the middle latitude phase it shows two strongly pronounced maxima.

The subject of our study are the long-living (two and more rotations solar filaments, published in the "Cartes synoptiques de la chromosphere solaire et catalogues des filaments" of the Meudon observatory for the period 1931-1963. Taking the coordinates of the centers of such filaments in consecutive rotations, we compute the displacement taking into consideration its direction. For the corresponding samples of (L) and (- L) we define the medians Me(L) and Me(- L) and compute the quantities:

$$A = n^{+}/n \; . \; Me(\ L)$$
$$B = n^{-}/n \; . \; Me(-\ L)$$
$$C = A + B.$$

The quantity C obtained by this algorithm characterizes the preferd direction in the longitudial displacement of the filaments in given 10° latitude zone and one year time interval. It is interesting to note the existence of the north-south asymmetry of the horizontal displacement of the filaments as a function of the heliographic latitude and the time. There is a rapid increase of the value of C at high latitudes during about three consecutive years. This effect is strongly pronounced for the cycle No 18.

So, the filament activity has latitude dependent effect varies with the phase of the solar cycle.

ON SOME STATISTICAL AND MORPHOLOGICAL CHARACTERISTICS OF THE QUIESCENT PROMINENCES OF SOLAR ACTIVITY CYCLE N 21

M.Sh.Gigolashvili, I.S.Iluridze
Abastumani Astrophysical Observatory, Georgian SSR Academy of Sciences, 383762 Abastumani, Mt. Kano'bili, Georgia, USSR

ABSTRACT

The observational data obtained at Abastumani Astrophysical Observatory of Georgian SSR Academy of Sciences, the data published in Monthly Bulletins "Solar Data" and Rome Astronomical Observatory Bulletin "Photographic Journal of the Sun" have been used. Long life-time objects maintaining their form and structural peculiarities were chosen for investigation. The filaments observed during less than 3 days were not considered.

The latitude distribution of quiescent prominences in time is studied both for each year of the solar activity cycle N 21 and for the whole one in 1975-1985. It was found that during the whole cycle the latitude distribution of quiescent prominences is of the pattern specific to both quiescent and active prominences at the moment of the solar activity minimum according to Waldmaier in 1933-1943. It should be noted that there is not such a similarity in yearly distribution during the whole solar activity cycle.

For revealing the sizes specific to quiescent prominences observed just on the solar limb, the differences $\Delta\varphi$ between the heliographic latitudes of the extremely northern and southern points were measured. The dependence of $\Delta\varphi$ on the prominence number is studied. It is seen that most frequently there are prominences with $\Delta\varphi \approx 3°$. Then, there is a large group with a wide interval of $\Delta\varphi$, the average of which $\Delta\varphi \approx 15°$, though two subgroups can be distinguished with their $\Delta\varphi$ about 10° and 19°. There is a small numer of prominences with $\Delta\varphi \approx 25°$.

It can be suggested that the prominences nearly parallel to the solar equator, belong to the first group. The prominences of the second group form a comparatively wide angle to the equator. The prominences almost meridional to the solar surface belong to the most scanty third group. The above dependence, reflects and includes both the lifetime of filaments and the effect of the solar differential rotation on the location on the solar surface.

MOTION OF HIGH LATITUDE SOLAR MICROWAVE SOURCES AND COMPARISON WITH POLAR PROMINENCES

S.URPO, S.POHJOLAINEN, H.TERÄSRANTA
Metsähovi Radio Research Station, Helsinki University of Technology, 02150 Espoo, Finland

B.VRSNAK, V.RUZDJAK, R.BRAJSA
Hvar Observatory, University of Zagreb, Faculty of Geodesy, 41000 Zagreb, Yugoslavia

A.SCHRÖLL
Solar Observatory Kanzelhöhe, University of Graz, 9521 Treffen, Austria

ABSTRACT[1]

Solar microwave sources at high solar latitudes have been observed with a 14 m radio telescope at the Metsähovi Radio Research Station in Finland. Several periods for observations were organized in 1986-1989 in order to detect sources close to the north and south pole of the Sun. Measurements at 22 and 37 GHz (wavelengths 14 and 8 mm respectively) have revealed the existence of high temperature and low temperature regions (relative to the quiet Sun level) at latitudes 50-80 degrees. The motions of these regions have been studied and compared with optical measurements of polar prominences. The temperature enhancement at 37 GHz is typically 100-400 K above the quiet Sun level (7800 K) at that frequency. Although in most cases temperature depression in a low temperature area amounts 50-300 K, at 37 GHz, the temperature drop in the low temperature area which was observed in July 1982 was as low as 900 K. The results of the radio measurements of the Sun at 22 and 37 GHz on high solar latitudes imply that high temperature areas correspond to polar faculae while low temperature areas correspond to polar prominences. The principal cause of the observed lower temperature area is the absorbtion by the filament.

[1]The full length paper is published simultaneously in the Hvar Observatory Bulletin.

POLAR CROWN FILAMENTS AND SOLAR DIFFERENTIAL ROTATION AT HIGH LATITUDES

R. Brajša [1], B. Vršnak [1], V. Ruždjak [1] and A. Schroll [2]

[1] Hvar Observatory, Faculty of Geodesy, University of Zagreb, Yugoslavia

[2] Sonnenobservatorium Kanzelhöhe der Universität Graz, Treffen, Austria

Rotation rates of 124 polar crown filaments are determined. Filaments are traced on Hα filtergrams taken at Sonnenobservatorium Kanzelhöhe, University of Graz, Austria, in the period 1979-9987, covering heliographic latitude ranges between 35°-80°. The lifetime of tracers varied from 1 to 6 days. Special care was taken to measure the positions of the footpoints of prominences, and the correction for an average height of 0.5%, 1% and 1.5% of the Solar radius was applied. The results were connected to those of Adams and Tang (1977) (prominences in latitutde range 0°-35°), and the least square fit to the function

$$\omega = A + B \sin^2 \emptyset + C \sin^4 \emptyset \qquad (1)$$

gives the values of the parameters A,B and C which are compared to other measurements in the table. In the above expression ω is the sidereal angular rotation rate in deg/day and $\emptyset$ is the latitude.

Parameters of eq. (1). Our errors in determination of the parameters A, B and C are 0.15, 0.9 and 0.9 respectively.

Method/tracer	A	B	C	Ref.
photosph./spec.	13.76	- 1.74	- 2.19	Howard & Harvey (1970)
filaments	14.42	- 1.40	- 1.33	d'Azambuja[2] (1948)
magn. fields	14.37	- 2.30	- 1.62	Snodgrass (1983)
Polar faculae	13.11	- 1.65	- 0.22	Makarova & Solonsky (1987)
pol.high temp. reg. at 37 GHz	11.55	+ 0.05	- 1.69	Urpo et al. (1989)
fil. h/R=0.5%	14.46	- 0.33	- 2.74	present work
fil. h/R=1%	14.45	- 0.11	- 3.69	present work
fil. h/R=1.5%	14.40	+ 0.86	- 5.64	present work

The paper in full length is published in the Hvar Observatory Bulletin.

Hvar Reference Atmosphere of Quiescent Prominences

Oddbjørn Engvold[1], *Tadashi Hirayama*[2], *Jean Louis Leroy*[3]
Eric R. Priest[4], *and Einar Tandberg-Hanssen*[5]

1 Introduction

It has become clear recently that prominences are not quiescent at all; they are highly dynamic and inhomogeneous. Many exciting new results about prominences were presented and discussed at this meeting.

Most quiescent prominences at high latitude have inverted magnetic polarity relative to the polarity expected from observations of photospheric fields. In comparison, inverted and normal polarities are more equally represented among active-region prominences. The magnetic fields threading quiescent prominences are both twisted and sheared. Both magntic field strength and plasma density appear to increase with height. At the same time as direct measurements suggest that the magnetic fields are basically horizontal, one observes vertical as well as horizontal mass motions in the range ± 5 $km\ s^{-1}$. Oscillatory motions are often present with periods in the range from about 3 to 60 minutes. The shorter periods could be local magnetohydrodynamic waves, whereas the longer periods might represent *MHD* oscillations of the medium-scale structures themselves. The role of the magnetic fields for prominence support was discussed in detail at the meeting. Large twisted magnetic loops may provide the "dip" in the field required by both normal and inverse magneto-stactic models for prominences, but momentum transfer to the gas via dissipating Alfvén waves seems on the other hand to be in good accordance with observations, i.e. it forecasts a vertical magnetic field which looks more like the observed morphology of prominences than the horizontal magnetic vector derived from polarimetric analysis. Prominence matter is concentrated in fine threads and knots. There is strong evidence that the threads are too small to be resolved with the best solar telescopes today. Estimates of the so-called filling factor give values from 0.1 to 10^{-6}, although 0.1 - 0.01 is more likely.

We now know more in broad terms than we did before this meeting, about the fundamental questions of their formation, support, and interaction with the surrounding solar atmosphere, bur many details remain to be worked out. Against this background, it was agreed that it could be helpful to put together a list of typical parameter values that one seems to be reasonable sure about. Such a list is given below. We also comment briefly on the general characteristics of quiescent prominences.

2 General Characteristics of Quiescent Prominences

The structures pertaining to prominences may for practical purposes be divided into *(i) large-scale, (ii) medium-scale,* and (iii) *small-scale.*

The *large-scale structures* are the coronal helmet streamers and their associated coronal cavities, or filament channels, which overlie the polarity reversal lines of photospheric magnetic fields. Filament channels are seemingly endlessly wrapped around the Sun. The typical widths and heights of helmet structures are 60 000 km and 50 - 100 000 km, respectively, with lifetimes from weeks to months. Figure 1 is a sketch showing a helmet streamer and filament channel, and Figure 2 is a more detailed theoretical model of the same.

[1] University of Oslo, Norway
[2] National Astronomical Observatory, Japan
[3] Observatoire de Pic-du-Midi, France
[4] University of St. Andrews, Scotland
[5] MSFC, Alabama, USA

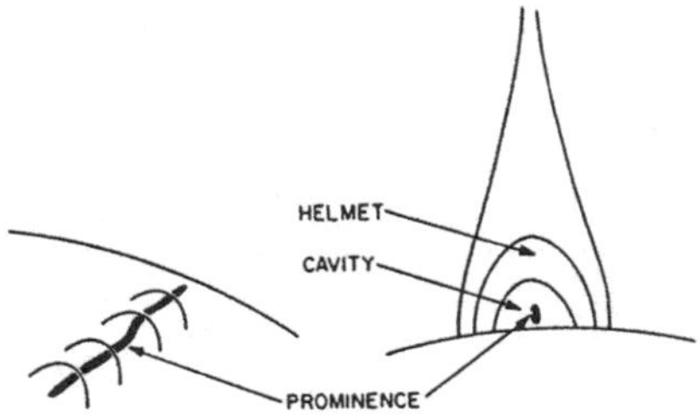

Figure 1: Topology of typical prominence-corona helmet configuration, showing on the left a disc view and whereas a limb view and on the right a limb view (Pneuman 1968)

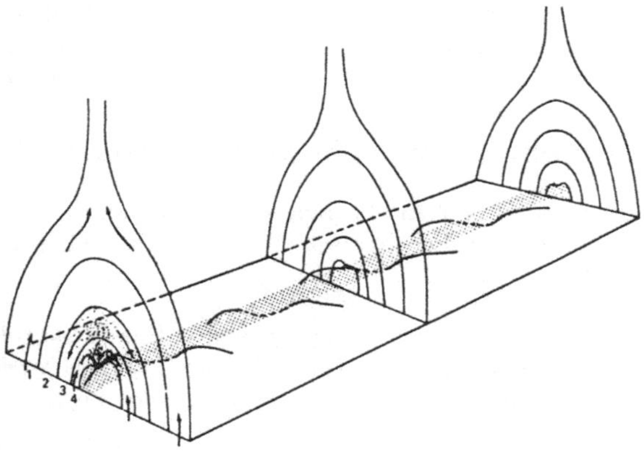

Figure 2: A 3-D description of a coronal streamer and its inner structures. The magnetic fields for the coronal cavity and the quiescent prominence are depicted as a global dipole and a sheared local field The direction of anticipated plasma flows is represented by arrows (An et al. 1985)

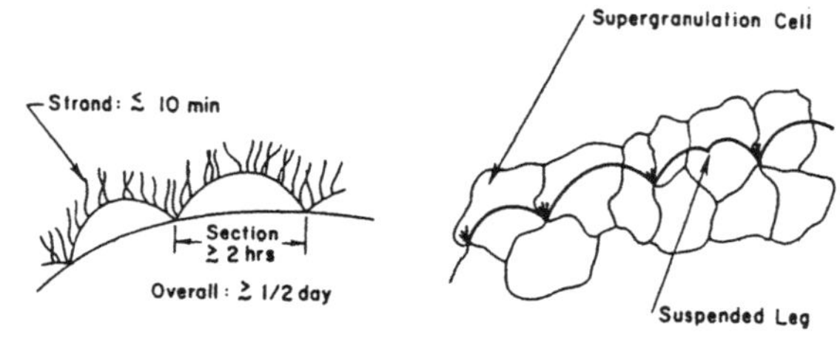

(a) (b)

Figure 3: (a) Lifetimes of prominence features and (b) the assumed location of prominence legs with respect to supergranulation cell boundaries (Forbes 1986)

The main body, "legs", and segments of a quiescent prominence (i.e. the volume within the cavity region which is filled with 6 - 8 000 K material), change their forms in the course of 2 to 12 hours and are refered to as *medium-scale* structures. Their dimensions are typically around 30 000 km, which interestingly is similar to the diameter of a supergranulation cell.

Bright threads and knots, which are the building blocks of quiescent prominences, constitute the prominence *small-scale structures.* The observed dimensions range from the current resolution limit of the best solar telescopes ($\frac{1}{3}$ arcseconds) up to a few arc seconds. Their lifetimes are a few minutes ($\leq$10 minutes), depending on size and brightness. Time series and Doppler measurements give flow velocities in the range ± 5 $km\ s^{-1}$. Small bright knots have been seen to move with velocities 15-25 $km\ s^{-1}$.

3 The Hvar Reference Atmosphere of Quiescent Prominences

Table 1 gives typical observed values of physical parameters in solar prominences, and values derived from prominence modelling. Details are found in papers of this proceedings and in the references below.

TABLE 1

Observed values for plasma parameters of quiescent prominences. One should notice that the values given in the table refer essentially to prominences observed at the limb, i.e. to relatively high prominences, and these may therefore not be regarded as representative for low quiescent prominences of active latitudes.

	Prominence		P-C Transition Region
	Central part	Edges	
T_e (K)	4 300 - 8 500	8 000 - 12 000	10^4 – 10^6
ξ_t ($km\ s^{-1}$)	3 - 8	10 - 20	30
n_e (cm^{-3})	10^{10} – 10^{11}	$10^{9.6}$	$3\ 10^{10}$ – 10^8
P_g ($dyn\ cm^{-2}$)	0.1 - 1	~0.02	~0.2
X	0.2 - 0.9		
B (gauss)	4 - 20		
V ($km\ s^{-1}$)	±5		~ 10

(Comments to table on next page)

T_e : One finds that $T_e \simeq T_{kin}$ in quiescent prominences. The most frequently observed value is $T_e \simeq$7500 K in the central parts of quiescent prominences.

B : The magnetic field strength in quiescent prominences is most frequently found to be around 8 gauss. Higher values ($\simeq$20 gauss) are frequently observed in the low prominences of active latitudes. The tabulated values indicate the commonly observed spread in B. Some observers find values as high as 30 gauss.

$X = \frac{n_{HI}}{n_H}$ is the degree of ionization of hydrogen derived from modelling of prominence plasma.

References

An, C-H, Suess, S T, Tandberg-Hanssen, E, and Steinolfson, R S: 1985 Solar Phys. **102**, 165

Hirayama, T: 1985, Solar Phys. **100**, 415

Pneuman, G W: 1968, Solar Phys. **3**, 578

Priest, E R (ed.): 1989, *Dynamics and Structure of Quiescent Solar Prominences*, Kluwer Academic Publ., Dordrecht

Tandberg-Hanssen, E: 1974, *Solar Prominences*, D Reidel, Dordrecht

PLASMA PARAMETERS IN QUIESCENT PROMINENCES

Eberhart Jensen and Jun Elin Wiik
Institute of Theoretical Astrophysics
University of Oslo, Norway

Introduction

From a set of four basic parameters; temperature, $T(T_e = T_i)$, mass density, ρ, degree of ionization, x, and magnetic field strength, B, parameters are derived that characterize the properties of the plasma in quiescent prominences.

Transport coefficients are of particular interest as the prominence plasma is far from homogeneous. Connected with the fine structure steep gradients must exist both in temperature, density and in the magnetic field strength. As the time constants in the fine structure are of the order of minutes the coupling to the magnetic field will lead to induced currents and ohmic losses.

With a moderate filling factor of the plasma threads, the prominence plasma will be "porous" to radiation. This property, which changes with time and must be very different in different parts of a prominence, will influence the degree of ionization, which is the most poorly known parameter in our set.

Specification of the damping lengths for Alfvén-waves, fast mode and acoustic waves at the end of our table requires knowledge of a fifth parameter, the period of the waves, denoted by τ.

The "background" medium that surrounds the cool prominence fine structure elements must have corona-like properties probably modified in varying degree by the presence of prominence matter. This second constituent has been ignored in the following.

Our basic parameters are known from observations with considerable difference in accuracy. We have used the intervals indicated in "The Hvar Reference Atmosphere of a Quiescent Prominence" (These proceedings) as our source. We thus obtain two values of the parameters in our table, representing an upper and a lower limit. For some parameters the resulting interval comes out to be rather liberal, reflecting considerable uncertainty in our knowledge. For others, such as gas pressure additional constraints are imposed. We have carried through our calculations for a pure hydrogen plasma only.

Dimensions are indicated where cgs units are not used.

Parameter	Central Part	Edge
Temperature, $T_e (=T_i)$	4300 - 8500	8000 - 12000
Mass density, ρ	2×10^{-14} - 2×10^{-12}	7×10^{-14}
Degree of ionization, $x=N_e/N$	0.8 - 0.1	0.1
Magnetic field strength, B	4 - 20	4 - 20
Electron density, $N_e = N_i$	$(1-10)\times 10^{10}$	4×10^{9}
Neutral H-density, N_n	2×10^{9} - 10^{12}	4×10^{10}
Total H-density, $N=N_n+N_i$	10^{10} - 10^{12}	4×10^{10}
Mean molecular weight, $\mu = \frac{1}{1+x}$	0.6 - 0.9	0.9
Electron pressure, $P_e = N_e kT$	6×10^{-3} - 10^{-1}	$(4-7)\times10^{-3}$
Gas pressure, $P_g = \frac{NkT}{\mu}$	0.1 - 2	2×10^{-2}
Magnetic pressure, $\frac{B^2}{8\pi}$	0.6 - 20	0.6 - 20
$\beta = \frac{8\pi P_g}{B^2}$	6×10^{-3} - 3	$(1-30)\times10^{-3}$
Velocity of sound, $V_s = (\frac{\gamma P_g}{\rho})^{1/2}$	$(8-10)\times10^{5}$	10^{6}
Alfvén-velocity, $V_A = \frac{B}{(4\pi\rho)^{1/2}}$	$(0.8-40)\times10^{6}$	$(4-20)\times10^{6}$
Thermal velocities, $V_e = (\frac{3kT}{m_e})^{1/2}$	$(4-6)\times10^{7}$	$(6-7)\times10^{7}$
$V_i = (\frac{3kT}{m_i})^{1/2}$	$(1-1.4)\times10^{6}$	$(1.4-2)\times10^{6}$
Plasma frequency, $\omega_{PL} = (\frac{4\pi^2 e^2 N_e}{m_e})^{1/2}$	$(6-20)\times10^{9}$	4×10^{9}
Gyro frequencies, $\omega_e = \frac{eB}{m_e c}$	$(0.7-4)\times10^{8}$	$(0.7-4)\times10^{8}$
$\omega_i = \frac{eB}{m_i c}$	$(0.4-2)\times10^{5}$	$(0.4-2)\times10^{5}$

Parameter	Central part	Edge
Gyro radii, $r_e = \frac{c(3km_e)^{1/2}}{e} \frac{T^{1/2}}{B}$	0.1 - 0.9	0.2 - 1.0
$r_i = \frac{c(3km_i)^{1/2}}{e} \frac{T^{1/2}}{B}$	5 - 40	7 - 40
Debye length, $\lambda_D = (\frac{kT}{4\pi e^2 N_e})^{1/2}$	$(1-6)x10^{-3}$	10^{-2}
Plasma parameter, $G=(N_e \lambda_D{}^3)^{-1}$	$(4-30)x10^{-4}$	$(1-3)x10^{-4}$
Coulomb logarithm, $\ln(12\pi G^{-1})$	10	10
Collision frequency, $\nu_{ei} = \frac{4(2\pi)^{1/2} e^4 N_i \ln\Lambda}{3(m_e)^{1/2} k^{3/2} T^{3/2}}$	$(5-100)x10^5$	$(1-2)x10^5$
Electrical conductivities, $\sigma_\parallel = \frac{e^2 N_e}{m_e \nu_{ei}}$	$(2x6)x10^2$ mho/m	$(5-8)x10^2$mho/m
$\sigma_\perp = \frac{\nu_{ei}{}^2}{\nu_{ei}{}^2+\omega_e{}^2} \sigma_\parallel$	$2x10^{-3}$ -6 mho/m	$(1-50)x10^{-4}$mho/m
$\sigma_H = \frac{\nu_{ei}\omega_e}{\nu_{ei}{}^2+\omega_e{}^2} \sigma_\parallel$	0.8 - 40 mho/m	0.3-2 mho/m
Thermal conductivities, $\kappa_\parallel = \frac{5k P_e}{2m_e \nu_{ei}}$	$(2-10)x10^3$	$(7-20)x10^3$
$\kappa_\perp = \frac{\nu_{ei}{}^2}{\nu_{ei}{}^2+\omega_e{}^2} \kappa_\parallel$	$2x10^{-2}$- 60	$(3-80)x10^{-3}$
$\kappa_H = \frac{\nu_{ei}\omega_e}{\nu_{ei}{}^2+\omega_e{}^2} \kappa_\parallel$	6 - 600	5 - 40

Parameter	Central part	Edge
Thermometric conductivities,		
$\lambda_{\parallel} = \frac{\kappa_{\parallel}}{\rho C_v}$	$8\text{x}10^6 - 2\text{x}10^9$	$(8-20)\text{x}10^8$
$\lambda_{\perp} = \frac{\kappa_{\perp}}{\rho C_v}$	$7\text{x}10^2 - 10^6$	$(0.2-7)\text{x}10^3$
$\lambda_H = \frac{\kappa_H}{\rho C_v}$	$3\text{x}10^5 - 2\text{x}10^7$	$(0.5-4)\text{x}10^6$
Ohmic damping length for A-waves, τ = 10-100s; $L_{A,ohm} = \frac{2}{\pi}\frac{\sigma}{c^2} V_A^3 \tau^2$	$7\text{x}10^{10} - 2\text{x}10^{18}$	$5\text{x}10^{13} - 10^{18}$
Damping lengths for non-linear waves, with periods τ=10-100s and flux density, $F = 5\text{x}10^5$		
Alfvén, $L_A = 8\text{x}10^{-5} \frac{\tau B^4}{\rho F}$	$2\text{x}10^5 - 10^{11}$	$6\text{x}10^6 - 4\text{x}10^{10}$
Fast mode, $L_f = 4.8\text{x}10^{-3} \frac{\tau B^{5/2}}{\rho^{3/4} F^{1/2}}$	$10^6 - 2\text{x}10^{10}$	$(2-900)\text{x}10^7$
Acoustic, $L_s = 2.3\text{x}10^{-4} \tau P_g^{1/2} V_s^{3/2}$	$5\text{x}10^5 - 5\text{x}10^7$	$(2-20)\text{x}10^6$

References

Braginskii, S.I., 1965, Reviews of Plasma Physics, 1, Consultants Bureau, New York, ed. M.A. Leontovich, p. 205.
Krall, N.A. and Trivelpiece, A.W., 1973, Principles of Plasma Physics, McCraw-Hill Book Company
Lang, K.R., 1974, Astrophysical Formulae, Springer Verlag.
Mitchner, M. and Kruger, C.H., 1973, Partially Ionized Gases, John Wiley & Sons.
Spitzer, L., 1967, Physics of Fully Ionized Gases, Interscience Publishers.
Wentzel, D.G., 1977, Solar Physics, 52, 163.

SUMMARY of IAU Colloquium 117, DYNAMICS OF PROMINENCES

by Eberhart Jensen

Some months ago I received 23 abstracts of papers to be presented at this meeting from Einar. Ordered after topics we get this ranking list;

1.	MHD-configuration, stability, preflare instability, forcefree magnetic ropes	10
2.	Formation	4
3.	Connection to flares	3
4.	Thermal instabilities	2
5.	Loops	2
6.	Determination of magnetic fields	1
7.	Instability of internal gravity waves	1

Listening to these contributions at the meeting, together with some 50 others I noticed that quite a few of them had chaged a good deal compared to the abstracts I received. So judging only from this modest sample, we may conclude that progress is fast in this field.

The sun has been called a gigantic plasma physics laboratory. At first sight one might think that prominences ought to be the ideal objects for study. Their physics must be easy to understand because we can see right through them! If we fail to understand what prominences are, where resolution goes down to a couple of hundred kilometers, what about stars and stellar astronomers? Looking at prominences we obviously have a tremendous advantage, and a unique opportunity to learn astrophysics.

Just now we experience a breakthrough in resolution, where fine detail may be seen down to less than 0.5 seconds of arc. It turns out that thin loops and threads are the structure elements that prominences are made of.

Reliable quantitative changes with time in the fine structure can now be obtained by image processing and sophisticated soft ware for de-stretching.

To understand the physics of prominences and to use this insight to make models that are not too far from the real thing, theoreticians need boundary conditions and other constraints.

At this meeting the fundamental parameters in prominences have been updated. But filling factors are difficult to specify, so both mass density and magnetic field strengths are probably on the low side as applied to the fine structure.

The birth place for prominences, the filament channels are fundamental for our understanding. At this meeting we have learned about the characteristic orientation of fibrils, the converging of magnetic flux and other signatures that proceed the formation of filaments. We have seen movies illustrating what is going on in great detail.

We have also learned that prominence feet are the locations where mass exchange with the chromosphere takes place. So we meet the old problem again - that of specifying an injection mechanism. We have heard about "the big foot" that filaments revolve around, and where impulsive heating starts. In addition to the ordered flow of matter through the feet we have the turbulent velocity field with velocities of 20 km/s or more, as we saw in the thriller movie yesterday. We also saw a velocity distribution where energy apparently was being pumped into the high-velocity tail prior to an eruption.

Dramatic examples of explosive events were described in great details. A couple of papers dealt with loop prominences - the only case where the physics is understood as far as I can see. However, here the crucial parameter, the magnetic field plays only a passive role.

With all these observations so full of beautiful dynamics and hap-penings in velocity space - how come that so much effort is devoted to the study of static configurations? They do not exist on the sun at all!

That is just a fact, some will call it a sad fact, but what can you expect on the foggy surface of a ball of hot gas? Why not make $V \neq 0$ and let it vary in time in a stochastic way. That is what the observations prescribe.

This reminds me of the first rule that should regulate the life of a solar physicist, "Never waste good seeing". Oddbjørn and I did just that, and here you see the results. From a beautiful drawing made 115 years ago, with all the fine details we now can take pictures of, we obtained a static slab model, just by taking a photo well out of focus.

But to be serious, we had some nice papers on modelling of fine structure in the form of threads, both dressed and undressed, and with convincing solutions of the transfer equation for this type of models.

I could continue for a long time, but there is no point in repeating the whole meeting.

This wonderful symposium has given us great inspiration for continued efforts in the study of one of the most exciting phenomena nature has to offer - solar prominences.

Lecture Notes in Mathematics

Vol. 1236: Stochastic Partial Differential Equations and Applications. Proceedings, 1985. Edited by G. Da Prato and L. Tubaro. V, 257 pages. 1987.

Vol. 1237: Rational Approximation and its Applications in Mathematics and Physics. Proceedings, 1985. Edited by J. Gilewicz, M. Pindor and W. Siemaszko. XII, 350 pages. 1987.

Vol. 1250: Stochastic Processes – Mathematics and Physics II. Proceedings 1985. Edited by S. Albeverio, Ph. Blanchard and L. Streit. VI, 359 pages. 1987.

Vol. 1251: Differential Geometric Methods in Mathematical Physics. Proceedings, 1985. Edited by P.L. García and A. Pérez-Rendón. VII, 300 pages. 1987.

Vol. 1255: Differential Geometry and Differential Equations. Proceedings, 1985. Edited by C. Gu, M. Berger and R.L. Bryant. XII, 243 pages. 1987.

Vol. 1256: Pseudo-Differential Operators. Proceedings, 1986. Edited by H.O. Cordes, B. Gramsch and H. Widom. X, 479 pages. 1987.

Vol. 1258: J. Weidmann, Spectral Theory of Ordinary Differential Operators. VI, 303 pages. 1987.

Vol. 1260: N.H. Pavel, Nonlinear Evolution Operators and Semigroups. VI, 285 pages. 1987.

Vol. 1263: V.L. Hansen (Ed.), Differential Geometry. Proceedings, 1985. XI, 288 pages. 1987.

Vol. 1265: W. Van Assche, Asymptotics for Orthogonal Polynomials. VI, 201 pages. 1987.

Vol. 1267: J. Lindenstrauss, V.D. Milman (Eds.), Geometrical Aspects of Functional Analysis. Seminar. VII, 212 pages. 1987.

Vol. 1269: M. Shiota, Nash Manifolds. VI, 223 pages. 1987.

Vol. 1270: C. Carasso, P.-A. Raviart, D. Serre (Eds.), Nonlinear Hyperbolic Problems. Proceedings, 1986. XV, 341 pages. 1987.

Vol. 1272: M.S. Livšic, L.L. Waksman, Commuting Nonselfadjoint Operators in Hilbert Space. III, 115 pages. 1987.

Vol. 1273: G.-M. Greuel, G. Trautmann (Eds.), Singularities, Representation of Algebras, and Vector Bundles. Proceedings, 1985. XIV, 383 pages. 1987.

Vol. 1275: C.A. Berenstein (Ed.), Complex Analysis I. Proceedings, 1985–86. XV, 331 pages. 1987.

Vol. 1276: C.A. Berenstein (Ed.), Complex Analysis II. Proceedings, 1985–86. IX, 320 pages. 1987.

Vol. 1277: C.A. Berenstein (Ed.), Complex Analysis III. Proceedings, 1985–86. X, 350 pages. 1987.

Vol. 1283: S. Mardešić, J. Segal (Eds.), Geometric Topology and Shape Theory. Proceedings, 1986. V, 261 pages. 1987.

Vol. 1285: I.W. Knowles, Y. Saitō (Eds.), Differential Equations and Mathematical Physics. Proceedings, 1986. XVI, 499 pages. 1987.

Vol. 1287: E.B. Saff (Ed.), Approximation Theory, Tampa. Proceedings, 1985–1986. V, 228 pages. 1987.

Vol. 1288: Yu. L. Rodin, Generalized Analytic Functions on Riemann Surfaces. V, 128 pages, 1987.

Vol. 1294: M. Queffélec, Substitution Dynamical Systems – Spectral Analysis. XIII, 240 pages. 1987.

Vol. 1299: S. Watanabe, Yu.V. Prokhorov (Eds.), Probability Theory and Mathematical Statistics. Proceedings, 1986. VIII, 589 pages. 1988.

Vol. 1300: G.B. Seligman, Constructions of Lie Algebras and their Modules. VI, 190 pages. 1988.

Vol. 1302: M. Cwikel, J. Peetre, Y. Sagher, H. Wallin (Eds.), Function Spaces and Applications. Proceedings, 1986. VI, 445 pages. 1988.

Vol. 1303: L. Accardi, W. von Waldenfels (Eds.), Quantum Probability and Applications III. Proceedings, 1987. VI, 373 pages. 1988.

Lecture Notes in Physics

Vol. 340: M. Peshkin, A. Tonomura, The Aharonov-Bohm Effect. VI, 152 pages. 1989.

Vol. 341: E.F. Milone (Ed.), Infrared Extinction and Standardization. Proceedings, 1988. III, 79 pages. 1989.

Vol. 342: M. Balabane, P. Lochak, C. Sulem (Eds.), Integrable Systems and Applications. Proceedings, 1988. VII, 342 pages. 1989.

Vol. 343: M. Month, S. Turner (Eds.), Frontiers of Particle Beams; Observation, Diagnosis and Correction. Proceedings, 1988. IX, 509 pages. 1989.

Vol. 344: M. Rascle, D. Serre, M. Slemrod (Eds.), PDEs and Continuum Models of Phase Transitions. Proceedings, 1988. VII, 229 pages. 1989.

Vol. 345: H. Holden, A. Jensen (Eds.), Schrödinger Operators. Proceedings, 1988. V, 458 pages. 1989.

Vol. 346: D. Lüst, S. Theisen, Lectures on String Theory. VII, 346 pages. 1989.

Vol. 347: Ph. Blanchard, J.-P. Dias, J. Stubbe (Eds.), New Methods and Results in Non-linear Field Equations. Proceedings. 1987. VII, 133 pages. 1989.

Vol. 348: A. Bohm, M. Gadella, Dirac Kets, Gamow Vectors and Gel'fand Triplets. VII, 119 pages. 1989.

Vol. 349: Y.A. Kubyshin, J.M. Mourão, G. Rudolph, I.P. Volobujev, Dimensional Reduction of Gauge Theories, Spontaneous Compactification and Model Building. X, 80 pages. 1989.

Vol. 350: G. Tenorio-Tagle, M. Moles, J. Melnick (Eds.), Structure and Dynamics of the Interstellar Medium. Proceedings, 1989. XXI, 537 pages. 1989.

Vol. 351: A. Dervieux, B. Larrouturou (Eds.), Numerical Combustion. Proceedings, 1989. VII, 481 pages. 1989.

Vol. 352: K.B. Wolf (Ed.), Lie Methods in Optics II. Proceedings, 1988. XII, 197 pages. 1989.

Vol. 353: M. Barthes, J. Léon (Eds.), Nonlinear Coherent Structures. Proceedings, 1989. X, 280 pages. 1990.

Vol. 354: J. De Coninck, F. Dunlop (Eds.), Wetting Phenomena. Proceedings, 1988. VII, 112 pages. 1990.

Vol. 355: R. Lima, L. Streit, R. Vilela Mendes (Eds.), Dynamics and Stochastic Processes. Proceedings, 1988. V, 298 pages. 1990.

Vol. 356: R. Wehrse (Ed.), Accuracy of Element Abundances from Stellar Atmospheres. Proceedings, 1988. V, 109 pages. 1990.

Vol. 357: J.L. Lumley (Ed.), Whither Turbulence? Turbulence at the Crossroads. Proceedings, 1989. V, 535 pages. 1990.

Vol. 358: J. Kovalevsky, Astrométrie moderne. VI, 294 pages. 1990.

Vol. 359: K. Kirchgässner (Ed.), Problems Involving Change of Type. Proceedings, 1988. XII, 207 pages. 1990.

Vol. 360: Y. Mellier, B. Fort, G. Soucail (Eds.), Gravitational Lensing. Proceedings 1989. XV, 315 pages. 1990.

Vol. 361: P. Gaigg, W. Kummer, M. Schweda (Eds.), Physical and Nonstandard Ganges. Proceedings, 1989. IX, 310 pages. 1990.

Vol. 362: N.E. Kassim, K.W. Weiler (Eds.), Low Frequency Astrophysics from Space. Proceedings. XII, 280 pages. 1990.

Vol. 363: V. Ruždjak, E. Tandberg-Hanssen (Eds.), Dynamics of Quiescent Prominences. Proceedings, 1989. XI, 304 pages. 1990.

GPSR Compliance

The European Union's (EU) General Product Safety Regulation (GPSR) is a set of rules that requires consumer products to be safe and our obligations to ensure this.

If you have any concerns about our products, you can contact us on ProductSafety@springernature.com

In case Publisher is established outside the EU, the EU authorized representative is:

Springer Nature Customer Service Center GmbH
Europaplatz 3
69115 Heidelberg, Germany

Batch number: 10640288

Printed by Printforce, the Netherlands